GENETIC PROGRAMMING THEORY AND PRACTICE

GENETIC PROGRAMMING SERIES

Series Editor

John Koza
Stanford University

Also in the series:

GENETIC PROGRAMMING AND DATA STRUCTURES: Genetic Programming + Data Structures = Automatic Programming! *William B. Langdon*; ISBN: 0-7923-8135-1

AUTOMATIC RE-ENGINEERING OF SOFTWARE USING GENETIC PROGRAMMING, *Conor Ryan;* ISBN: 0-7923-8653-1

DATA MINING USING GRAMMAR BASED GENETIC PROGRAMMING AND APPLICATIONS, *Man Leung Wong and Kwong Sak Leung;* ISBN: 0-7923-7746-X

GRAMMATICAL EVOLUTION: ***Evolutionary Automatic Programming in an Arbitrary Language***, *Michael O'Neill and Conor Ryan;* ISBN: 1-4020-7444-1

GENETIC PROGRAMMING IV: ***Routine Human-Computer Machine Intelligence***, John *R. Koza, Martin A. Keane, Matthew J. Streeter, William Mydlowec, Jessen Yu, Guido Lanza;* ISBN 1-4020-7446-8

The cover art was created by Leslie Sobel in Photoshop from an original photomicrograph of plant cells and genetic programming code. More of Sobel's artwork can be seen at www.lesliesobel.com

GENETIC PROGRAMMING THEORY AND PRACTICE

edited by

Rick Riolo
Center for the Study of Complex Systems
University of Michigan

Bill Worzel
Genetics Squared

KLUWER ACADEMIC PUBLISHERS
Boston / Dordrecht / New York / London

Distributors for North, Central and South America:
Kluwer Academic Publishers
101 Philip Drive; Assinippi Park
Norwell, Massachusetts 02061 USA
Telephone (781) 871-6600
Fax (781) 681-9045
E-Mail: kluwer@wkap.com

Distributors for all other countries:
Kluwer Academic Publishers Group
Post Office Box 322
3300 AH Dordrecht, THE NETHERLANDS
Telephone 31 78 6576 000
Fax 31 78 6576 254
E-Mail: services@wkap.nl

Electronic Services <http://www.wkap.nl>

Library of Congress Cataloging-in-Publication Data

GENETIC PROGRAMMING THEORY AND PRACTICE
Edited by Rick Riolo and Bill Worzel
ISBN: 1-4020-7581-2

Printed on acid-free paper. Printed in the United States of America.

Contents

Contributing Authors

Wolfgang Banzhaf is Associate Professor of Computer Science at the University of Dortmund, Germany.

Jason M. Daida is an Associate Research Scientist in the Space Physics Research Laboratory at The University of Michigan, Ann Arbor.

Joseph A. Driscoll is an Assistant Professor in the Department of Computer Science at Middle Tennessee State University.

Stephen Freeland is an Assistant Professor of Bioinformatics within the Biology Department of UMBC (University of Maryland, Baltimore County).

David E. Goldberg is the Jerry S. Dobrovolny Distinguished Professor in Entrepreneurial Engineering and Director of the Illinois Genetic Algorithms Laboratory at the University of Illinois at Urbana-Champaign.

Erik D. Goodman is Professor of Electrical and Computer Engineering and of Mechanical Engineering at Michigan State University.

David Hill is a graduate student in the Department of Civil and Environmental Engineering at the University of Illinois at Urbana-Champaign.

Daniel Howard is a QinetiQ Fellow and Head of the Software Evolution Centre at QinetiQ in Malvern, UK.

Jianjun Hu is a Ph.D student in Computer Science at Michigan State University.

Martin A. Keane is a consultant to the gaming industry and works with Genetic Programming Inc.

Mark Kotanchek is the group leader for Applied Math and Physics within Physical Sciences Research and Development of the Dow Chemical Company, Midland, MI, USA.

Arthur K. Kordon is a Research and Development Leader in the Applied Math and Physics Group within the Physical Sciences Research and Development of the Dow Chemical Company.

John Koza is Consulting Professor at Stanford University in the Biomedical Informatics Program in the Department of Medicine and in the Department of Electrical Engineering.

Duncan MacLean is a founder of and Chief Engineer for Genetics Squared, Inc, a computational discovery company working in the pharmaceutical industry.

Miguel Nicolau is a researcher and part-time PhD student at the University of Limerick, Ireland.

David A. Ostrowski works in the Business Systems Research group at the Ford Research Laboratory. Dr. Ostrowski is also an adjunct faculty member in the Department of Computer Science at the University of Michigan-Dearborn.

Una-May O'Reilly is a research member of the Living Machines and Humanoid Robotics group in the Computer Science and Artificial Intelligence Laboratory at the Massachusetts Institute of Technology.

Robert G. Reynolds is a professor in the Department of Computer Science at Wayne State University and the director of the Artificial Intelligence Laboratory. He is also an Associate Adjunct Research Scientist with the Museum of Anthropology and a faculty member of the Center for the Study of Complex Systems at the University of Michigan-Ann Arbor.

Rick Riolo is Director of the Computer Lab and Associate Research Scientist in the Center for the Study of Complex Systems at the University of Michigan.

Conor Ryan is a senior lecturer and University Fellow at the University of Limerick, Ireland.

Kumara Sastry is a member of Illinois Genetic Algorithms Laboratory and a graduate student of Material Science and Engineering at the University of Illinois at Urbana-Champaign.

Kisung Seo is Visiting Assistant Professor of ECE at Michigan State University and also Assistant Professor of EE at Seokyoung University in Seoul.

Guido Smits is a Research and Development Leader in the Applied Math and Physics group within Physical Sciences Research and Development of Dow Benelux, Terneuzen, NV.

Terence Soule is a Professor at the University of Idaho, where is a member of the Computer Science Department and of the Bioinformatics and Computational Biology Program.

Lee Spector is Dean of the School of Cognitive Science and Associate Professor of Computer Science at Hampshire College in Amherst, Massachusetts.

Matthew Streeter is a Systems and Research Programmer at Genetic Programming, Inc. in Mountain View, California.

Dave Wilkinson is a geophysicist in the Seismic Analysis and Reservoir Property Estimation Team at ChevronTexaco Exploration and Production Technology Company.

Bill Worzel is the President and co-founder of Genetics Squared, Inc, a computational discovery company working in the pharmaceutical industry.

Deyi Xie is a geologist in the Seismic Analysis and Reservoir Property Estimation Team at ChevronTexaco Exploration and Production Technology Company.

Tina Yu is a computer scientist in the Mathematical Modeling Team at ChevronTexaco Information Technology Company.

Anjun Zhou is a principal of the Advanced Research Center at State Street Global Advisors.

Preface

The work described in the chapters in this book was first presented at a workshop titled "Genetic Program, Theory and Practice," organized by the Center for the Study of Complex Systems at the University of Michigan, Ann Arbor, 15-17 May 2003. The goal of this workshop was to start an exchange of research results and ideas between those who focus on Genetic Programming (GP) theory and those who focus on the application of GP to various real-world problems. In order to facilitate these interactions, the number of talks and participants was small, and the time for discussion was large. Further, participants were asked to review each other chapters *before* the workshop, with each chapter reviewed by both a "theory" and an "applications" researcher. Those reviewer comments, as well as discussion at the workshop, are reflected in the chapters presented in this book. Additional information about the Workshop, addendum's to chapters, and a site for continuing discussions by participants and by others can be found at http://cscs.umich.edu:8000/GPTP-2003/ .

We thank all the Workshop participants for making the Workshop an exciting and productive three days. In particular we thank all the authors, without whose hard work and creative talents Workshops and books like this one would not be possible. We also thank the keynote speakers: Dr. John Holland, Dr. Stephen Freeland, Ms. Lynne Ellyn and Mr. Tom Chase who delivered three thought provoking speeches that inspired a great deal of discussion among the participants.

The Workshop received support from several sources, including:

- The Center for the Study of Complex Systems (CSCS);
- State Street Global Advisors, Boston, MA;
- Christopher T. May, RedQueen Capital Management; and
- DTE Energy Foundation, Michigan.

We thank all of the contributors for their kind and generous support for the Workshop and for GP research in general.

A number of people made key contributions to running the Workshop and assisting the attendees while they were in Ann Arbor. Foremost among them was

Howard Oishi, assisted by Jesse Buck and Mike Charter. We thank them all for running yet another very successful workshop. We also thank those who helped with reading and copy-editing chapters, including Jim Odell, Bill Rand, Bill Tozier, and Eric Wollesen. Jesse Buck also put in many hard hours working on assembling this book. Melissa Fearon's editorial efforts were invaluable, from the initial plans for the book through its final publication. Thanks also to Deborah Doherty of Kluwer for helping with various technical publishing issues. Finally, we thank Carl Simon, Director of CSCS, for his support for this endeavor from its very inception, and to James Odell for serving on the Organizing Committee and helping out in various ways throughout.

RICK RIOLO AND BILL WORZEL

Foreword

I was very happy and excited when I was invited to participate to the workshop on Genetic Programming (GP) Theory and Practice organised by the Center for the Study of Complex Systems (CSCS) held at the University of Michigan, Ann Arbor, in May 2003. The focus of this workshop was explicitly on how theory and practice should interact and what they can teach each other: an important and timely topic, particularly in the light of the many, but still very recent and not widely understood, developments seen in GP theory the last few years.

The participants' list was impressive, including many of the most influential and respected researchers in genetic programming and evolutionary computation. I was really looking forward to attend this workshop and learn and interact with them all, when "disaster" struck: I discovered that our new baby would be born only days before the workshop, which made it impossible for me to leave my family and attend the event. To punish me for my desertion the organisers of the event, Rick Riolo and Bill Worzel, very kindly asked me whether I would write a foreword for this volume gathering the proceedings of the workshop — an invitation I was very glad to accept.

I am very impressed by the quality of the papers presented at the workshop and now printed in this volume, both in terms of the novelty and importance of theoretical results and the variety and credibility of real-world applications of genetic programming. In terms of both contributions to science and inspiration for future practical applications this workshop appears to have been a great success. Knowing the depth and breadth of many of the participants I have no doubts that the discussion of all contributions has been lively and enormously productive.

Many events take place regularly in evolutionary computation. However, very few are specifically devoted to genetic programming, and definitely none focuses on the "theory meets practice" idea behind this workshop, which is an urgent topic, not just for genetic programming, but for the whole field of genetic and evolutionary computation. I hope this event will be the first of a series, where theoreticians and practitioners will be able to exchange and generate new ideas, look back critically at what has been achieved and develop a vision of the things to come, thereby guiding and inspiring future research in genetic programming.

I am sure I will learn a lot from studying this volume in detail, and I am sure many others, practitioners and theoreticians alike, can gain much from doing the same — Thank you Rick and Bill for having put together such a high profile and inspiring event.

Professor Ricardo Poli
Department of Computer Science
University of Essex, May 1999

Chapter 1

GENETIC PROGRAMMING: THEORY AND PRACTICE

Bill Worzel[1] and Rick Riolo[2]
[1]*Genetics Squared, Inc., Milan, MI 48160,* [2]*Center for the Study of Complex Systems, University of Michigan*

Key words: Genetic programming, theoretical biology, diversity, bloat, population dynamics, particulate genes,

1. GP as an Art Struggling To Become A Craft

Genetic Programming (GP) is a young art that is just beginning to make its way into applications programming in a serious way. The dynamics of the GP process is poorly understood with many serious questions remaining. People applying GP to real-world problems have relied more on intuition than theory, experience more than mathematics. To reach the next stage in its development, GP theory and practice must both advance. Theory must inform practice and practice must test theory.

To address these challenges to GP, the Center for the Study of Complex Systems (CSCS) of the University of Michigan organized a workshop titled "Genetic Programming: Theory and Practice" in Ann Arbor from May 15th to May 17th, 2003. The Workshop, sponsored by DTE Energy, State Street Global Advisors, Red Queen Capital Management and the CSCS, brought together many of the leading practitioners with the leading theorists to explore the relationship between those two approaches to GP research.

Over the three days of the Workshop, alternating presentations by GP theorists and practitioners highlighted the differences between the two camps. The applications researchers, led by researchers at Dow Chemical,

The Ford Motor Company, State Street Advisors, DTE Energy and Texaco, showed many diverse and powerful applications of GP that collectively had an impact on decisions and systems involving billions of dollars for their respective companies. Smaller companies such as Genetics Squared and researchers spinning GP out of academic institutions highlighted what the future may hold for GP as they sought to apply GP to areas such as bioinformatics and analog circuit design. The collective attitude of this group was: "Success is its own justification."

The attendees from the theoretical camp took a very different approach, using simple models to study the dynamics of GP and to identify important parameters and limitations. Issues such as bloat, population size, building blocks and diffusion models were raised, discussed and debated. This group had a stern view of the future of GP, counseling discipline, simplicity and fundamental theory while never losing sight of the highly complex nature of Genetic Programming.

While at times the two groups seemed to be speaking on different subjects, the lively exchanges during discussion periods, lunches and breaks, demonstrated the shared community and desire by the participants to carry GP to the next level. There was definitely a sense that GP is on the verge of growing, adding sophistication in approach and diversity of application.

This book is an integral part of the GP Theory and Practice Workshop, with one chapter for each talk at the workshop. In this chapter we give a brief overview of the chapters, emphasizing a few of the themes and issues that either recur across chapters or that we feel are critical for the future direction of GP research.

2. Survey of Theory

John Holland, in the first day's keynote speech, summed up the value of theory with the straightforward comment that: "Theory tells us where to look!" In light of the Workshop's talks and discussions, it might be added that "Practice tells us when we're looking in the wrong place!" Holland went on to discuss the importance of models, especially models that exaggerate certain characteristics of a system since not all aspects of an environment are equally important. The ability to select and create a 'grammar' for the manipulation of fundamental elements is a key aspect to creating a powerful, general model. Over-specification using excessive numbers of variables is antithetical to successful modeling. Genetic programming is often used to discover which are the key variables in a problem and generate rules for combining the selected variables. However Holland's keynote implicitly challenged GP investigators to apply GP to the problem of discovering coherent groups of rules that would constitute a formal grammar.

Stephen Freeland delivered the keynote speech for the second day of the Workshop starting with a survey of evolutionary theory and the information theoretical aspects of codon mapping. In his speech he raised a key point for evolutionary computing: When is the adoption of biological models appropriate for evolutionary computing and when is it pointlessly slavish? He identified three areas of natural evolution that he felt were particularly appropriate for the Evolutionary Computation (EC) community:

- The use of particulate genes and their implication for population genetics
- The importance of an adaptive genetic code
- The dichotomy of genotype and phenotype

Freeland also suggested that the freedom of evolutionary computing from the constraints of the natural world allow us to "experiment" with various aspects of evolution and could lead to a better understanding of evolutionary theory. He called for an ongoing collaboration between EC programmers and theoretical biologists that would enrich both communities. Freeland has graciously contributed a chapter to this book (Chapter 19) that describes many of the ideas covered in his keynote speech and outlines possible areas for future collaboration.

For a long time, the question of whether the underlying theory of Genetic Algorithms (GAs) could be extended to GP has been debated. On the one hand, there seemed to be good reason to believe the GA model should inform GP behavior. On the other hand, it was easy to come up with reasons why this should not be so.

In the last several years, Poli, Langdon and others, have come forward with mathematical proofs that extend the schema theory to GP. At the same time Daida and others have presented results that are hard to explain using the schema theory alone. Does this invalidate the schema theory or does it suggest that more work needs to be done? Both viewpoints were presented in this workshop.

Beyond the schema debate, work on questions such as the role of building blocks, population size, the impact of operator selection, and diversity were raised and discussed in the context of understanding GP mechanics and how these factors influence the success of a GP search.

Lee Spector, in Chapter 2, surveys a bit of the history and development of the field of genetic programming, identifying several key points for future theoretical development. One of his main conclusions is that the GP community should not only continue to borrow ideas from biology (as suggested by Freeland) but that it should also look into borrowing biological *theory* to explain GP phenomena. He identifies four main areas for future theoretical work:

- The impact of code modularization on existing theory and the extension of theory to describe this impact;
- The relationship of "environment" to the code and the code's "development" in response to its environment;
- A theory of diversity and dynamical GP where the evolutionary mechanisms "evolve" in situ; and
- The implicit "topology" of GP systems as expressed by such mechanisms as multi-deme or niching systems.

It is worth noting that Spector has been investigating some of these issues in a "virtual world" environment by using his PushGP system to evolve agent controllers - something that harkens back to Holland's keynote speech and the importance of developing rules generated by independent agents.

In Chapter 4 Wolfgang Banzhaf describes a GP system modeled after regulatory pathways in natural genetic systems. Included in this approach are promoter and suppressors and a complete network of "protein-gene" interactions - perhaps moving GP closer to cellular automata than conventional programming systems. Banzhaf also set out to explore the issues of response to an environment - almost as if he was answering Spector's call! The theoretical implications of this model include a clear demonstration of the value of neutral mutation, sensitivity to critical points in the "environment" (i.e., the problem space) and the natural development of encapsulated behaviors. The model demonstrates significant behavioral differences from "standard GP" that are intriguing and may offer a more robust system for solution development.

Jason Daida outlines a diffusion model for the progress of GP in Chapter 7. In particular he presents data that shows that in the "standard GP" system that does not use architecture altering operators, there is a highly constrained area where most search will take place, leaving an area several orders of magnitude larger that will almost never be reached. His suggestions for escaping this structural constraint include the importance of Automatically Defined Functions (ADFs) and other structure altering operators and large populations that give the greatest scope for discovery. His call for a "meta-GP" echoes Freeland's comments on the importance of genotype-phenotype mapping as such a system can alter the structure during execution due to the "environment" it runs in. One important implication of this work is that it shows that while the Schema Theorem may be proven for GP, it is not the entire story in describing the dynamics of effective GP systems.

In Chapter 8 Justinian Rosca arrives at a similar conclusion regarding diffusion and GP dynamics by constructing a simple model that ties code growth, and in particular the appearance of ineffective code sequences ("bloat") to the tendency of trees to grow by diffusion. Surprisingly he's able to use this to show that code growth is *not* exponential in most cases but rather linear.

Of all the participants in the GP Workshop, David Goldberg argued most passionately that GP and other evolutionary algorithms, while drawing heavily from biological models, should still be viewed as an engineering discipline. Drawing on his experience in Genetic Algorithms, Sastry, O'Reilly, Goldberg and Hill in Chapter 9 identify the ability to create an adequate supply of building blocks in a population as a necessary condition for GP success. This chapter gives a preliminary formula for estimating building block supply for a given problem. In Chapter 13 Sastry and Goldberg proceed on to outline a methodology for "competent genetic programming," again arguing for an engineering sensibility to the process of developing successful GP systems. In particular they argue that the engineering principle of design decomposition must be brought to GP. The early identification of building blocks is key in their strategy followed by a focused refinement and combination of these blocks in a coherent solution. Once the building blocks have been identified, GP can be used more efficiently to find good solutions.

In Chapter 11 Bill Langdon describes an intriguing application of GP to the realm of reversible computing. While Langdon shows a certain skepticism of the entire area of reversible computing, he demonstrates that GP can be an effective method for the design of reversible computing systems - a potentially significant factor in the future development of both reversible computing systems and quantum computers. This may suggest that GP will be an important tool for future computing systems which, while theoretically powerful, are difficult to design using normal techniques.

In Chapter 12 Conor Ryan and Miguel Nicolau reverse the direction of the flow of history by taking the lessons of GP and applying them to Genetic Algorithms. They outline a system of grammar construction using GP-like trees that are mapped into GAs, thus sidestepping the issues of position that has such a strong influence over the success of a GA. In this they exploit the already known tendency of GP to arrive quickly at preliminary structural components and to begin the diffusion process of adding the "details" at lower levels.

A great deal of ink has been spilled over the years describing the selection of operators in GP. In Chapter 16 Terence Soule uses a linear representation to avoid issues related to tree growth in order to explore the impact of different operators on code bloat. Soule concludes that operators with global effects create pressure for code growth in order to surround critical sections with ineffective code (so-called "introns"). These introns prevent crossover from destroying effective code segments. Meanwhile local-effect operators create code that is more likely to remain effective even if they are broken up during crossover so bloat is less likely to occur.

3. Survey of Applications

From simple beginnings, GP has begun to enter the leading edge of commercial applications, particularly in areas where there are difficult problems that require novel solutions. Perhaps the flood of data in recent years and the fact that computing power has become a common commodity is contributing to the growth of GP in the applications world. For example, functions of the type generated by GP describing complex data are preferable to statistical or even "black-box" solutions. To generate such functions it is not unusual to have to evaluate several million programs in the course of a single GP run, so the growth of the "Beowulf" style cluster computers provides an accessible engine for large scale evolutionary computing on "real world" problems.

Applications researchers using GP have their own stories to tell. Confronted with unusual problems they have developed their own versions of GP and rules-of-thumb to get the best results. Sometimes these rules fit well into GP theory, sometimes they do not, but their work has always been driven by success, not theory. The applications chapters in this book describe some of the problems encountered by GP practitioners, the ways they solved the problems and the modifications to the GP algorithm they developed along the way.

Lynne Ellyn and Tom Chase of DTE Energy delivered the keynote at the start of the third day. They described the growing need for a computing solution that could integrate large bodies of disparate data. They also highlighted a need common to many corporations: Computing solutions must be accepted by corporate management so they must both be comprehensible and sensible. For this reason "black box" solutions such as those provided by neural nets are not desirable for decision support software and at the same time, solutions must "make sense" for them to be accepted in a corporate environment. This is a theme that was echoed in one way or another by many of the GP practitioners at the Workshop.

In Chapter 3, Driscoll, Worzel and MacLean describe using GP to analyze biological data produced by "gene chips" that map the gene expression patterns associated with cellular behavior. They describe applying GP to produce diagnostic and prognostic classification rules for better diagnostic decisions leading to better treatments for diseases. Much of their work centers on producing robust results from a limited sample set having many variables in the form of gene expression levels for thousands of genes. In wrestling with this problem they developed two hybrid systems that combine GP with separate optimization methods to refine the GP results.

In Chapter 5, David Ostrowski and Robert Reynolds describe an interesting combination of traditional software engineering techniques and genetic programming. By making the comparison of traditional software development with genetic programming, Ostrowski and Reynolds are able to use "white box" and "black box" testing methods to create a GP solution that has both a desirable design structure and good performance characteristics. This approach is being used by the Ford Motor Company to developing pricing strategies that fit within a framework of acceptable behavior. In integrates the cultural behaviors of the marketplace with optimization criteria to get the best return on investment. As such it will influence millions of dollars of revenue generated by one of the world's largest companies.

The range of papers at the Workshop covered everything from the highly theoretical to nuts-and-bolts practical. In Chapter 6, Goodman, Hu and Seo describe a highly practical system with significant theoretical implications. They propose a system of sub-populations where each sub-population is based on fitness. As an individual improves its fitness it "graduates" to the next level. This solves the problem of maintaining diversity while still having a focused search for very fit individuals.

Genetic Programming has in common with other evolutionary algorithms the combination of exploration and exploitation of a search space. Despite the value of exploration, it has repeatedly been shown that encapsulating code greatly improves GP efficiency. In Chapter 10, Daniel Howard takes a new look at exploration, comparing it with lateral thinking where alternatives are explicitly studied in search of new solutions. The system described selects sub-trees for further encapsulation based on the frequency of their appearance in the population and their persistence across generations. This approach is compared with standard Automatically Defined Functions (ADFs) and was shown to have a slight advantage in computational efficiency.

In his first day's keynote speech, John Holland discussed the importance of "existence proofs," that is, a proof that something can occur. In computer science this usually means proving that a system or technique can do something previously considered undoable. John Koza has spent much of the last decade working to prove that GP could produce a novel design of an analog circuit. In 2002 he described GP produced designs that were completely novel and yet outperformed the best known human designs. In Chapter 14, Koza, Streeter and Keane summarize the long road to this success and the important milestones and features of GP needed to reach it.

Dow Chemical has probably made the longest, most sustained use of GP in an industrial setting. From the mid-1990s on, they have used GP for symbolic regression, particularly in the area of modeling chemical processes. Over the years they have integrated GP into a tool kit that includes neural nets and support vector machines (SVMs). In Chapter 15, Kotanchek, Smits and Kordon describe an integrated environment where GP is used in conjunction with a number of other empirical techniques, in part as to improve confidence in a solution by getting confirming results from several tools. Thus, while the transparency of GP solutions is useful, the confidence of having similar results developed using a diverse set of tools is important when developing control systems for $1-billion chemical reactors.

In Chapter 17, Tina Yu, Dave Wilkinson and Deyi Xie describe the use of GP to evaluate the development potential of an oilfield. This is a crucial decision on which hundreds of millions of dollars are staked. The data this decision is based on is composed of a mix of different types of data including seismic reflection data and well and drilling logs. GP's ability to work with highly diverse data and to integrate it into a coherent solution proves to be important. As with Dow's approach it is necessary to combine the GP system with other systems including a fuzzy-modeling and a fuzzy inference system to assess a field's potential yield. The results produced show the flexibility of the GP system used. The system used, *Disciplicus* is one of the few commercially available GP software packages but it shows the growing maturity of GP software.

Pioneers in new technologies must be prepared to prove their techniques many times over before they gain acceptance. State Street Global Advisors (SSGA) have millions of dollars at stake and have traditionally used hand-crafted models in making investment decisions. To keep at the forefront of their industry SSGA has established a policy of reviewing their models every few years and creating new models that were then tested head-to-head with the old models using historical data.

In Chapter 18 Anjun Zhou describes the process of bringing genetic programming into this competition. She encountered stiff resistance to the idea and developed a tiered approach that would allow her to create not only effective models but comprehensible models that would "make sense" to the investment team. In order to demonstrate that a machine-derived solution could be effective, she not only had to out-perform the old model, but because of the resistance to GP, the GP-derived model had to beat out the two new hand-crafted models. Not only did she win the competition but the GP solution outperformed its nearest competitor by 150 basis points - a huge margin of success. Would-be investors, be aware that Zhou presents one of the "also-ran" solutions in this chapter as the selected model is proprietary and is being used by State Street Advisors to manage over $100 million of emerging market investments!

4. The Relationship Between Theory and Practice

If GP is to move beyond an art, or even a craft, then there must be sound theory to inform practitioners. If theory is to be reliable and useful, it must be supported by the experience of practitioners. The goal of this Workshop was to explore how theory should inform practice and how practice supports (or contradicts) theory, and it will be interesting to see how this Workshop influences the development of GP.

Many interesting results were discussed at the Workshop and are described in this book. Driscoll et al, discovered GP behaviors that can be traced to overfitting, suggesting that the best results in difficult problems are reached earliest with comparatively simple programs. This suggests that the Building-Block Hypothesis (BBH) may be correct as the main structure of a program is reached early in the evolutionary process and the fine details arrived at incrementally over many subsequent generations. On the other hand, Goodman and Hu developed a new system of sub-populations (or demes) that, while based on the BBH, actually suggests that the structural limits in GP described by Daida are correct. Soule and Goldberg made specific recommendations on operator selection and population sizing. These guidelines will no doubt be put to the test as practitioners continue to push the envelope of what GP can do and to map out new terrain in commercial applications of GP.

5. Summary

The goal of the GP Theory and Practice Workshop was to bring theorists and practitioners together to expand the dialogue between them. The chapters in this book represent some of those interactions, in that each chapter was reviewed prior to the Workshop by two other attendees, one theory and one applications researcher. The results were tentative but intriguing with lively discussions, food for thought and a great deal of bridge building. Perhaps most interesting was the dialogue between Freeland and most of the other participants. While most GP people have contacts among theoretical biologists, there have never been any formal ties between the two groups. The lively and almost constant exchange between Freeland and the other participants suggests that both disciplines have something to gain from collaboration and it is likely that there will be papers exploring this connection in the future. It is one of the odd things about evolutionary computing systems that, biologically inspired, the practice of the art often resembles field scientists observing the behavior of a living creature: keeping "field notebooks" and watching the response to different "stimuli" in the form of changed parameters. Perhaps the gulf between theory and practice is also borrowed from biology and the tension between them is natural. If so then hopefully this tension will drive as much creativity as it has in the sciences.

Chapter 2

AN ESSAY CONCERNING HUMAN UNDERSTANDING OF GENETIC PROGRAMMING

Lee Spector
Cognitive Science
Hampshire College
Amherst, MA 01002 USA
lspector@hampshire.edu

Abstract This chapter presents a personal perspective on the relation between theory and practice in genetic programming. It posits that genetic programming practice (including both applications and technique enhancements) is moving toward biology and that it should continue to do so. It suggests as a consequence that future-oriented genetic programming theory (*mathematical* theory, developed to help analyze, understand, and predict system behavior) should also borrow, increasingly, from biology. It presents specific challenges for theory *vis-à-vis* recent technique enhancements, and briefly discusses possibilities for new forms of theory that will be relevant to the leading edge of genetic programming practice.

Keywords: biology, development, representation, diversification, phylogeography, visualization.

1. Theory and Practice

Researchers in the field of genetic and evolutionary computation borrow features of biological systems to produce adaptive problem-solving technologies. For the earliest such technologies, only the most general, skeletal features of biological genetics (as then understood) were appropriated. For example, many early systems represented candidate solutions as linearly structured, positionally encoded genomes, borrowing loosely from the structure of DNA discovered by Watson and Crick in 1953. In addition, these systems used algorithms for mutation and recombination based loosely on DNA mechanics, and schemes for population-level adaptation based on an abstraction of Darwinian natural selection.

What has the role of *theory* been in this endeavor historically? By "theory" here I mean *mathematical* theory, developed to help analyze, understand, and perhaps predict the behavior of genetic and evolutionary computation systems in quantitative terms. Under this definition theory has been essential in the field from its earliest days; for example, Holland's seminal monograph on genetic algorithms, first published in 1975, was largely theoretical, and introduced the formal notion of "schemata" which continues to drive much of the theory in the field today (Holland, 1992). In much of the early work theory and practice occurred together and directly motivated one another.

Genetic programming, by which I mean the specialization of "genetic and evolutionary computation" to systems that evolve executable computer programs, arose later, in the late 1980's and early 1990's. In the succeeding years, genetic programming theory and practice (in which I include both applications-oriented work and development of new techniques) have both undergone steady progress, although not always in a coordinated ways.

With the basic theory already established, system builders and applications engineers felt free to experiment with variations ("hack"), without paying much attention to mathematical theory. For the theorists, on the other hand, there was much work to be done in clarifying and developing the mathematics, irrespective of the activities of the hackers. As a result, there has been a divergence. Understandably, theory has tended to lag behind practice; it is easier to hack systems than it is to integrate the resulting hacks into a rigorous mathematical framework for the purposes of analysis and prediction.

For example, recent advances in genetic programming schema theory have broadened its applicability to include variable-length and tree-structured genomes, along with popular mutation and crossover algorithms (see, for example, Poli, 2001). This was a substantial and successful mathematical undertaking, but for all of its successes it captures a state of the art that is already about 10 years old. During the intervening decade practice has flowered, producing much more complex and capable systems for which there is not yet a strong theoretical framework.

One example of the gap between theory and practice is *automatically defined functions*, a technique extension developed in 1994 that has since gained widespread acceptance, and is believed by many to be essential for the application of genetic programming to difficult, large-scale problems (Koza, 1994). There is not yet a mathematical theory of genetic programming that provides any real leverage in understanding how and why automatically defined functions are helpful. Meanwhile, practice has moved beyond "simple" automatically defined functions, to systems with *architecture-altering operations* (Koza, 1995) or other mechanisms that allow the number of automatically defined functions and other aspects of modular architecture to evolve as a system runs (Spector and Robinson, 2002).

Should theoreticians therefore turn their attention to extending schema theory to handle automatically defined functions and architecture-altering operations? I do not think that is likely to be the most productive move. Automatically defined functions and architecture-altering operations are only the tip of the iceberg. Among the other innovations that are playing an increasingly important role in genetic programming practice are developmental processes (at genetic, morphological, and behavioral levels), evolved representations, structured "deme" geography, and coevolution. The prospect of extending current theory to cover these complications seems daunting indeed, and of course the innovations in practice continue to accumulate at a rapid clip.

None of this should be taken to dispute either the utility of early, "simple" genetic programming systems for certain classes of problems, or the utility of the newly enhanced theoretical apparatus for understanding and improving these systems. But practice is moving ahead rapidly, and this may be an opportune time to take stock of the situation and to contemplate the possibilities for new forms of theory that will be relevant to the leading edge of practice.

2. To Life!

Can we characterize the specific directions in which genetic programming practice is moving? This would be a tall order, as genetic programming is a field of rapid and sprawling innovation. The on-line genetic programming bibliography[1] contains 3,068 entries as of this writing, the great majority of which refer to publications within the last ten years. A high percentage of publications in the field describe innovative enhancements to the basic technique; several collections have a roughly tripartite division into theory, technique enhancements, and applications. Even if only something like 10% of the publications in the last ten years describe innovations in genetic programming practice, and I believe this is a conservative estimate, there have nonetheless been several hundred extensions to the technique since the field first blossomed in the early 1990's. Clearly, a systematic survey and analysis of the "direction" in which all of these innovations are pushing the field would be a major undertaking (and beyond the scope of this chapter).

Nonetheless, one can make some rough generalizations. Innovations in technique are usually made by researchers who find that their systems hit some sort of limit before they can solve the problems put to them. When this happens the researchers often turn to other fields for inspiration.

[1] http://liinwww.ira.uka.de/bibliography/Ai/genetic.programming.html

If the perceived limit is one of raw computational resources (memory or computing time) then researchers often turn to other areas of computer science for solutions, for example by optimizing representations or procedures (as in Keijzer, 1996), modifying the techniques to take better advantage of available hardware and software architectures (Spector and Stoffel, 1996; Nordin, Banzhaf, and Francone, 1999; Poli and Langdon, 1999), or adding features derived from software engineering (e.g. modules, as in Koza, 1994; Spector, 1996).

If, on the other hand, the perceived limit is in the adaptive capacity of the technology, then researchers often turn to the science that studies the best known examples of adaptive systems: biology. For example, recent advances in genetic programming technique use mechanisms derived from DNA dynamics (Koza, 1995; Hansen, 2003), learning mechanisms in neural networks (Teller, 1999; Downing, 2001), immune systems (Nikolaev, Iba, and Slavov, 1999), regulatory networks (Banzhaf, 2003), and biological gene expression processes (Keller and Banzhaf, 2001; Ferreira, 2001; Wu and Garibay, 2002).

As a result, a significant portion of the leading edge of the field continues to head toward biology. Even in work that appears quite distant from biology on the surface, one often finds arguments that aspects of the proposed techniques were motivated by some hither-to under-appreciated feature of biological systems. Insofar as the adaptive power of biological systems still vastly outstrips that of any currently existing human-engineered computational system this is quite reasonable, and it is a trend that we should expect to continue.

3. Life Evolving

Biology, however, is a moving target. Conceptions of the fundamental mechanisms of biological evolution have changed repeatedly and dramatically since the 1960s and 1970s, when the core concepts of genetic and evolutionary computation were established. One often-cited example concerns the role of symbiosis and other "mutualisms"; originally thought to have been curious exceptions to ordinary biological processes, these phenomena are now widely thought to have played central roles in the development of multicellular life (Maynard Smith and Szathmáry, 1999), with some researchers arguing that they are major sources of evolutionary novelty on par with natural selection (Margulis, 2000).

Indeed, the very notion of a "gene," which has suffered dramatic semantic shifts in its 94-year history[2], now appears to be deeply problematic. Evelyn Fox Keller, after describing several recently discovered complications to the traditional picture of gene expression and gene-protein interaction, writes:

[2]The term "genetics" was introduced by William Bateson in 1906, and the term "gene" was introduced three years later by Wilhelm Johannsen (Keller, 2000).

> Techniques and data from sequence analysis have led to the identification not only of split genes but also of repeated genes, overlapping genes, cryptic DNA, antisense transcription, nested genes, and multiple promotors (allowing transcription to be initiated at alternative sites according to variable criteria). All of these variations immeasurably confound the task of defining the gene as a structural unit. (Keller, 2000, p. 67)

So, if we are to continue to move in biological directions, we might expect to see the notion of "gene" that is implicit in our genetic representations to undergo significant alterations. Indeed, we might also expect the boundaries between the "genetic" representations and the other components of our systems to continue to blur. For example, Keller argues that the biological mechanisms that generate variation are themselves under genetic control, and that the evolutionary processes are thereby themselves products of evolution:

> The critical dependence of genetic stability on proofreading and repair enzymes may have come as a great surprise, but more surprising yet was the discovery of "repair" mechanisms that sacrifice fidelity in order to ensure the continuation of the replication process itself—and hence the survival of the cell. Far from reducing error, such mechanisms actively generate variations in nucleotide sequence; moreover, it appears that when and where they come into play is itself under genetic control. (Keller, 2000, p. 32)
>
> That regulation of genetic stability and mutability is a feature of all living systems in now widely accepted. (Keller, 2000, p. 35)
>
> Recent laboratory studies of bacterial evolution provide further confirmation, lending support to the notion that organisms have evolved mechanisms for their own "evolvability." (Keller, 2000, p. 37)

Mechanisms related to these ideas have been used in genetic and evolutionary computation for some time (for example, in *evolution strategies* (Bäck and Schwefel, 1995), and in discussions of the evolution of evolvability (Altenberg, 1994)). The lesson to draw from Keller's observations, and from the expected further "biologicalization" of genetic programming hypothesized in this chapter, is that these sorts of mechanisms are likely to play an increasingly important role in the field. To be relevant to future genetic programming practice, therefore, genetic programming theory should be applicable to, and informative with respect to, systems with these sorts of extensions.

4. Challenges

In this section I would like to highlight a few specific challenges for genetic programming theory based on recent practice. These are not meant to be exhaustive, and they are drawn mostly from my own work (for no better reason than the fact that these are the challenges with which I am most familiar). Nonetheless I believe that they can provide useful pointers to opportunities for future-oriented theory.

Evolved Code Structure

As noted above, genetic programming systems now routinely include mechanisms for the evolution of *automatically defined functions* (Koza, 1994) or *automatically defined macros* (Spector, 1996). Other mechanisms for evolved modularity have also been developed (for example, Kinnear, 1994). Because several studies have shown that these features can significantly extend the reach of genetic programming, useful theory ought to provide firm mathematical leverage with respect to questions about how and when automatic modularization can enhance the adaptive powers of genetic programming systems.

The essential difficulty here is that the semantic structure of a modular program can be related to its syntactic structure in complex ways. For example a smaller program may execute more primitive operations than a larger program, even without considering loops, if the smaller program makes more significant use of modularity. Changes to a program will likely have effects of a magnitude that is correlated not only with the "depth" at which the changes occur, but also with the relation between the changes and the program's modular architecture; for example a change within module will probably have an impact that is proportional to the number of times that the module is called.

The challenge is more severe, but also more necessary, for systems in which the "architecture" of evolving programs is itself subject to variation and natural selection. By "architecture" I mean the number of modules and the nature of their interfaces to each other and to the "main program" (or the "result-producing branch" in Koza's terminology). Koza's system of architecture-altering operations employs new genetic operators (in addition to standard crossover and mutation) to permit the addition and deletion of modules, along with other architectural changes (Koza, 1995). These new genetic operators were explicitly derived from biological gene duplication processes. In my own PushGP system, architecture can evolve without such mechanisms; modules are created and executed by means of code-manipulation instructions that are always available, and the action of ordinary genetic operators (like crossover and mutation) can, by changing sequences of such instructions, change the architecture of the overall program (Spector and Robinson, 2002).

Evolved Developmental Processes

"Development" occurs at several levels in biology. As mentioned above, genetic mechanisms (in which we should include not only DNA but also RNA, enzymes, and proteins involved in gene expression, at least) themselves undergo complex forms of environmentally-mediated development in the normal course of cell activity. In addition, complex developmental processes at a larger (morphological) level of aggregation mediate the long transition from embryo to adult in multicellular organisms. Finally, behavioral/psychological developmental processes are often critical in producing adults capable of completing the life cycle of complex life forms.

Many of these developmental processes have already been incorporated, at some level of abstraction, into genetic programming systems. I have already mentioned several of the systems that use some form of "gene expression" development. In my own PushGP system programs are routinely self-manipulating, and the sequence of primitive instructions that is eventually executed can be very difficult to predict from the surface structure of the genome (program) prior to "development."

Morphological development has also appeared in various forms, usually when the execution of an evolved program builds a secondary structure which is then responsible for the desired problem-solving behavior. This is analogous to the construction, by a biological genome, of a body which is then responsible for behavior in the world (although this is really a flawed analogy, as indicated by the quotes from Keller above). Examples include Koza's work on the evolution of programs that build electrical circuits (Koza *et al.*, 1999) and my own work on the evolution of programs that build quantum gate arrays (Barnum, Bernstein, and Spector, 2000).

Behavioral or psychological development has been incorporated in many systems that evolve structures that subsequently undergo "learning" processes as they interact with their problem environments. The most salient examples of this approach are systems that build neural networks that are then trained on problems from their target domains using standard neural network learning algorithms; note that many of these systems incorporate both morphological and behavioral development.

Evolved Diversification

As indicated in the quotes from Keller above, in biology the mechanisms of diversification are themselves under genetic control and it appears clear that the mechanisms of evolution have thereby themselves evolved.

Within genetic and evolutionary computation there is a long tradition of systems in which mutation rates are genetically encoded. Within genetic programming more specifically there have been several explorations of "Meta-GP," in which not only the rates but also the algorithms for diversification are genetically encoded and can therefore evolve (Schmidhuber, 1987; Edmonds, 2001). In these systems co-evolving populations of program-manipulation programs are used in lieu of traditional genetic operators to produce the offspring of the individuals in the primary (problem-solving) population.

In my own recent work on "autoconstructive evolution" with the Push programming language, programs construct their own offspring using code-manipulation instructions that are available in the instruction set along with problem-related primitives. In these systems the "genetic operators" are just the parts of individuals (possibly interleaved with other parts of individuals) that produce offspring — when a program is run it may do something that confers fitness (for example, providing correct answers to symbolic regression fitness cases, or navigating an agent toward food, etc.) *and* it may also produce a child. The code for the child may be produced in any computable way by the parent, possibly using the code of the parent, the code of other individuals, randomly generated code, etc. The Pushpop system grafted these concepts onto a fairly traditional genetic programming algorithm (Spector and Robinson, 2002; Spector, 2002). In more recent work my colleagues and I have built autoconstructive evolution systems embedded in 3D virtual worlds (in Jon Klein's BREVE system (Klein, 2002), the latest development versions of which include a Push language interpreter; see (Spector itet al., 2003)).

Geography

Geography plays a critical role in evolutionary biological theory. Most theories of speciation rely on notions of geographic isolation, location-based ecological niches, and clinal variation of species characteristics across gradients of temperature and other environmental features. Indeed, the interaction between geography and evolutionary processes has been an area of exploding interest in the last decade, referred to by some as "phylogeography" (Avise, 2000).

These concepts had little direct influence on early work in genetic and evolutionary computation, although the concept of "demes" (local breeding populations, between which migration occasionally occurs) provides a coarse-grain analog that has had considerable impact (see e.g. Fernández *et al.*, 2003). More recently, however, concepts from artificial life models (in which spatial distribution is often modeled explicitly) have influenced work in genetic programming. One example is work by Punch and Rand combining genetic programming with elements from Holland's Echo architecture and Brooks's subsumption architecture (Punch and Rand, 2000). Another example is my work on evolution within spatially continuous 3D virtual worlds (Spector itet al., 2003). In such environments, many other biological concepts that had previously played little or no role in genetic and evolutionary computation also come into play. For example, in many ALife-derived systems different individuals have different life expectancies, and their reproductive behaviors may change based on age and environmental circumstances.

One might expect, based on the importance that concepts of spatial distribution and geography have had in biological theory, that the influence of these concepts on genetic programming practice will continue to increase.

5. To Theory!

What kinds of mathematical theory would help us to analyze, understand, and predict the behavior of genetic programming systems that exhibit the challenges outlined above? There is a sense in which current theory already "handles" all of these challenges; for example even with all of the complications raised above, a genetic programming system can still be viewed as a markov chain (Poli, Rowe, and McPhee, 2001) and related theoretical apparatus can still be applied. But this is an unsatisfactory answer for at least two reasons. First, while it is conceivable that existing theoretical frameworks could be extended to meet these challenges this would require considerable effort and time (Poli, personal communication). Second, I suspect that the resulting theories, if generalized to handle such radical extensions to the basic technique, will be so general that they will cease to have explanatory power.

Are there alternatives? I believe that there are. The obvious place to look for mathematical theory that might be applicable to a field heading toward biology is biology itself. While relations between evolutionary biology and mathematics have not always been harmonious (see for example Keller, 2002), there have nonetheless emerged many mathematical theories of aspects of biological evolution. These have often provided valuable insights into the nature and dynamics of evolving populations even when they ignored seemingly important aspects of biological systems (such as development, which was ignored in early population biology) and even when they were based on naive models of the molecular foundations of genetics (such as the assumption that individual traits are determined by individual genes).

Biologists have invented mathematical or quasi-mathematical theories of diversification, development, and the relations between genetics and geography. The current explosion of work in genomics is producing a wealth of mathematical theory concerning DNA "code structure" and gene expression (see for example Karp, 2002). There seems, therefore, to be a great deal of theory available for transplantation into genetic programming, suitable for addressing the specific challenges outlined above.

What differentiates mathematical theory in biology from the existing mathematical theory in genetic programming? I believe that the primary difference is that biologists (even mathematical biologists) begin with large scale phenomena (populations of complex organisms) and work downward to the mechanisms out of which these phenomena are built, while genetic programming theorists begin with a set of low level mechanisms and work upward to the large scale phenomena exhibited by our systems.

In genetic programming we have the luxury of *knowing* the exact nature of the mechanisms since we construct them, while in biology many of the mechanisms are still unknown. But this luxury is fleeting and perhaps illusive as well. As discussed above, the mechanisms used in genetic programming are now in considerable flux. In addition, while we may have complete knowledge of the mechanisms at the lowest level of abstraction this may not be the level at which the most fruitful theories can be built. For example, if a system incorporates developmental processes and adaptive representations then it is possible that mechanisms at a somewhat higher level, for example "body plans" and "regulator genes" will be essential to understanding its behavior. It is not obvious that such mechanisms will emerge from bottom-up theories based on, for example, a schema analysis of programs containing code-manipulation instructions.

Biology, for its part, has always had the luxury of truly interesting large scale phenomena to study, and biologists have built mathematical frameworks capable of providing useful generalizations of complex adaptive systems even in the face of ignorance about the ways in which those systems are implemented. This is, by itself, a good reason for genetic programming theorists to look to biological theory for inspiration. An additional reason is the fact that many of the new features being added to genetic programming systems originate in biology.

One example of the theoretical move advocated here, already underway in the community, is the empirical study of diversity and diversification in genetic programming systems. The study of diversity has deep roots in biology, which ought to provide context for these studies. In addition, many of the relevant questions can be asked in a way that transcends implementation details. Measures of diversity have been adopted by several researchers (a nice survey is available in Burke, Gustafson, and Kendall, 2002). In addition, researchers have begun to explore relations between diversification and adaptation; one of my studies, for example, showed that adaptive populations of endogenously diversifying Pushpop programs are reliably diverse (Spector, 2002). But this is only a start.

6. Prospects

It is possible that biological tools will not actually help us to understand our genetic programming systems. One possibility is that mathematical models that are sufficiently rich to capture the most important features of complex adaptive evolutionary systems, whether organic or technological, will be too complex or abstract to help *humans* understand how and why the systems behave as they do. This is a depressing thought, but I have little more to say about it one way or the other.

Another possibility is that the "right" theory for genetic programming (however extended) will bear little or no similarity to current theories in biology. One reader of a draft of this essay agreed with the critique of current genetic programming theory but felt that biology had unfortunately little to offer; for example, she found in her own work that most of the mathematical apparatus of population genetics, while useful in analyzing simple systems, failed to scale up to the complexity found in applied genetic programming work. As noted earlier, however, biology is in the midst of major transitions and there are bound to be substantial theoretical innovations in the coming years. My contention is simply that wherever biology goes, genetic programming should follow.

Yet another possibility is that the right theory for genetic programming will bear no resemblance to current *or future* theories in biology. After all, the critics will say, genetic programming is a problem solving technology, while biological evolution is not; the similarities are superficial and should remain so. If so, then so much the worse for genetic programming theory; it will have to borrow its fundamentals from other fields (perhaps thermodynamics; Adami, 1997) or invent them from whole cloth.

In any event, it is probably worth noting that if the goal is *human understanding* of genetic programming then there may also be alternatives to mathematics. One alternative is understanding via visualization; often the human visual system can discover high-level, emergent properties for which we have, as yet, no formal theory. Genetic programming researchers have always used the traditional forms of scientific visualization (fitness graphs, statistical plots, etc.), but if the conjecture of this chapter is correct, and genetic programming is indeed moving toward biology, then other forms of visualization suggest themselves. In particular, insofar as future systems, like biological systems, exist and function in 3D (virtual) spaces, direct observation of 3D geometry as it evolves over time — a sort of virtual microscope for virtual ecosystems — may provide essential insights. This approach has already proven valuable in the field of artificial life (e.g. Spector and Klein, 2002) and it might be expected to have more relevance to genetic programming the closer genetic programming moves to biology.

Acknowledgments

Apologies to John Locke for the title. Riccardo Poli provided inspirational comments, though he bears no responsibility for the uses to which the inspiration has been put. Bill Langdon, Una-May O'Reilly, and Anjun Zhou provided thoughtful reviews that raised more good questions and issues than I could address here. This effort was supported by the Defense Advanced Research Projects Agency (DARPA) and Air Force Research Laboratory, Air Force Materiel Command, USAF, under agreement number F30502-00-2-0611, and by NSF grant EIA-0216344. Support was also provided by Hampshire College for the Institute for Computational Intelligence.

References

Adami, C. (1998). *An Introduction to Artificial Life*. Berlin: Springer Verlag.

Altenberg, L. (1994). "The Evolution of Evolvability in Genetic Programming." In *Advances in Genetic Programming*, Kinnear, K.E. Jr. (Ed.), pp. 47-74. The MIT Press.

Avise, J.C. (2000). *Phylogeography: The History and Formation of Species*. Harvard University Press.

Bäck, T. and Schwefel, H.-P. (1995). "Evolution Strategies I: Variants and their computational implementation." In *Genetic Algorithms in Engineering and Computer Science*, P. Cuest, *et al.* (Eds.). John Wiley & Sons Ltd.

Banzhaf, W. (2003). Artificial Regulatory Networks and Genetic Programming. In *Genetic Programming, Theory and Practice*, Rick L. Riolo and Bill Worzel (Eds.). Kluwer.

Barnum, H., Bernstein, H.J. and Spector, L. (2000). Quantum circuits for OR and AND of ORs. *Journal of Physics A: Mathematical and General, 33(45):* 8047–8057.

Burke, E., Gustafson, S. and Kendall, G. (2002). A Survey And Analysis Of Diversity Measures in Genetic Programming. In *Proceedings of the Genetic and Evolutionary Computation Conference (GECCO 2002)*, W. B. Langdon, *et al.* (Eds.), pp. 716–723. San Francisco, CA: Morgan Kaufmann.

Downing, K.L. (2001). Reinforced Genetic Programming. *Genetic Programming and Evolvable Machines 2(3):* 259–288.

Edmonds, B. (2001). Meta-Genetic Programming: Co-evolving the Operators of Variation. *Elektrik, the Turkish Journal of Electrical Engineering and Computer Sciences 9(1):* 13–29.

Fernández, F., Tomassini, M. and Vanneschi, L. (2003). An Empirical Study of Multipopulation Genetic Programming. *Genetic Programming and Evolvable Machines 4(1):* 21–51.

Ferreira, C. (2001). Gene Expression Programming: A New Adaptive Algorithm for Solving Problems. *Complex Systems 13(2).*

Hansen, J.V. (2003). Genetic Programming Experiments with Standard and Homologous Crossover Methods. *Genetic Programming and Evolvable Machines 4(1):* 53–66.

Holland, J.H. (1992). *Adaptation in Natural and Artificial Systems,* second edition. The MIT Press, Cambridge, MA.

Karp, R.M. (2002). Mathematical Challenges from Genomics and Molecular Biology. *Notices of the AMS 49(5):* 544–553.

Keijzer, M. (1996). Efficiently Representing Populations in Genetic Programming. In *Advances in Genetic Programming 2*, Angeline, J., and Kinnear, K.E. Jr. (Eds.), pp. 259–278. The MIT Press.

Keller, E.F. (2000). *The Century of the Gene*. Harvard University Press.

Keller, E.F. (2002). *Making Sense of Life: Explaining Biological Development with Models, Metaphors, and Machines.* Harvard University Press.

Keller, R.E. and Banzhaf, W. (2001). Evolution of Genetic Code on a Hard Problem. In *Proceedings of the Genetic and Evolutionary Computation Conference (GECCO 2001)*, pp. 50–56. Morgan Kaufmann.

Kinnear, K.E. Jr. (1994). Alternatives in Automatic Function Definition: A Comparison of Performance. In *Advances in Genetic Programming*, Kinnear, K.E. Jr. (Ed.), pp. 119–141. The MIT Press.

Klein, J. (2002). BREVE: a 3D Environment for the Simulation of Decentralized Systems and Artificial Life. In *Proceedings of Artificial Life VIII, The 8th International Conference on the Simulation and Synthesis of Living Systems*, pp. 329–334. The MIT Press.

Koza, John R. (1994). *Genetic Programming II: Automatic Discovery of Reusable Programs.* The MIT Press, Cambridge, MA, USA.

Koza, J. R. (1995). Gene Duplication to Enable Genetic Programming to Concurrently Evolve Both the Architecture and Work-Performing Steps of a Computer Program. In *IJCAI-95 Proceedings of the Fourteenth International Joint Conference on Artificial Intelligence*, pp. 734–740. San Francisco, CA: Morgan Kaufmann.

Koza, J.R., Andre, D., Bennett, F.H. III and Keane, M. (1999). *Genetic Programming 3: Darwinian Invention and Problem Solving*. Morgan Kaufman, San Francisco, CA, USA.

Margulis, L. (2000). *Symbiotic Planet*. Basic Books.

Maynard Smith, J., and Szathmáry, E. (1999). *The Origins of Life*. Oxford University Press.

Nikolaev, N.I., Iba, H. and Slavov, V. (1999). Inductive Genetic Programming with Immune Network Dynamics. In *Advances in Genetic Programming 3*, L. Spector *et al.* (Eds), pp. 355–376. The MIT Press.

Nordin, P., Banzhaf, W. and Francone, F.D. (1999). Efficient Evolution of Machine Code for CISC Architectures using Instruction Blocks and Homologous Crossover. In *Advances in Genetic Programming 3*, L. Spector *et al.*. (Eds.), pp. 275–299. The MIT Press, Cambridge, MA, USA.

Poli, R. (2001). General Schema Theory for Genetic Programming with Subtree-Swapping Crossover. In *Genetic Programming, Proceedings of EuroGP 2001*, J.F. Miller *et al.* (Eds.), pp. 143–159. Springer Verlag.

Poli, R., and Langdon, W.B. (1999). Sub-machine-code Genetic Programming. In *Advances in Genetic Programming 3*, Spector, L., *et al.* (Eds.), pp. 301–323. The MIT Press.

Poli, R., Rowe, J.E. and McPhee, N.F. (2001). Markov Chain Models for GP and Variable-length GAs with Homologous Crossover. In *Proceedings of the Genetic and Evolutionary Computation Conference (GECCO 2001)*, pp. 112–119. San Francisco, CA: Morgan Kaufmann.

Punch, W.F. and Rand, W.M. (2000). GP+Echo+Subsumption = Improved Problem Solving. In *Proceedings of the Genetic and Evolutionary Computation Conference (GECCO 2000)*, pp. 411–418. San Francisco, CA: Morgan Kaufmann.

Schmidhuber, J. (1987). Evolutionary principles in self-referential learning. Diploma thesis, Institut für Informatik, Technische Universität München.

Spector, L. and Robinson, A. (2002). Genetic Programming and Autoconstructive Evolution with the Push Programming Language. *Genetic Programming and Evolvable Machines 3(1)*: 7–40.

Spector, L. and Klein, J. (2002). Evolutionary Dynamics Discovered via Visualization in the BREVE Simulation Environment. In *Workshop Proceedings of the 8th International Conference on the Simulation and Synthesis of Living Systems*, pp. 163–170. Sydney, Australia: University of New South Wales.

Spector, L., Klein, J., Perry, C., and Feinstein, M. (2003). Emergence of Collective Behavior in Evolving Populations of Flying Agents. In *Proceedings of the Genetic and Evolutionary Computation Conference (GECCO-2003)*, Cantu-Paz, E., *et al.* (Eds.), pp. 61-73. Springer-Verlag.

Spector, L. and Stoffel, K. (1996). Ontogenetic programming. In *Genetic Programming 1996: Proceedings of the First Annual Conference* (Cambridge, MA, 28–31 July 1996), J. R. Koza (Eds.), pp. 394–399 The MIT Press.

Teller, A. (1999). The Internal Reinforcement of Evolving Algorithms. In *Advances in Genetic Programming 3*, L. Spector *et al.* (Eds.), pp.325–354. The MIT Press.

Wu, A.S. and Garibay, I. (2002). The Proportional Genetic Algorithm: Gene Expression in a Genetic Algorithm. *Genetic Programming and Evolvable Machines 3(2)*: 157–192.

Chapter 3

CLASSIFICATION OF GENE EXPRESSION DATA WITH GENETIC PROGRAMMING

Joseph A. Driscoll[1], Bill Worzel[2] and Duncan MacLean[2]
[1]*Middle Tennessee State University, Murfreesboro, TN 37132*
[2]*Genetics Squared, Inc., Milan, MI 48160*

Abstract: This paper summarizes the use of a genetic programming (GP) system to develop classification rules for gene expression data that hold promise for the development of new molecular diagnostics. This work focuses on discovering simple, accurate rules that diagnose diseases based on changes of gene expression profiles within a diseased cell. GP is shown to be a useful technique for discovering classification rules in a supervised learning mode where the biological genotype is paired with a biological phenotype such as a disease state. In the process of developing these rules, it is necessary to devise new techniques for establishing fitness and interpreting the results of evolutionary runs because of the large number of independent variables and the comparatively small number of samples. These techniques are described and issues of overfitting caused by small sample sizes and the behavior of the GP system when variables are missing from the samples are discussed.

Key words: Genetic programming; Classification; Molecular diagnostics

1. Introduction

This paper summarizes the use of a genetic programming (GP) system to develop classification rules for gene expression data that hold promise for the development of new molecular diagnostics. Molecular diagnostics is the name given to the method of diagnosing diseases using the presence of certain DNA sequences or changes to the pattern of genes expressed within an entity. Specifically, this work focuses on discovering simple, accurate rules that diagnose diseases based on changes of gene expression profiles within a diseased cell. GP is shown to be a general system for discovering classification rules in a supervised learning mode where the genotype is paired with a phenotype such as a disease state. Using GP, simple, concise rules were developed which clearly identified samples as belonging to previously identified diseases or disease sub-classes that were established using existing pathology techniques. The GP-derived rules offer the possibility of more reliable diagnostics as well as informing prognostic and treatment decisions. Work is underway to develop commercial diagnostics using this approach.

In the process of developing these rules it was necessary to develop new techniques for establishing fitness and interpreting the results of evolutionary runs because of the large number of independent variables and the comparatively small number of sample classes. Two specific cases are described and the results and methods summarized.

1.1 Gene Chips

The invention of cDNA microarrays and oligonucleotide arrays has opened a new window into understanding cellular behavior. These "gene chips" have allowed researchers to discover a great deal about cellular mechanics by monitoring the changes in gene expression caused by diseases or by therapeutic agents. This technology is built on the fact that for every new protein needed for a biological process, a new copy of the DNA sequence is made from the gene that describes a desired protein. Because DNA is double-stranded, if the DNA for a gene is separated from its complementary strand, it can be "combined" or hybridized with a tagged strand of complementary DNA (cDNA) in order to detect the amount and nature of expressed genes. By measuring the number of tagged strands present in a sample, the expression level of a gene can be measured. A description of this technology may be found in (Gerhold et al., 1999).

Gene chips have been used to detect and analyze diseases by comparing the gene expression profiles of diseased tissue with that of healthy tissue or by comparing the gene expression profiles between different disease types. See for example (Khan et al., 2001) for a description of the use of gene chips to classify different types of cancer tumor tissues that are difficult to diagnose using traditional pathology techniques. Because the genes that are part of a disease signature are not known in advance and because gene chips can have thousands or tens of thousands of genes on them, they are useful for screening samples to find those genes that are associated with a particular disease or disease sub-type. But one sticking point in the use of gene chips to diagnose diseases is that they are expensive and the analysis of the data they produce is time consuming. Cells are noisy and complex with much going on that has nothing to do with the disease being studied and gene chips are sensitive to differences in sample preparation. Various statistical and machine learning techniques have been applied to this problem, but they all produce results of varying accuracy and comprehensibility and the results often require tens or hundreds of genes to be monitored. If a method could be discovered that simplifies the results produced by gene chips and was more understandable to biologists, then other, cheaper and more reliable techniques such as quantitative PCR (Polymerase Chain Reaction) could be used to diagnose diseases simply and accurately.

1.2 Disease Classification

The main use of gene chips today is by researchers who are trying to understand the cellular mechanics of an entity. Gene chips have been used extensively to understand what characterizes a physical expression of a process at the genetic level and, in particular, what characterizes a disease. To do this, tissue samples taken from individuals with a disease are analyzed using gene chips. The resulting gene expression profiles are compared with profiles from healthy samples or from different disease types (e.g., different types of cancer) and conclusions are drawn about the cellular mechanisms particular to a disease.

Currently, the most common way gene expression-based diagnostics are developed is by using hierarchical cluster analysis to discover genes that are consistently expressed at a higher or lower level than in the other samples. The genes with a high variance of expression level, often numbering in the hundreds, are visualized and selected on the basis of association with biological processes thought to be of significance in the disease under study. The selected genes are then studied on a functional basis, and one or more genes are selected as possible therapeutic targets or as biomarkers for the disease.

Recently, machine learning techniques have been applied to the problem of sample classification and, in particular, neural nets and support vector machines have begun to gain currency in bioinformatics circles. The Shipp et al. study described in Section 3.2 used several different approaches including weighted voting among genes identified using cluster analysis, K-Nearest Neighbor analysis and SVM and so provides an interesting summary of existing methods.

2. Methods

The work described here uses genetic programming to discover diagnostic rules that will accurately classify samples into different diseases using any technology that can reliably measure the expression levels of the genes used in the rules. In its typical implementation, a training set of disease samples for which gene expression profiles have been produced using gene chips are analyzed to produce a Boolean expression describing whether a given sample belongs to a given class. The rule is then checked against a test set of samples to evaluate its ability to generalize to unknown samples. A generic example of this kind of rule is shown below:

IF (GENE 162 * GENE 2324 > GENE 636) THEN 'CANCER TYPE A'

Here the training has produced a rule where the expression level of a gene, here titled GENE 162 is multiplied by the expression level of GENE 2324 and the product is compared with the expression level of GENE 636. If the expression level of the product of this multiplication is greater than the expression level of GENE 636, then the rule classifies the sample as belonging to 'CANCER TYPE A', a previously defined type of cancer.

To get these results, the genetic programming system is presented with gene chip data taken from known samples of both target disease and healthy or other disease tissues, and it evolves the predicate IF-clause for the target disease class. These evolved rules are initially developed using a training set of samples with the number of correctly classified samples being the measure of fitness of the candidate rule. The fitness measure used to evolve expressions may be varied so that more weight is given to rules producing fewer "false positive" errors (i.e., ones that identify healthy individuals as being diseased) or by giving more credit to rules that produce fewer "false negative" errors (i.e., ones that identify diseased individuals as healthy) depending on the proportion of samples in a data set. Similarly, the fitness measure may be varied for other reasons depending on the goals of a study. For example, a researcher may want to produce rules that use genes that code for proteins that are known to enter the bloodstream in preference to those that stay within the diseased cell.

In the work described here, generational GP was used with tournament selection. Fitness was calculated by the formula: 0 - (number of FP * FP weight) - (number of FN * FN weight) where FP is the number of False Positive misclassifications (i.e., where a sample is incorrectly classified as belonging to the target class) and FN is the number of False Negative misclassifications (i.e., where a sample is incorrectly classified as not belonging to the target class). A perfect rule will therefore have a fitness of 0, and each misclassification gives the fitness a larger negative value.

The use of FP and FN weighting was created to help compensate for disproportionate sample numbers for each class. There might be only 8-10 samples belonging to a given class out of a possible 80 samples, so a "just say no" strategy of denying that any samples belonged to the target class would create a rule that was accurate 85%-90% of the time. By penalizing the false negative rules disproportionately, the fitness would bias the search toward programs that were accurate at classifying the target class samples. As mentioned above this weighting can also be used to bias the search for other reasons external to the evolutionary process.

Elitism was used in that the two best individuals were always propagated into the next generation so that the most fit individuals in a generation were always at least as good as the best in the previous generation. The tableau in Table 3-3 shows an example of the various GP parameters used in one study.

An example of the value of using genetic programming to discover diagnostic rules may be seen in the use of genetic programming in analyzing the data set described in (Khan et al., 2001). In their work, Khan et al. describe developing a neural network-based system that could completely and accurately classify samples based on their gene expression profiles, but to do this they constructed a very complex mechanism that required 200 genes as inputs. Using genetic programming, the authors were able to evolve 4 rules, one for each cancer type that used a maximum of 3 genes per rule for a total of 10 genes used for all 4 rules.

2.1 n-fold cross validation

Ideally, a large data set would be available and would be partitioned into large training and test sets. However, in many cases, the total amount of data is very small, resulting in small training and test sets since it is more difficult to have a representative set of samples in a small training set, the system may not learn the most general classification concept. Similarly, a small test set does not allow a very thorough assessment of the generality of the learned concept. One method used to cope with small data sets is n-fold cross validation. This is an evaluation procedure, external to the GP system itself.

A data set can be divided into n equal partitions, called folds. One fold is selected as the test set, and the remaining n-1 folds collectively become the training set. These sets are then used in a full GP run. Next, another of the n folds is selected as a new test set, and again the remaining n-1 folds are used as the training set. A new GP run is performed, starting completely from scratch, using these sets. This process is used a total of n times, until each fold has served once as the test set. Performance is expressed via the total number of false positive, false negative, true positive, and true negative classifications encountered during the n runs.

If only a single run is performed (e.g., n=1 above), one may wonder if the results are truly representative. It is possible that the random distribution of samples into training and test sets was particularly "lucky", and so the system's classification performance is not very representative of how the system will perform, in general, on data sets in the domain. To combat this problem, a value of 10 or more is usually used for n. The collective performance of the system over this many runs provides a more realistic assessment of the system's performance in the domain.

2.2 Classification with floating point values

Because of the small number of samples available in some cases, it was necessary to develop another approach that would differentiate between programs of equal fitness. To do this, an approach was created that could provide a numeric measure of how well a program differentiated between samples of the target class and samples from other classes.

Instead of creating Boolean-valued rules, rules returning floating point values may be created. An accompanying threshold value allows the interpretation of the floating point output as a Boolean classification. Assume that each rule produces a floating point output. When a particular rule is applied to the *m* training instances, a vector of *m* floating point values is produced, one per training instance. Further assume that a floating point value called a slice point exists. Classification of the instances into two classes proceeds as follows: if a training instance produced a value greater than or equal to the slice point, then it is classified as being a member of the target class. Otherwise, it is classified as not being a member of the class.

The value of the slice point is different for each rule and is determined during training. To produce a set of candidate slice point values, each rule's floating point output values (one per training instance) are sorted into ascending order. The midpoints between these sorted values are the candidate slice points. Each candidate is tested by using it to classify the training instances. The slice point that results in the best classification (on the training set instances) is chosen and stored with the rule. It is later used when applying the rule to the blind test set.

By comparing the numerical distance between samples that fall on either side of the slice point, one can draw some conclusions about the "definiteness" of the classification. In other words, if a rule clearly separates samples according to class on either side of the slice point and the samples have the good degree of separation from one another, it is supposed that this is a better rule than one where samples from both classes cluster near the slice point.

This proved useful because in some cases a number of "perfect" classifiers was discovered (i.e., they classified all samples correctly) and it was necessary to develop a way to differentiate between the rules.

2.3 Combining multiple rules for a single classification

Because some situations arose where there were sub-classes within the main target class, it became necessary to develop programs that were sets of rules, each helping to identify a sub-class, i.e., a sub-grouping of samples within the target class that have different genotypic features from the other samples in the target class. To do this, a method was developed to evolve groups of rules that were combined and interpreted as a single Boolean classification value.

Assume that each individual in the evolutionary population is actually a set of Boolean rules. For a given sample, each rule has access to the data set inputs, and each produces a Boolean output. As a result, a set of Boolean values is produced, one per rule. The rules' outputs can be combined to produce a single Boolean answer which indicates class membership for the given instance.

Truth tables provide a useful way to understand how the multiple Boolean values can be interpreted as a single value. Assume that there is only a single rule in the individual. If its output is called X, then the four possible interpretations of this value are X, X' (i.e., not X), always 0 (i.e., false), and always 1 (i.e., true). This is summarized in Table 3-1.

Table 3-1. Truth table for an individual with one Boolean-valued rule

X	**X**	**X'**	**Always 1**	**Always 0**
0	**0**	**1**	**1**	**0**
1	**1**	**0**	**1**	**0**

Only one of these four interpretations will be used for the individual, and each individual can use different interpretations. The best interpretation (i.e., the one producing the best classification fitness) to use is determined during training and is stored for later use with the blind test set.

If an individual has two Boolean rules, then there are sixteen possible interpretations of the rule outputs, summarized in Table 3-2.

Table 3-2. Truth table for an individual with two Boolean-valued rules

X	Y	I_0	I_1	I_2	I_3	I_4	I_5	I_6	I_7	I_8	I_9	I_A	I_B	I_C	I_D	I_E	I_F
0	0	0	0	0	0	0	0	0	0	1	1	1	1	1	1	1	1
0	1	0	0	0	0	1	1	1	1	0	0	0	0	1	1	1	1
1	0	0	0	1	1	0	0	1	1	0	0	1	1	0	0	1	1
1	1	0	1	0	1	0	1	0	1	0	1	0	1	0	1	0	1

The outputs from the two rules, X and Y, serve as an index into Table 3-2; the four possible combinations of Boolean outputs correspond to the four rows in the truth table. The sixteen columns, labeled I_0 through I_F, represent possible ways of interpreting the two rules' values in order to produce a single Boolean output value. Training determines the optimal interpretation (i.e., the optimal column) for each rule, where optimality is determined by training set classification performance. As an example, assume that interpretation I_4 is chosen. If the rule outputs are X=0 and Y=1, then the overall Boolean result is 1. This value is found by finding the row in the table corresponding to the output values, and then looking in the I_4 column for the Boolean value.

3. Results

The ability of GP to produce simple, accurate classification rules was demonstrated with several data sets. These experiments, and their results, are described in the following sections.

3.1 Khan et al.

The data set used by Khan et al. contains gene expression measurements, obtained using cDNA microarrays, from four types of cancer cells: neuroblastoma (NB), rhabdomyosarcoma (RMS), the Ewing family of tumors (EWS), and Burkitt lymphomas (BL), a type of non-Hodgkin lymphoma. No non-cancerous tissue was used as the goal of the work was to train neural networks capable of accurately classifying tissue taken from a tumor as one of these four tumor types based on gene expression levels. This is desirable because the tumors are difficult to diagnose using standard histological techniques. If a more definitive approach could be developed, it would improve the diagnostic process.

The microarrays measured the expression levels of 6567 genes. This data was later filtered to remove samples below a minimum level of expression, leaving expression data for 2308 genes. A total of 88 cells was analyzed, each having an associated 2308 gene expression level measurements. Data for 63 of these cells was used as a training set, with the remaining 25 cells making up a blind test set. The task of the neural network classifier trained on this data was to accept gene expression values as input and, using this information, determine the class of the sample.

Principal Component Analysis (PCA) (Rao, 1964) was used to determine the ten most significant components of the data. For a given sample, these ten values were used as inputs to a neural network. The neural network had four outputs, corresponding to the four possible classes. Each output value could range in value from 0 to 1, with 0 indicating that the sample was not a member of the corresponding class and 1 indicating membership in the class. Intermediate values represented varying levels of confidence as to class membership.

Rather than train a single network for classification, a total of 3750 neural networks were trained, using the data in the training set. The entire set of 3750 networks was used as a single classifier. A sample's 10 inputs were presented to all of the networks, and the outputs for all networks averaged. This resulted in four average class membership values, one for each of the four classes. The class receiving the highest average value was chosen as the class for the input sample. If none of the average values was above a threshold, then this condition is interpreted as giving the input sample none of the four possible classifications (e.g., the sample is not from one of the four classes learned by the neural networks). With these methods, all training and blind test samples were correctly classified.

3.1.1 GP and the Khan data set

GP was used to produce classification rules for the Khan data set. The rules for each class were evolved against all other samples in the set (i.e., the population was considered as target/~target during the evolution of rules for a particular class). Since each class had a perfect classifier, there was no danger of having a sample identified as belonging to more than one class. To provide for a fair comparison of their system and the system described here, n-fold cross-validation was not used to generate multiple training and test sets. Instead the training and blind test sets used were the same as with Khan et al.'s work with 63 samples used for training, and 25 samples used as a blind test set. The task was the same also: to produce a classifier capable of correctly classifying cell samples based on their gene expression values. Instead of a set of neural networks, GP produced four IF-THEN classification rules, one for each class.

The entire Khan data set was used; no reduction of the number of genes (e.g., with PCA) was made. All 2308 expression measurements were available to GP. GP was able to automatically select which variables to use in the classification rules. Preprocessing methods that reduce the number of variables, such as with PCA, risk discarding subtle, crucial information in the data set that could be necessary for accurate classification. In particular, standard PCA is known to have difficulties when the data contains nonlinear relationships, such as those that exist in biological data.

The following table summarizes *typical* parameters used in the Khan runs. The actual parameters varied from run to run. The most variable parameter was population size.

Table 3-3. Typical GP Parameter Settings on Khan et al., data

Parameter	**Setting**
Objective	Evolve classification rules for gene expression data
Terminal Set	Gene expression values
Function Set	Boolean and floating point operators: <, >, <=, =>, =, AND, OR, NOT, *, /, +, -, SQR, EXP, SQRT
Evolution type	Generational
Selection Scheme	Tournament, size = 7
Population size	20000 total, 5000 per deme
Crossover probability	0.9
Mutation probability	0.1
Termination criteria	Perfect classifier on blind test set or max number of generations (200)
Initial tree depth	5
Initial node count	5
Maximum tree depth	7
Maximum node count	15
Number of demes	4
Deme migration frequency	Every generation
Deme migration percentage	5% of individuals
Fitness	Number of correct classifications on training set

The following rules perfectly classify all samples from both the training and blind test sets. In these rules, EXP(x) indicates the exponential function, EXP(x) = e^x. SQR(x) indicates the squaring operation, SQR(x) = x^2.

IF (GENE 1318 * GENE 1953 > GENE 2049) THEN 'EWS'
IF (GENE 845 > EXP(GENE 521)) THEN 'BL'
IF (GENE 141 > SQR(GENE 841)) THEN 'NB'
IF (GENE 1651 / GENE 1954 <= GENE 553) THEN 'RMS'

Clearly, the rules produced using GP are simpler than the system described in Khan et al., using fewer genes and producing as accurate results. Moreover, the use of GP is transparent in the sense that the mechanism used to classify them is available for inspection. This contrasts well with neural nets and support vector machines (SVMs). The simplicity of the rules is due to the relatively sharp limits to tree depth (see tableau above). This was typical of the approach used in other sets where depth was only added when complexity was needed to create effective classifiers.

There is a danger of overfitting or discovering spurious connections in a data set that has so many variables and such few samples. However, the fact that the test set was a single-blind set and the evolved rules were still perfect classifiers gives confidence in the rules' robustness.

In addition to the robustness of the rules when applied to the test set, a look at the recurrence of genes across runs shows a tendency for certain genes to recur with exceptionally high frequency. Table 3-4 shows that despite resetting the random seed each time, the top three recurring genes occur between two and four times more frequently than the next most common gene and the most recurrent gene appears in eight times as many rules as the 9th most recurrent gene. In Table 3-4, Raw Count is the number of occurrences of each gene, counting multiple uses in a single rule. Rule Count is the number of unique rules in which a gene appears. In other words, if a gene occurs 3 times in a single rule the Raw Count is 3 but the Rule Count is only 1.

Table 3-4. Frequency of gene occurrences in 500 separately evolved rules; Raw Count is the number of occurrences of a gene, Rule Count is the number of rules in which a gene appears.

Gene Description	Raw Count	Rule Count
profilin 2	262	196
glutathione S-transferase A4	203	163
hematopoietic cell-specific Lyn substrate 1	167	137
Wiskott-Aldrich syndrome (ecezema-thrombocytopenia)	63	53
Homo sapiens inducible protein mRNA, complete cds	37	35
dual specificity phosphatase 6	31	30
Fc fragment of IgG, receptor, transporter, alpha	30	30
thyroid hormone receptor interactor 6	29	25
proteasome (prosome, macropain) 26S subunit, ATPase, 6	27	23

Finally, a t-test of the correlation of the rules was performed. This was done in two ways: by randomizing sample labels of the training set and evolving rules to classify the randomly labeled samples and by randomizing the labels of the test set. The results of these tests are shown in Table 5.

Table 3-5. T-Test of best rules compared to rules evolved against randomly labeled samples

T-Test Summary	EWS	BL	NB	RMS
Randomized Test	3.86E-14	8.31E-09	7.93E-10	7.44E-1
Randomized Train	2.53E-09	1.08E-06	1.16E-05	3.05E-0

Combining the facts that the evolved "best rules" were robust when tested against a blind sample set, that the genes selected by GP for use were heavily biased toward certain genes and a t-test comparing the evolved rules against rules evolved for randomly labeled samples, it is likely that the evolved rules were not based on accidental combinations of gene expression levels.

3.2 Prognosis of Treatment Response in DLBCL

Shipp et al. (Shipp et al., 2002) study Diffuse Large B-Cell Lymphoma (DLBCL) using gene expression profiles and postulate a different cause for this disease than Alizadeh et al. (Alizadeh et al., 2000) had proposed. Using a number of different algorithms, the authors were able to correctly derive a prognosis for the disease in approximately 75%-80% of the cases, depending on the algorithm used. Though it is striking that all of the algorithms used misclassified almost the same set of samples, the authors of (Shipp et al., 2002) drew no conclusions from this fact.

Using GP to analyze the Shipp et al. samples, approximately the same level of accuracy in classifying the patient's responsiveness to treatment was achieved. But what was striking was that the GP results were significantly less accurate than results achieved when GP was applied to diagnostic problems. Moreover, the behavior of the GP system has led to the conclusion that there were some difficulties with the data, since every effort to improve the accuracy of results actually *decreased* it. Specifically, the best results were achieved by making short, shallow runs (i.e., few generations and shallow GP trees). Longer runs or deeper trees produced clear signs of overfitting with worse results on the blind test samples.

Furthermore, the samples that are misclassified using GP were almost identical to the samples misclassified by Shipp et al., which is significant given the very different approaches they used. Since it seemed unlikely that there were a large number of samples that were incorrectly labelled when classification was based on a simple, observable fact of responsiveness to treatment, it was concluded that there was a more fundamental underlying problem. The first hypothesis was that the misclassified subset might be a third class that was perhaps, "partially responsive". But attempts to classify by clustering this set were complete failures. It was concluded that that there is some other information needed to make a more accurate classification such as demographics or patient history.

4. Discussion

Several different approaches for classifying biological samples using GP are described here. These were developed out of necessity because of the unusual features of the data and other problems including:

- Small sample size in relation to the number of variables
- Missing data elements
- Classification errors

While working to resolve these issues, several questions have arisen about GP that are further discussed below.

4.1 How should small sample sets be used with GP?

Since the training and test sets differ in each of the n-folds, the results of n-fold cross validation indicate how well the learning algorithm performs on the particular learning task. This is a fairly standard approach used in supervised learning problems. However, how should the results be interpreted in stochastic approaches such as genetic programming? Clearly, if the fitness on the training data in a fold is significantly better than the fitness on the test data, it is an indication of overfitting (i.e., discovering programs that are too specialized to the unique, non general characteristics of the samples presented in the training sets).

However, if the median fitness across a number of folds is N and the fitness for one fold is 2 sigma (2Σ) more does that make that fold's candidate meaningless? After all, the aberrant run may happen because the GP system found a uniquely good solution from among a number of mediocre results, e.g. it reached a hard-to-find global maximum. Suppose the median fitness across a run (i.e., the median of all folds) is N but the median of the next run is more than 2Σ different. What does that mean about the programs evolved in that run?

This raises the question of whether there is a better way to use a limited data set. An attempt to introduce artificial noise to create new versions of the samples met with only moderate success. The goal of cross-fold validation is to use available data to increase confidence in the robustness of the learning algorithm in the particular domain. The slice algorithm described above was an attempt to assess the robustness by introducing another measure of classification: separation distance between classes.

Beyond the interaction between GP and existing data, is there a way to predict, for a number of independent variables, whether there are enough samples to get good solutions independent of the fitness level reached? In other words, a problem may be hard, so hard that the fitness level of the best solutions remain far from perfect. Is there a way to predict the robustness of a solution for a problem with N variables given that there are P samples available to learn from?

What are the pathologies of GP behavior and what do they indicate about the problem, the control parameters or the data? In the Shipp et al. study described above, the persistent overfitting led to a tentatively correct conclusion that information necessary to solve the problem was missing. In essence the only way to create accurate classifiers was to create programs that were tied to the unique features of the training sets. Simple "glances" at the data produced less accurate, but more robust results.

However, it is believed that this can also occur if training samples are misclassified and in at least some of the published gene chip data sets, this is known to be the case. In essence, misclassified samples lead GP into overfitting by causing it to adopt the characteristics of one class in the programs designed to classify another class. This can also occur in cases where there is a missing or "hidden" class unknown to the researchers who are assigning class labels to the samples. This has occurred in at least one related cancer set where one type of leukemia was shown through cluster analysis to actually consist of two classes: a dominant type and a smaller, rarer subtype. This is described in (Armstrong et al., 2001) where a subtype of acute lymphoblastic leukemia is defined that was unknown prior to the application of gene chips to the study of cancers.

4.2 Is diversity problem specific?

Beyond indications about the data or the problem space, some problem-independent behaviors should be apparent based on changes to the GP control parameters. Clearly having a very high mutation rate leads to random walk-like search behaviors since effective code fragments are destroyed by excessive mutation regardless of the problem space. But what about other control parameters and their relationship to a problem?

One parameter that was varied quite a bit in the studies described here is population size. This is a question that has been hotly debated in the GP literature. Intuitively, it would seem that larger populations, particularly if segregated into multiple demes, would give more "genetic" diversity leading to a larger parallel exploration of the problem space. However, several papers (Luke and Panait, 2002; McPhee, 1999) suggest that this is not the case. Instead population diversity narrows quickly, and in the final generations most or all of the individuals can be traced back to a very small number of individuals in the initial population. Yet in much of the recent work done by the authors on biological and chemical data, there has in fact been a great deal of apparent diversity within populations, especially across demes. It is suspected that this is because these problems have many diverse solutions so there is less pressure to focus on a limited set of genetic material that belongs to a significantly better "neighborhood" of solutions.

Curiously, despite the diversity of solutions discovered, there remain certain "themes" that recur across evolutionary runs. This is shown in the frequency table (Table 3-4) where the genes that appear most frequently in solutions across multiple separate runs are summarized. Note that 3 genes occur between 2 and 4 times more frequently in the solutions discovered across 500 runs. This suggests there is a genuine attraction for these genes *independent of the starting population* but not enough to preclude other solutions. Indeed, a casual glance at two classification rules (shown below) for neuroblastoma (class 'NB') shows how different two "perfect" classifiers can be: they use different genes, combined in different ways to classify the sample.

IF GENE 741 > (EXP(GENE 1478) + SQRT(GENE 1992)) THEN 'NB'
IF GENE 141 > SQR(GENE 841) THEN 'NB'

4.3 What are the implications of optimization done in conjunction with the genetic programming algorithm?

The optimization of the slice point choice (in the floating point case) and the Boolean truth table for combining multiple programs (in the Boolean case) can improve the fitness achieved in a run. More work needs to be done to quantify the benefit, but it was found that problems otherwise beyond the scope of the available computing resources became tractable.

On other occasions when tackling symbolic regression problems, a linear combination of programs with real valued outputs was used, the combination being found by a linear least-squares process. Optimization steps to improve the fitness of individuals are not unprecedented in the GP literature. (McKay,

1995) used a non-linear optimization method to find good values for the constants in an expression, and (Raidl, 1998) used linear optimization to optimize weights introduced to the top-level terms of a GP expression tree.

Clearly a "hybridized" GP system can be useful for some problems. Are these the equivalent of hybrid GA systems where the local search strength of stochastic hill-climbing is combined with the global search characteristics of a genetic algorithm?

5. Conclusions

The experience with these data sets shows that GP can be used to generate classification rules from gene chip data that may be the basis for commercially viable diagnostics. However, the small sample size relative to the large number of variables will continue to challenge our ability to produce robust rules.

The transparency of the rules produced gives an insight into the biological processes captured by the gene chips that is not available from other "black box" solutions, and the creation of algebraic rules that combine gene expression levels suggest interesting quantitative relationships between genes. Moreover the automatic selection of genes used and the repeatability of the appearance of key genes give access to potential biomarkers that would normally require much more effort to discover.

The flexibility of GP allowed adaptation to the requirements of different problems and data sets. This included the development of the slice point method, adjustment of fitness function relative to the proportion of target class samples to the entire sample set, and the ability to bias the search in specific directions as required.

Some of the problems and questions encountered included the need to develop a sense of a reliable sample-to-variable ratio to provide consistent results and overfitting due to missing or incorrectly classified data. Population diversity seemed to be quite high in all cases, perhaps because of the large number of optima in the form of good solutions. This is reminiscent of some genetic algorithm applications that have yielded many solutions to what were previously seen as fairly deterministic problems such as antenna design (Linden and Altshuler, 1999).

The significance of the results from a biological standpoint is currently under study, however there are some telling signs that there is biological relevance in the results. In particular, several of the genes used in the evolved rules were also identified by Khan et al. as being significant and some were associated with known disease processes. In the DLBCL work, the fact that four separate and very different techniques identified many of the same samples as misclassified, is suggestive, particularly when an independent analysis by Dr. Stephen Krawetz of Wayne State University suggested that the gene chip selected was lacking some key genes. In essence, a bottom-up analysis of the data and a top down analysis by a biologist looking at the gene pathways associated with the disease reached the same conclusion - which is at least suggestive that the GP had discovered biologically-based connections between the data.

Finally, in the studies described here and in five other studies of similar cancer data, the pattern of certain genes having a dominant presence across multiple GP runs was always seen. The genes identified in this way are being studied as possible biomarkers for diseases or disease states. The fact that GP gives access to such information sets it apart from other machine learning techniques and increases the usefulness of GP to biologists.

References

Bojarczuk, C. C., Lopes, H. S., and Freitas, A. A. (2001). Data mining with constrained-syntax genetic programming: applications to medical data sets. Intelligent Data Analysis in Medicine and Pharmacology (IDAMAP-2001)

Brown, M. P. S., Grundy, W. N., Lin, D., Cristianini, N., Sugnet, C., Furey, T. S., Manuel Ares, J. & Haussler, D. (1999). Support vector machine classification of microarray gene expression data. University of Santa Cruz Technical Report. UCSC-CRL-99-09 http://www.cse.ucsc.edu/research/compbio/genex/genex.ps.

Brown, M. P. S., Grundy, W. N., Lin, D., Cristianini, N., Sugnet, C., Furey, T. S., Manuel Ares, J. & Haussler, D. (1999). Supplemental data for "Knowledge-based analysis of microarray gene expression data by using support vector machines", available at http://www.cse.ucsc.edu/research/compbio/genex/.

Brown, M. P. S., Grundy, W. N., Lin, D., Cristianini, N., Sugnet, C., Furey, T. S., Manuel Ares, J. & Haussler, D. (2000). Knowledge-based analysis of microarray gene expression data by using support vector machines. *Proc. Natl. Acad. Sci.. (USA) 97:* 262-267

Eisen, M., Spellman, P., Brown, P., and Botstein, D. (1998). Supplemental data for "Cluster analysis and display of genome-wide expression patterns", *Proc.Nat.Acad.Sci. (USA) 95:* 14863-14868, available at http://rana.stanford.edu/clustering.

Eisen, M., Spellman, P., Brown, P., and Botstein, D. (1998). Cluster analysis and display of genome-wide expression patterns, *Proc.Nat.Acad.Sci. (USA) 95:* 14863-14868.

Gerhold, D., et al. (1999). DNA chips: Promising Toys have become Powerful Tools. Trends Biochem Sci.; 24(5):168-73

Khan, J. *et al.* (2001). Classification and diagnostic prediction of cancers using gene expression profiling and artificial neural networks. *Nat Med 7:* 673-679

Khan, J. *et al.* (2001). Supplementary information for Javed Khan, et. al, Nature Medicine; 7(6):673-679, http://www.nhgri.nih.gov/DIR/Microarray/Supplement/.

Linden, D. and Altshuler, E. (1999). Evolving Wire Antennas using Genetic Algorithm. Proceedings of the First NASA/DoD Workshop on Evolvable Hardware, 225-232, IEEE Computer Society, Los Alamitos, CA.

Luke, S. and Panait, L. (2002). Is the Perfect the Enemy of the Good? In *Proceedings of the Genetic and Evolutionary Computation Conference,* pp. 820-828, Morgan Kauffman, San Francisco, CA.

McKay, B., *et al.* (1995). Using a tree structured genetic algorithm to perform symbolic regression. In *First International Conference on Genetic Algorithms in Engineering Systems: Innovations and Applications*, A. M. S. Zalzala (Ed.); GALESIA, volume 414, pages 487--492, Sheffield, UK, 12-14 September. IEEE.

McPhee, N. F. and Hopper, N. J. (1999). Analysis of Genetic Diversity through Population History. In *Proceedings of the Genetic and Evolutionary Computation Conference*, pp. 1112-1120, Morgan Kauffman, San Francisco, CA.

MYGD: Munich Information center for Protein Sequences (MIPS) yeast genome database, http://www.mips.biochem.mpg.de/proj/yeast.

Raidl, G. R. (1998). A Hybrid GP Approach for Numerically Robust Symbolic Regression. In *Genetic Programming 1998: Proceedings of the Third Annual Conference,* J. R. Koza, *et al.* (Eds.), pp. 323-28. University of Wisconsin, Madison. San Francisco: Morgan Kaufmann Publishers.

Rao, C. R. (1964). The Use and Interpretation of Principal Component Analysis in Applied Research, *Sankya, Series A*: 26: 329-358

Tan, K. C., Tay, A., Lee, T. H., and Heng, C. M. (2002). Mining multiple comprehensible classification rules using genetic programming. In *Proceedings of the 2002 Congress on Evolutionary Computation CEC,* 1302-1307.

Teller, A. and Veloso, M. (1995). PADO: Learning Tree Structured Algorithms for Orchestration into an Object Recognition System. Technical Report CMU-CS-95-101, Carnegie Mellon University, Dept. of Computer Science.

Chapter 4

ARTIFICIAL REGULATORY NETWORKS AND GENETIC PROGRAMMING

W. Banzhaf
Department of Computer Science
University of Dortmund, D-44221 Dortmund, GERMANY
banzhaf@cs.uni-dortmund.de

Abstract An artificial regulatory network able to reproduce a number of phenomena found in natural genetic regulatory networks (such as heterochrony, evolution, stability and variety of network behavior) is proposed. The connection to a new genetic representation for Genetic Programming is outlined.

Keywords: Regulatory Networks, Artificial Evolution, Evolutionary Algorithms, Genetic Programming, Development, Heterochrony

1. Introduction

Artificial evolution has proved useful in the optimization of numerical parameters in long-standing combinatorial problems, as well as in structural design problems. The former are known under labels such as Genetic Algorithms (Holland, 1992; Goldberg, 1989a), Evolutionary Strategies (Rechenberg, 1994; Schwefel, 1995) and Evolutionary Programming (Fogel and Walsh, 1965; Fogel, 1995), the latter are mostly known as Genetic Programming (Koza, 1992; Banzhaf *et al.*, 1998).

One key problem of all search algorithms is the "curse of dimensionality" (Bellman, 1961). This expression refers to the exponential growth of the search space's volume as a function of dimensionality. Thus even if there is an evolutionary computation solution to a problem of given size, the same problem with increased dimension might be completely unsolvable. The entire hope of Evolutionary Computation rests with the fact that methods can be implemented which are not volume-based but path-based. This means that history-dependent (or other constraint-dependent) search operators scan only a small part of the search space around the present areas of probing. It is expected that it will be possible to approach good solutions with this method.

In biology, developmental processes restrict the path of evolution to only a tiny fraction of the search space of possible forms and behaviors in Nature (Gould, 2002). This fact was a point of debate among biologists for more than a century. Why would Nature restrict itself so radically, to the point of risking extinction of entire classes of organisms when changing conditions would dictate adaptations which were impossible to reach from a particular branch of the tree of life? We cannot but think that there is some advantage to this approach in natural search spaces. Perhaps the density of solutions is equal everywhere? Perhaps neutral variations[1] are easily achieved in natural evolutionary systems, so as to relax the constraints somewhat? Perhaps the pressure to achieve solutions quickly resulted in a somewhat stream-lined search process?

Evolutionary and developmental biology were separated for most of the 20th century. However, a recent convergence of principles in biology suggests the evolutionary computation community should also take a closer look at development. Since GP search spaces seem to be closer to natural search spaces (density of solutions, ubiquity of neutral variations), the time has come to explore developmental approaches, notably in connection with GP. A number of researchers have, over the last decade, explored some aspects of developmental approaches to GP, including the genotype-phenotype map (Altenberg, 1995; Banzhaf, 1994; Spector and Stoffel, 1996; Keller, 1996; O'Neill *et al.*, 2003; Koza *et al.*, 1999), the genetic code (Freeland, 2002; Keller and Banzhaf, 1999), the mechanisms of gene expression and regulation (see for example (Kargupta, 2002)).

Over recent years, a key insight of biologists studying development and evolution was that variety in the living world can be traced back to three genetic mechanisms, active both in development and evolution:

1 Interactions between the products of genes

2 Shifts in the timing of gene expression (heterochrony)

3 Shifts in the location of gene expression (spatial patterning)

If we take this conclusion seriously, and go to the root of its natural implementation, we arrive at regulatory networks. Nature uses regulatory networks as a means to set up and control these mechanisms. This way, Nature unfolds the patterns and shapes of organism morphologies and of their behavior. In addition, regulatory networks mediate between development and evolution, since many evolutionary effects can be followed through their regulatory causes.

[1]Neutral variations are changes in genotypes that do not lead to fitness changes of phenotypes.

How can we make use of these insights in artificial evolutionary systems? Previous work in the area is scattered. Eggenberger (Eggenberger, 1997) has studied the patterning of artificial 3D-morphologies. Reil (Reil, 1999) has set up an artificial genome and studied some consequences for artificial ontogeny. Kennedy (Kennedy, 2001) examined a model of gene expression and regulation in an artificial cellular organism. Bongard and Pfeifer have considered the relation between evolving artificial organisms and behavior (Bongard and Pfeifer, 2002).

This chapter presents a new model of an artificial regulatory network which should be useful for algorithms like GP employing structural evolution. The model is a simplification and abstraction of the key elements of protein-genome interaction. It is not yet connected to a semantics of structures. That will be the next step. At this point, we can merely study the inner workings of the model and outline uses for it in GP. Section 2 explains the overall view of the regulatory network model, section 3 views it from the static (structural) perspective, while section 4 looks at the dynamic perspective. Section 5 explains the concept of heterochronic control. Section 6 exemplifies the plasticity of such systems to evolutionary pressure, section 7 discusses stability of steady states and means of communication. Finally, section 8 summarizes the discussion and outlines future steps. Our entire discussion here is qualitative, because only sample networks showing typical behavior will be presented.

2. A new genetic representation based on artificial regulatory networks

Our regulatory network model consists of a bit string, the 'genome', and mobile information-carrying molecules, 'proteins', which are equipped with bit patterns for interaction with the genome. Together, they represent a theoretically closed world with a network of interactions between genome and proteins, and a dynamics determined by this network.

A mechanism for reading off genes and for producing proteins with particular bit-patterns is used often called a 'genotype-phenotype mapping.' Proteins are able to wander about and to interact with any pattern on the genome, notably with 'regulatory sites' located upstream from genes. By attaching to these special sites, they can influence the production of (other) proteins. We observe the production of proteins and the dynamics of their concentration changes as a result of the interplay between all the interactions taking place simultaneously.

The genome is implemented as a sequence of 32-bit (integer) numbers. The length of the sequences, L_G, determines the length of the genome and is frequently used as a parameter. A particular START pattern, the 'promoter', is used to signal the beginning of a gene on the bit string (analogous to an open reading frame (ORF) on DNA), starting at the next integer. The signal used is arbitrary and was chosen as 'XYZ01010101', with XYZ arbitrary bytes and the one-byte pattern which in a genome generated by randomly choosing '0's and '1's will appear with a probability of $2^{-8} \approx 0.0039 = 0.39\%$. Genes follow the promoter and have a fixed length of $l_g = 5$ 32-bit integers resulting in an expressed bit pattern of 160 bits for each gene.

Upstream from the promoter site there are two special sites. One enhancer site and one inhibitor site both are of length 32 bits. Attachment of proteins to these sites will result in changes in protein production of the corresponding gene. It is assumed that a very low production of proteins takes place if both sides are unoccupied. Usually, however, there will be proteins around to influence expression rate of a particular gene, and we shall look at that in more detail later. In this simple model, we restrict ourselves to just one regulatory site for expression and one for suppression of proteins. This is a radical simplification with regard to natural genomes, where 5-10 regulatory sites that might even be occupied by complexes of proteins are the rule.

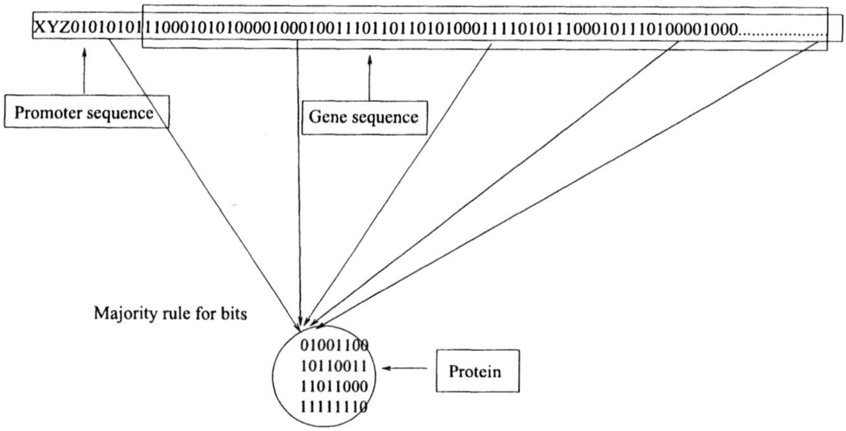

Figure 4-1. The genotype-phenotype mapping. Proteins are produced from genes via the genotype-phenotype mapping function.

In this model, we disregard the transcription process completely. Further, there are no introns, no RNA-like mobile elements and no translation procedure resulting in a different alphabet for proteins. Instead, proteins consist of bit patterns of a particular type: Each protein is a 32-bit number resulting from a many-to-one mapping of its gene: On each bit position in the gene's integers the majority rule is applied so as to arrive at one bit for the protein. In the case of a tie (not possible with an odd number for l_g), it is resolved by chance (see Figure 4-1).

Proteins can now be examined as to how they 'match' the genome : Each bit pattern of a protein can be compared to the genome pattern. The comparison is implemented by an XOR operation which results in a "1" if both patterns are complementary. Thus, complementarity between genome and protein bit patterns - and therefore their match - is determined by the number of bits set resulting from this XOR operation. In general, it can be expected that a Gaussian distribution results when measuring the match between a protein and all the bit sequences of a random genome. Notably, there are a few high-matching and a few low-matching positions and many average-matching ones on the genome.

3. Static view

Table 4-1 gives three examples of genomes with increasing size. We list the number of genes which roughly follows the 0.39% rule, the maximum match between resulting proteins and their genome at any location, and the number of times such a maximum match has been found. As can be seen the number of proteins with maximum match remains about the same, but their specificity increases.

Table 4-1. Sample genomes of increasing size. Specificity of highest matches increase.

Genome length L_G	Number of genes	Max. match	Freq. max. match
1,000	3	25	3
10,000	37	28	4
10,0000	409	30	3

Viewed from a protein's point of view, which scans the genome for good matching sites, Figure 4-2 depicts a typical situation: There is a wide variety of matching degrees, with very few high and low peaks. Average match is 16 bits just in the middle between 0 (min) and 32 bits (max), and statistically a Gaussian distribution emerges.

Let's change perspective and look from the genome's point of view, and more specifically, from the point of view of regulation sites. A number of proteins are produced and floating by, with some providing better matches to the site, other proteins providing worse matches. In principle, each protein has the potential to interact with each regulatory site, and the degree of matching will determine the probability of occupation of a certain site with a certain protein. The situation is depicted in Figure 4-3.

Because proteins are competing for attachment to regulatory sites, the probability of occupation with a particular protein is dependent on the degree of matching of all other proteins to this site. It is therefore necessary to normalize the degree of matching between the various proteins.

Under the simplifying assumption that the occupation of two regulatory sites per gene modulate the expression of the corresponding protein, a network of interactions between genes and proteins can be deduced, which can be parametrized by strength of match. Figure 4-4 shows a sample network, taken from an example genome with 32 genes / proteins.

Although no evolution has taken place (recall these genomes are generated by randomly drawing bits), the network of interactions shows a highly structured view of the resulting interactions. Despite this impression, however, the networks must be considered very complex (in terms of layers vs. participating nodes) and a deep hierarchy of interactions is visible.

A different picture emerges if we slightly change the strategy of generating our random genome. Figure 4-5 shows another genome which was generated by 'growing' a genome from a single 32-bit integer number. A series of length duplications, followed by mutations was applied to arrive at a genome of the same length $L_G = 10,000$. As can be seen from the figure, the result is a much shallower hierarchy, with particular master genes holding sway over an entire set of other genes, and other genes merely connected to their master gene.

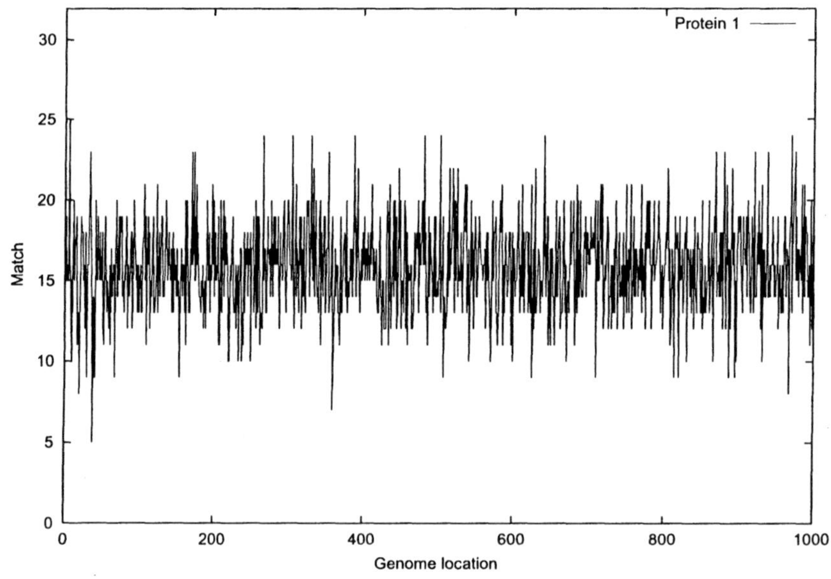

Figure 4-2. Matches of a sample protein with its genome. The genome was generated randomly, and matches are visibly distributed randomly as well.

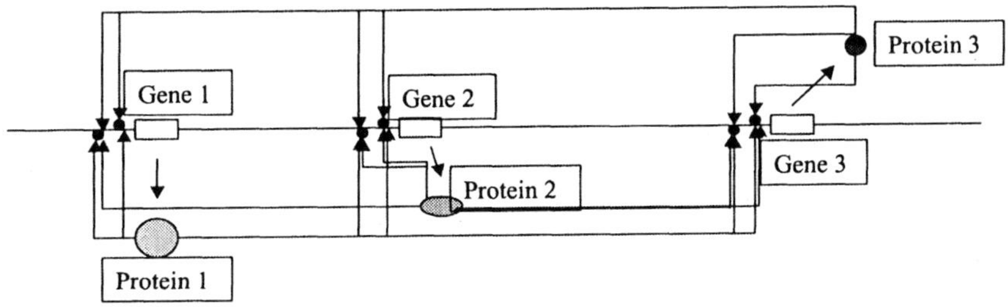

Figure 4-3. Interaction between genome and proteins, as seen from the regulatory sites. Each protein matches more or less to all regulatory sites.

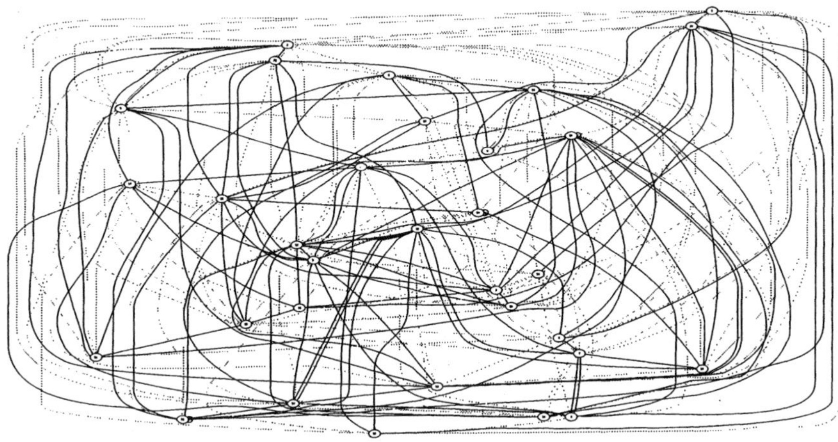

Figure 4-4. Network of interactions resulting from matches of proteins with the genome for a larger example. Depicted are 9 out of 32 proteins and their matches with the regulatory sites of all 32 genes of this example ($L_G = 10,000$). Black: Enhancing interactions; Gray: Inhibiting interactions.

As has been observed in natural genome organization (Dasson *et al.*, 2000), shallow hierarchies, up to the point of modularity, are a hallmark of biological organisms. It is interesting to note that a simple process of duplication and divergence suffices to reach a similar state, even from a random genome.

4. Dynamic view

Our discussion will now leave the static picture and concentrate on dynamics of the interaction network. A match between protein and regulatory site of a gene leads to activation or inhibition of protein production of the corresponding gene. Generally, the influence of a protein i with $i = 1, ..., n_p$ on an enhancer/inhibitor

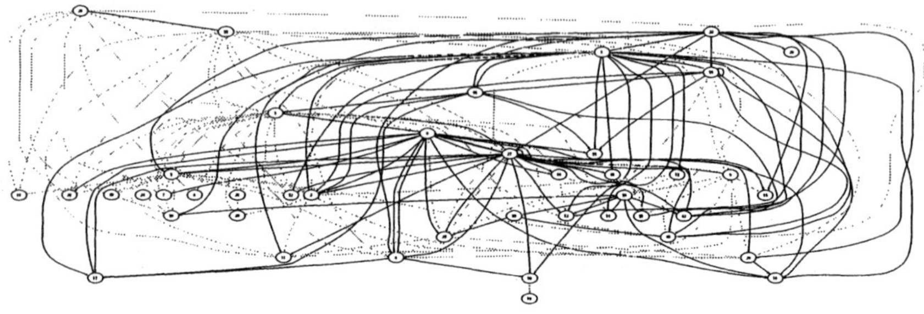

Figure 4-5. Network of interactions resulting from matches of proteins with a genome generated through a growth algorithm: A series of genome length duplications is followed by diversifying mutations ($L_G = 10,000$). Matching parameter is at higher values for arriving at a network complexity of approximately the same size as before. Black: Enhancing interactions; Gray: Inhibiting interactions.

site is exponential in the number of matching bits, $exp(\beta(u_i - u_{max}))$ where u_{max} is the maximum match achievable.

The concentration of protein molecules c_j of protein j modulates this strength to produce the following excitatory/inhibitory signals for the production of protein i:

$$e_i = \frac{1}{N}\sum_j c_j e^{\beta(u_j^+ - \bar{u}_{max}^+)} \quad (4.1)$$

$$in_i = \frac{1}{N}\sum_j c_j e^{\beta(u_j^- - \bar{u}_{max}^-)} \quad (4.2)$$

where a scaling was done as to have a maximum match for the best matching protein, both in excitatory and inhibitory signals.

Given these signals, protein i is produced via the following differential equation

$$\frac{dc_i}{dt} = \delta(e_i - in_i)c_i - \Phi \quad (4.3)$$

A flow term assures that concentrations remain in the simplex, $\sum_i c_i = 1$, resulting in competition between sites for proteins.

If we look at the dynamics of concentration changes of proteins, starting from a state of equal concentration that reflects the native low-level expression of all genes, we can observe some proteins increase their level of concentration, then fall again, with usually one being left over. Thus, a typical dynamic system behavior can be seen, well known under the name 'point attractor' in dynamical systems theory (Hirsch and Smale, 1997).

For different random genomes (different number of genes, matching etc) the dynamics is remarkably different. There are cases of longer and shorter time scales, there are complicated and simple dynamics. Figures 4-6 to 4-9 show four different dynamics resulting from four different genomes.

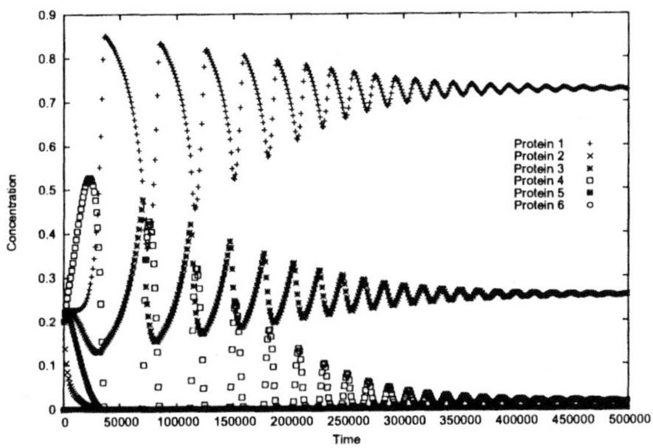

Figure 4-6. Time development of protein concentrations. Different dynamical systems are realized by different genomes (see later figures). Here, a dampened (nonlinear) oscillator type of dynamics is exhibited.

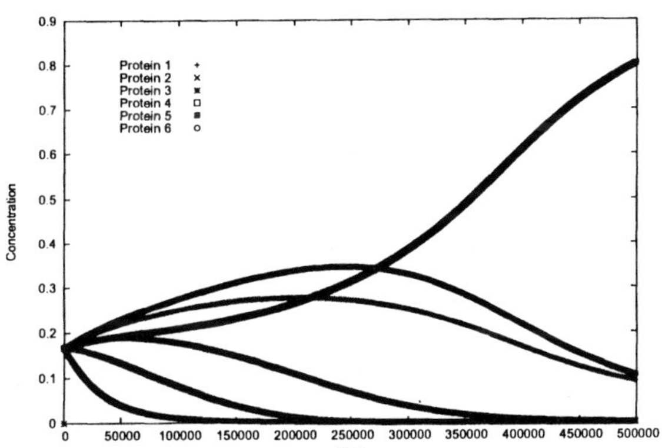

Figure 4-7. Time development of protein concentrations. Slow and smooth development of concentrations due to the particular bit patterns–and thus coupling strengths–between genes and proteins.

It should be noted that this richness of dynamics is merely a result of different genomes of the same length, with different protein patterns resulting in different matching and regulation results. No development or evolution has yet been put in place.

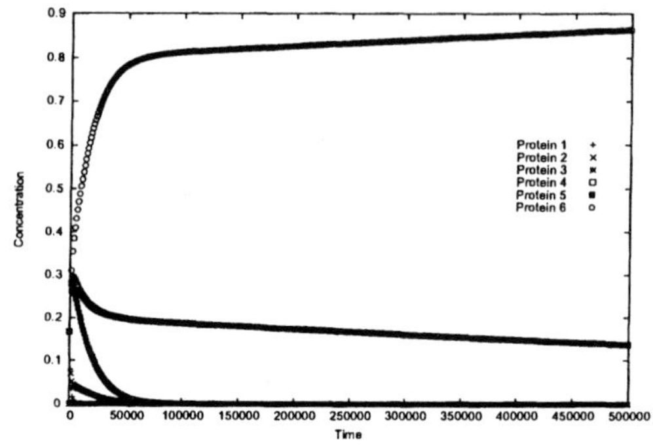

Figure 4-8. More examples of dynamics: Quick settlement into a point attractor.

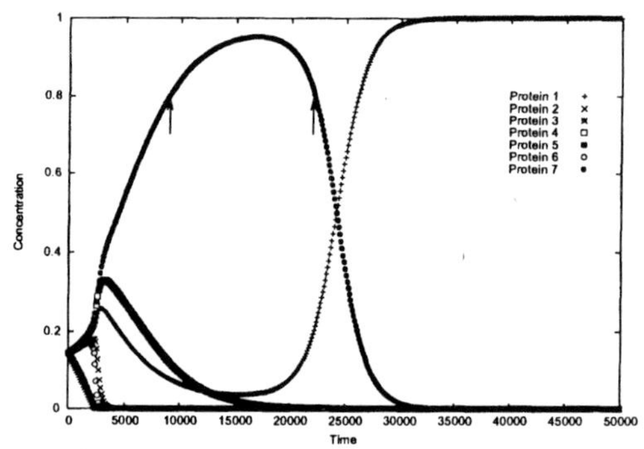

Figure 4-9. Extended transition phase with one protein achieving high values of concentration and a subsequent switch to expression of one other gene.

5. Heterochronic control

If we look at this from the perspective of how many proteins are above or below a certain production threshold, we can observe the turning on or turning off of genes (on/off could be set equal to $x \times 2$ or $x \times 1/2$ of initial production, or it could be based on an absolute concentration value). This translates into a timing of onset/offset of gene production. Figure 4-9, right, for instance, shows the timing of the start and end of concentrations above 0.8 for protein 7 (arrows), $t_{on} = 9,000, t_{off} = 22,000$.

Changing the degree of matching between regulatory sites and proteins by one or two bits can result in dramatic changes in the dynamics, but it need not. Sometimes there are no changes at all and we have a neutral variation. Some example changes that actually varied the expression are shown in Figure 4-10 and 4-11. It is interesting to note that variations in patterns are translated by the

ARN into time variations, similar to what has been observed in natural GRNs (Davidson, 2001)

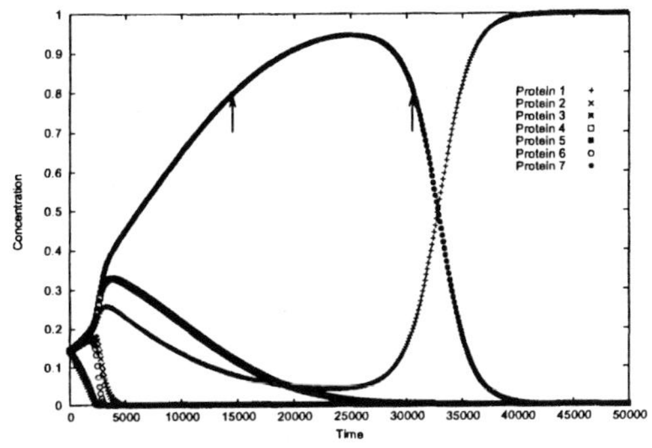

Figure 4-10. Genome of Figure 4-9. Degree of matching between protein 7 and inhibitory site to gene 4 changed by one bit. Timing of expression of protein 7 changes substantially: $t_{on} = 14,500, t_{off} = 30,500$.

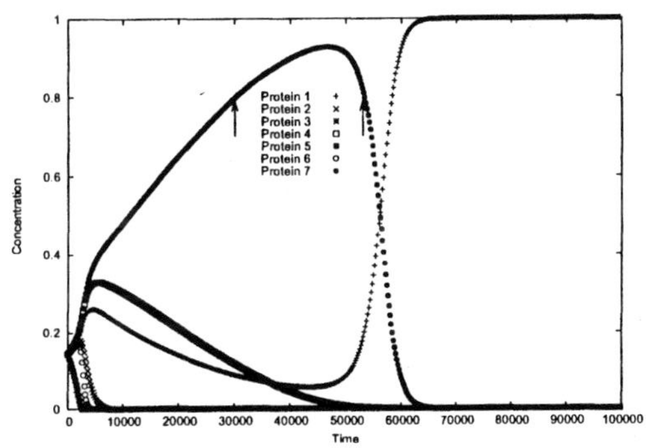

Figure 4-11. Genome of Figure 4-9. Degree of matching between protein 7 and inhibitory site to gene 4 increased by another bit, timing changes even further $t_{on} = 30,000, t_{off} = 53,000$.

Heterochrony, i.e. a variation in the timing of onset or offset of certain genes are heavily used in development for generating particular structural effects (McKinney and McNamara, 1991; McKinney, 1999). As we can see by comparing Figures 4-10 and 4-11, and 4-9, small changes cause small effects. The same principle could be also of use in modeling "physiological" reactions, for instance under the control of external factors exceeding certain threshold values.

Interestingly, the range of possible changes is partitioned logarithmically, due to the change of occupation probability, that is depending on an exponentiated matching difference between proteins and DNA bit-patterns. This can be seen most easily, if we put all concentration curves of protein 1 and 7 from the proceeding experiments into one plot, see Figure 4-12. We can clearly see the range of changes expanding with further additions of bit flips.

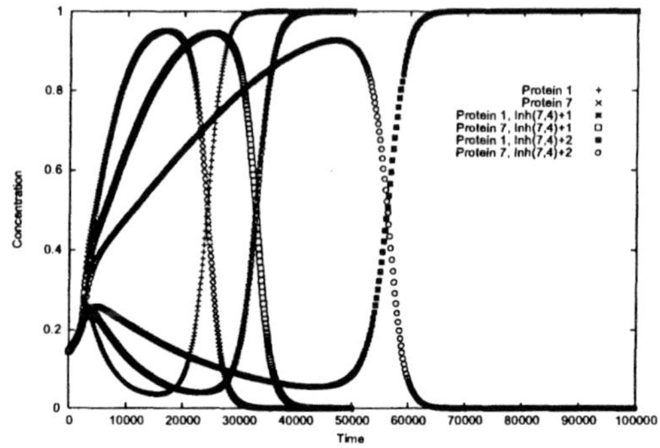

Figure 4-12. Genome of Figure 4-7. Degree of matching between protein 7 and inhibitory site to gene 4 changed progressively. Timing of expression of protein 7 changes in increasing step sizes.

6. Evolution

The most important question to be addressed with such a model is whether it would be possible to define arbitrary target states and evolve the genome/protein network toward this target state. Our first results in a typical simulation are shown in Figure 4-13. It shows the progress of a network in approaching the target concentration of a particular protein, here protein 6. As we can see, the evolutionary process quickly converges towards this target state. It must be emphasized that the very simplest way of doing evolution was used here, a $(1 + \lambda)$ evolution strategy, with $\lambda = 1$ (Rechenberg, 1994). Various experiments were performed with the same genome (not shown here), allowing evolution of other concentration levels for other proteins. We can see from the figure, that steep declines in the deviation (error) curve are followed by apparent stagnation periods. These stagnation periods are, however, accompanied by continued changes in the genome under evolution. It is merely the mapping of the genome that does not show many consequences of these variations. By construction we designed a system with many neutral pathways. Evolutionary progress is thus interrupted superficially, but goes on in genomes due to neutral steps.

This can be seen if we consider the changes in concentration levels of all proteins at $t = 100$ in Figure 4-14. Here we can discover that all protein concentrations change over time, with many stagnation periods for all proteins.

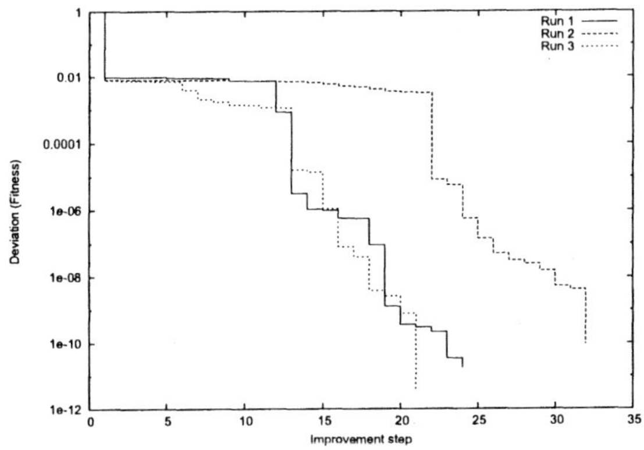

Figure 4-13. Evolution at work: 3 different runs of a $(1 + \lambda)$ strategy to arrive at a prespecified concentration of one particular protein: $c_6 = 0.085$ at time $t = 100$.

Huge steps are sometimes shown by certain proteins, which are not reflected in the fitness of an individual, due to the focus on measuring only the deviation from $c_6 = 0.085$ for fitness.

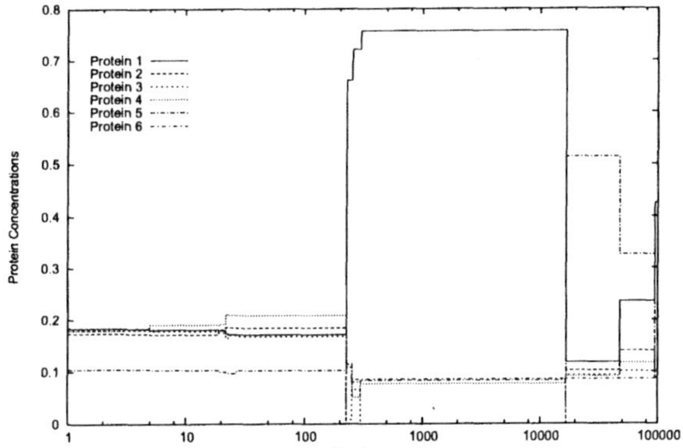

Figure 4-14. Evolution at work: Same run as in Figure 4-13, with all protein conentrations protocolled at $t = 100$. As can be seen, protein concentration of selected protein 6 meanders towards goal state $c_6 = 0.085$, whereas other protein concentrations pass through huge swings.

When comparing the figures for heterochronic control one cannot but have the impression of rather small variations between the different expression patterns. Changes in exon content could bring about more effects. Figures 4-15 and 4-16 shows two 1-bit mutations in the expressed part of gene 4 and 1, respectively.

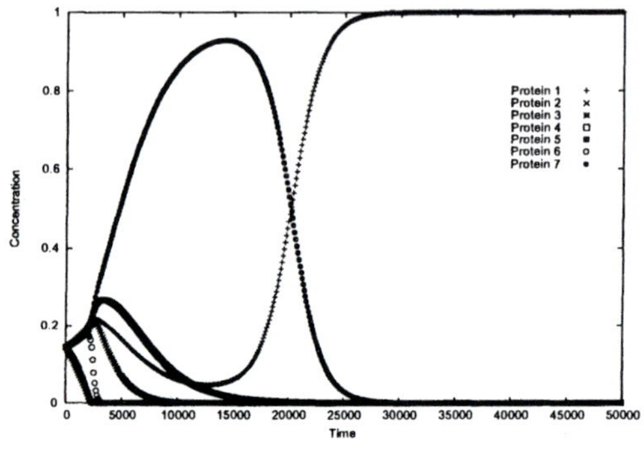

Figure 4-15. One-bit mutations in the expressed part of gene 4. Mutation does not change the phenotype.

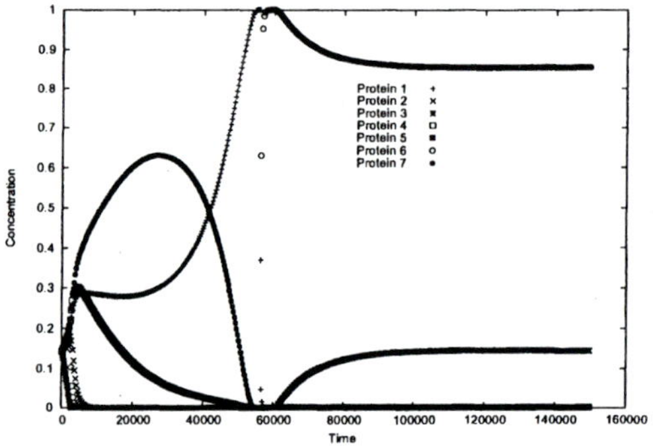

Figure 4-16. One-bit mutations in the expressed part of gene 1. Mutation radically changes the phenotype. A completely new behavior, including the dominance of another protein is visible in the right figure.

Differences are striking, though still some similiarity (particularly in the earlier iterations) remains. Around iteration 60,000, however, a radical switch occurs in the behavior that could not be observed without the mutation. This gives an indication how novelty might be generated in such a system: by slight changes in patterns, entirely new 'phases' appear in the phenotype.

Here we have not been concerned with diffusion and spatial variation, another fruitful area when examining development in biology. Much is already known about the early pre-structuring of organisms and the subsequent unfolding of these patterns into real organs and structures. This process, however, requires that cells are able to communicate with each other, which in turn manifests itself in a common language.

7. Stability and Communication

What would be easier than using the mobile elements producable by cells, to let them carry meaning? This theme is the subject of the present section. We report on a few experiments with proteins that have been added or removed from a network, thus simulating the import from or export to other cells.

Two question will be asked: (i) Is the regulatory network providing a stable environment, so that the export of protein does not perturb the behavior of the network? (ii) Is the regulatory network sensitive enough to change its behavior upon the impingement of protein from the outside?

It is clear that both requirements are somewhat contradictory. Nevertheless, both questions can be answered in the affirmative. Some proteins are very stable, regardless of whether one adds or removes them; others are sensitive to addition, still others to removal (see Figures 4-17-4-22 for examples, percentages given in protein change per time unit). Obviously, the network provides again a very rich behavioral environment, where various features can be subject to selection.

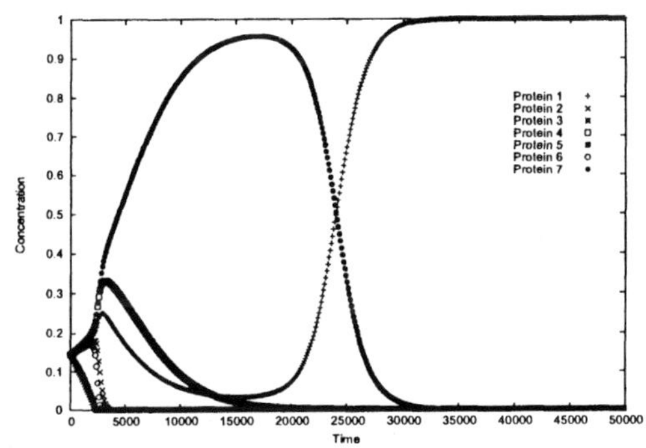

Figure 4-17. Removal of protein produced by the network. Removal of protein 1, 10%. There is only a slight adaption of the dynamics, no real change in the behavior visible.

8. Summary and Perspectives

In this contribution we have shown that a simple model for artificial regulatory networks can be formulated which captures essential features of natural genetic regulatory networks. Although we have only shown qualitative results, the difference in behavior of these networks from usual genetic representations can be seen already from the few examples shown here.

Our next steps are to move from an analyis of qualitative behavior to quantifying certain features like stability and evolvability. Even more interesting, however, is to press ahead and find a proper connection of this type of genetic

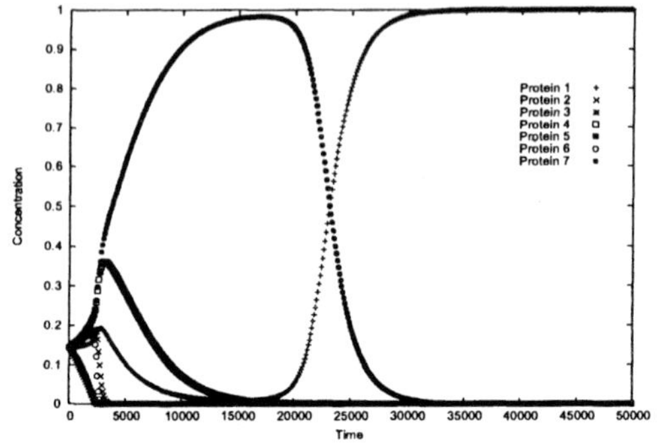

Figure 4-18. Removal of protein produced by the network. Removal of protein 1, 100%. . There is only a slight adaption of the dynamics, no real change in the behavior visible.

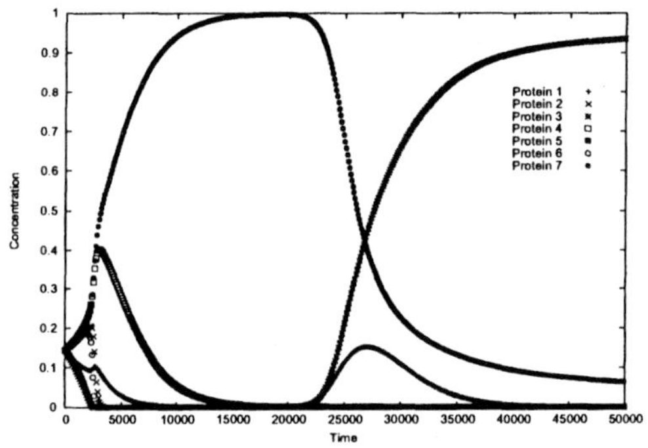

Figure 4-19. Removal of protein produced by the network. Removal of protein 1, 300%. There is only a slight adaption of the dynamics when removing the protein.

representation to genetic programming. Here we can only outline our present thinking in this direction.

The proteins used here as a means for regulation might carry a second role, namely that of agents or objects behaving in the outside world. This would mean that a translational mechanism would be put into place that translates the protein pattern into another representation useful in the 'outside' world. A plain method, for instance, would be to interpret the 32 bits of the protein as an instruction for a 32-bit processor. There are some GP-systems able to digest bit patterns of arbitrary type and generating useful behavior from it (see (Banzhaf *et al.*, 1998) and (Nordin *et al.*, 1999)). Concentrations could be used to fix an order for the sequence of instructions to be executed. Much more complex relationships between the information carried by the protein and program constructs are reasonable and will be introduced in due time.

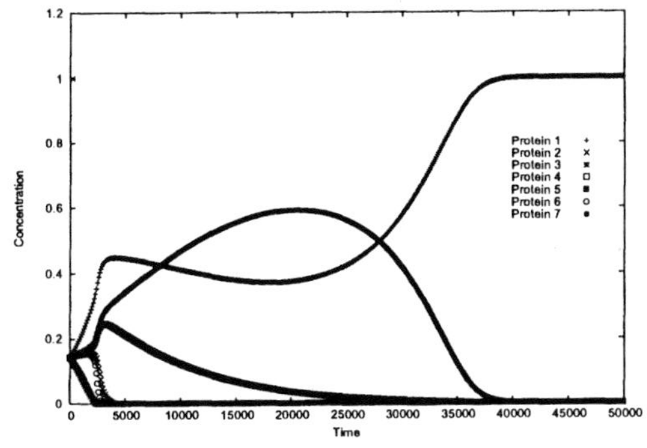

Figure 4-20. Addition of protein produced by the network. Addition of protein 1, 200 %. There is a stronger reaction of the dynamics when adding the protein.

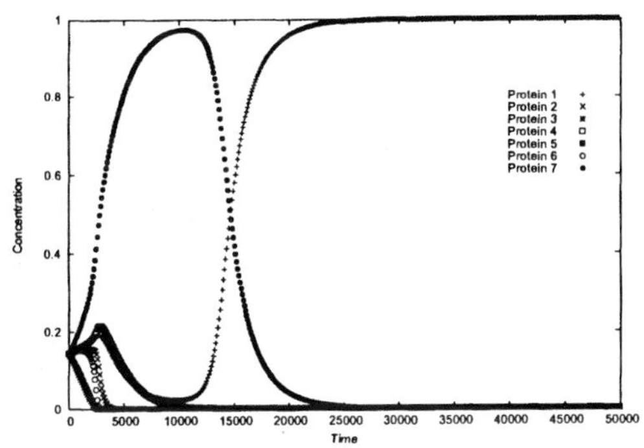

Figure 4-21. Addition of protein produced by the network. Addition of protein 7, 200 %. The opposite from Figure 4-20: There is only a slight adaption of the dynamics when adding the protein..

Acknowledgement

The author gratefully acknowledges a sabbatical stay at the Institute for Genomics and Bioinformatics at UC Irvine, where part of the ideas that lead to this work were born. Especially he wants to acknowledge the hospitality of its director, Prof. Pierre Baldi, and of its manager, Mrs. Ann Marie Walker.

References

Altenberg, L. (1995). Genome growth and the evolution of the genotype-phenotype map. In *Evolution and Biocomputation*, W. Banzhaf and F. H. Eeckman (Eds.), pp. 205–259. Springer-Verlag, Berlin, Germany.

Banzhaf, W. (1994). Genotype-phenotype-mapping and neutral variation – A case study in genetic programming. In *Parallel Problem Solving from Nature III* (Jerusalem, 9-14 Oct. 1994),

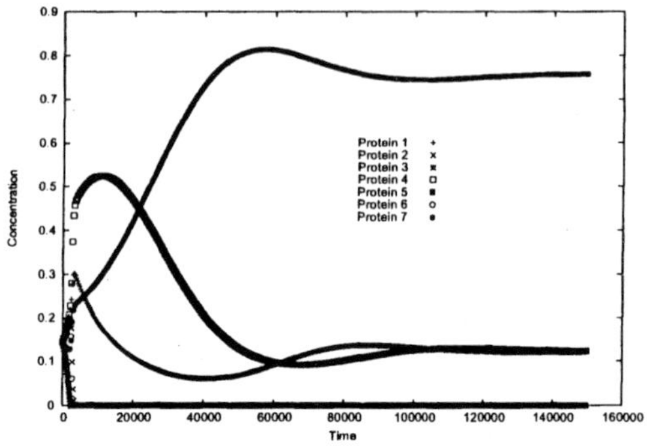

Figure 4-22. Removal of protein produced by the network. Removal of protein 7, 200 %. Note the opposite effect from Figure 4-19: there is a strong reaction when removing the protein.

Y. Davidor, H.-P. Schwefel, and R. Männer (Eds.), pp. 322–332. Vol. 866 of *Lecture Notes in Computer Science (LNCS)*. Springer.

Banzhaf, W., Nordin, P., Keller, R. and Francone, F. (1998). *Genetic Programming - An Introduction*. Morgan Kaufmann, San Francisco, CA.

Bellman, R. (1961). *Adaptive Control Processes: A Guided Tour*. Princeton University Press, Princeton, NJ.

Bongard, J. and Pfeifer, R. (2002). Behavioral selection pressure generates hierarchical genetic regulatory networks. In *Proceedings of the Genetic and Evolutionary Computation Conference*, W. B. Langdon et al. (Ed.), Morgan Kaufmann.

Dassow, G., Meir, E., Munro, E. and Odell, G. (2000). The segment polarity network is a robust developmental module. *Nature 406:* 188–192.

Davidson, E. (2001). *Genomic Regulatory Systems*. Academic Press, San Diego, CA.

Eggenberger, P. (1997). Evolving morphologies of simulated 3d organisms based on differential gene expression. In *Proceedings of the 4th European Conference on Artificial Life* (1997), I. Harvey and P. Husbands (Eds.), pp. 205–213. Springer Verlag.

Fogel, D. (1995). *Evolutionary Computation*. IEEE Press, New York.

Fogel, L., Owens, A. and Walsh, M. (1965). Artificial intelligence through a simulation of evolution. In *Biophysics and Cybernetic Systems*, M. Maxfield, A. Callahan, and L. Fogel (Eds.), pp. 131–155.

Freeland, S. J. (2002). The darwinian genetic code: An adaptation for adapting? *Genetic Programming and Evolvable Machines 3*: 113–127.

Goldberg, D. (1989). *Genetic Algorithms in Search, Optimization and Machine Learning*. Addison Wesley, Reading MA.

Gould, S. (2002). *The Structure of Evolutionary Theory*. Belknap Press of Harvard University Press, Cambridge, MA.

Hirsch, M. and Smale, S. (1997). *Differential Equations, Dynamical Systems and Linear Algebra*. Academic Press, Reading, MA.

Holland, J.H. (1992). *Adaptation in Natural and Artificial Systems*, second edition. The MIT Press, Cambridge, MA.

Kargupta, H. (2002). Editorial: Special Issue on Computation in Gene Expression. *Genetic Programming and Evolvable Machines 3*: 111–112.

Keller, R. and Banzhaf, W. (1996). Genetic programming using genotype-phenotype mapping from linear genomes into linear phenotypes. In *Genetic Programming 1996: Proceedings of the First Annual Conference* (Stanford University, CA, USA, July 1996), J. R. Koza, D. E. Goldberg, D. B. Fogel, and R. L. Riolo (Eds.), pp. 116–122. MIT Press.

Keller, R. and Banzhaf, W. (1999). The evolution of genetic code in genetic programming. In *Proceedings of the Genetic and Evolutionary Computation Conference (GECCO 1999)* (Orlando, Florida, USA, July 1999), W. Banzhaf *et al.* (Eds.), pp. 1077–1086. Morgan Kaufmann.

Kennedy, P. and Osborn, T. (2001). A model of gene expression and regulation in an artificial cellular organism. *Complex Systems 13(1)*.

Koza, John (1992). *Genetic Programming: On the Programming of Computers by Natural Selection*. MIT Press, Cambridge, MA, USA.

Koza, J.R., Andre, D., Bennett, F.H. III and Keane, M. (1999). *Genetic Programming 3: Darwinian Invention and Problem Solving*. Morgan Kaufman, San Francisco, CA, USA.

McKinney, M. (1999). Heterochrony: Beyond words. *Paleobiology 25*: 149–153.

McKinney, M. and McNamara, K. (1991). *Heterochrony: The Evolution of Ontogeny*. Plenum Press, New York ,NY.

Nordin, P., Banzhaf, W. and Francone, F.D. (1999). Efficient Evolution of Machine Code for CISC Architectures using Instruction Blocks and Homologous Crossover. In *Advances in Genetic Programming 3*, L. Spector *et al.*. (Eds.), pp. 275–299. The MIT Press, Cambridge, MA, USA.

O'Neill, M., Ryan, C., Keijzer, M. and Cattolico, M. (2003). Crossover in grammatical evolution. *Genetic Programming and Evolvable Machines 4:* 67–93.

Rechenberg, I. (1994). *Evolutionsstrategie "93*. Frommann Verlag, Stuttgart.

Reil, T. (1999). Dynamics of gene expression in an artificial genome - implications for biological and artificial ontogeny. In *Proceedings of the 5th European Conference on Artificial Life*, D. Floreano *et al.*. (Eds.), pp. 457–466. Springer.

Schwefel, H.-P. (1995). *Evolution and Optimum Seeking*. Sixth-Generation Computer Technology Series. John Wiley & Sons, Inc., New York.

Spector, L. and Stoffel, K. (1996). Ontogenetic programming. In *Genetic Programming 1996: Proceedings of the First Annual Conference* (Cambridge, MA, 28–31 July 1996), J. R. Koza (Eds.), pp. 394–399 The MIT Press.

Chapter 5

USING SOFTWARE ENGINEERING KNOWLEDGE TO DRIVE GENETIC PROGRAM DESIGN USING CULTURAL ALGORITHMS

EXPLOITING THE SYNERGY OF SOFTWARE ENGINEERING KNOWLEDGE IN EVOLUTIONARY DESIGN

David A. Ostrowski[1] and Robert G. Reynolds[2]

[1]*Ford Motor Company Scientific Research Laboratories,Dearborn, Michigan 48121*

[2]*Dept. of Computer Science, Wayne State University, Detroit, Michigan 48202*

Abstract: In this paper, we use Cultural Algorithms as a framework in which to embed a white and black box testing strategy for designing and testing large-scale GP programs. The model consists of two populations, one supports white box testing of a genetic programming system and the other supports black box testing. The two populations communicate by sending information to a shared belief space. This allows a potential synergy between the two activities. Next, we exploit this synergy in order to evolve an OEM pricing strategy in a complex agent-based market environment. The new pricing strategy generated over $2 million dollars in revenue during the assessment period and outperformed the previous optimal strategy.

Key words: Genetic Programming, Cultural Algorithms, Hybrid Genetic Programming Environments, Agent-Based Modeling, OEM strategy evolution, Black box testing, White Box Testing

1. Introduction

Software Engineering methodologies have demonstrated their importance in the efficient solution of software for complex real-world problems. The process of designing and testing programs can be described as the heuristic search through the space of possible programs (Simon, 1986). Heuristic search is often performed using evolutionary computation methods in situations of high dimensionality (Koza, 1990). Software design methods when directly applied to evolutionary computation practices can reveal detailed information with regards to program constraints that can be exploited by genetic programming systems.

A strong relationship exists between software design and testing. Software testing practices reinforce and verify the design process by determining program faults through the identification of knowledge that can allow the programmer to identify the causes and relate them back to a specification. Testing allows the programmer to update the software specification and refine it. Two complementary approaches to software testing are white and black box testing (Pressman, 1987). White box testing is used to investigate a programs structure, which includes the examination of program logic and control flow. Black box testing allows the examination of a program from a purely functional perspective in terms of the production of input output pairs, with no consideration as to how they were specifically generated. In the context of program testing, black box testing is applied first with the goal of determining whether the programs performance matches the requirements. When a lack of fit is identified, a white box approach is used to directly relate the behavior back to the specification in the program.

These techniques can also be applied in the context of software design in order to produce design knowledge. We view the program construction process as being supported by alternating phases of white and black box testing. Initially, the white box testing approach is informally applied in order to support the initial construction of a prototype. Here, we are examining a program structure or effectively modifying the structure in response to its performance. Once a program's structure has been developed to a suitable level of performance, it is then given over to the black box testing process. The goal of black box testing is to identify any faults within the existing program representation in terms of the programs requirements. Once faults have been identified, they can in turn be used to infer new tests for the white box testing phase. This alternating sequence of white and black box testing parallels what human programmers do during program development.

In this paper we will investigate the application of this approach to the automated design of genetic programs. In particular, the impact that knowledge generated by black and white box testing, individually and together, has on expediting the genetic programming process will be examined. More generally, we are interested on the interplay between traditional software design practices and the automatic development process.

Is the information traditionally produced by software design practices for human developers also useful in guiding the designs produced by genetic programming systems?

1.1 Cultural Algorithms as a Framework Within Which to Collect and Apply Software Engineering Knowledge.

Cultural Algorithms enhance the evolutionary process by the application of a belief structure to the traditional evolutionary population (Reynolds, 1979). This second structure emulates symbolic cultural evolution as opposed to biological evolution. Here this symbolic knowledge base can maintain beliefs about a program's syntax and semantics. This knowledge is influenced by, and in turn influences, future populations of genetic programs in a manner similar to the preservation of culture among human populations. This enhanced version of evolutionary computation is suitable for application to program design since the knowledge maintained by the integration of the two software development activities can assist in the overall software design process. Zannoni and Reynolds (Zannoni, 1994; Zannoni, 1997) demonstrated the use of Cultural Algorithms to speedup the Genetic Programming development process by collecting knowledge that related program syntax with program semantics based upon the performance for a given training set. This we consider to be an example of "white-box" knowledge that associated sub-trees in a genetic program with individuals that performed well on a given training set. We will build upon their approach by adding to the belief space knowledge produced by the application of another software development activity, black-box testing.

Black-box testing is used to identify mismatches in the functional and non-functional specification for a program and its behavior. Then, new test cases are generated that reflect those deficiencies. These new cases are integrated with those used during the white box testing phase. Software metrics are used to measure non-functional aspects of a requirements specification that focus on design quality. The Cultural Algorithm approach that we are going to employ here uses two Cultural Algorithms chained together: one for white box testing and the second for black box testing. They will each utilize a shared belief space representing implicit and explicit constraints on the developing program. It is hypothesized here that this knowledge is necessary to guide the design of solutions to complex multi-dimensional engineering problems.

This paper is structured in the following manner. In Section 2, we briefly discuss the complementary types of knowledge that can be generated by white box and black box testing respectively. Examples that illustrate how such knowledge can be used to guide Genetic Program Search are given for each. However, in previous approaches either specific black box or white box knowledge is used in an application, but not in explicit conjunction with each other. We postulate that the explicit coordination of both types of testing will produce symbiotic effects on the search process. That is, while each type of knowledge can in fact improve search efficiency on its own, what happens when they are explicitly used together to direct search? These approaches are explicitly chained together within a Cultural Algorithm framework in order to investigate this hypothesis.

Section 3 discusses Cultural Algorithms as a framework in which to integrate the two approaches and describes the Dual Cultural Algorithm Genetic Programming system (DCAGP). In Section 4 we use the system to evolve a pricing strategy for an Original Equipment Manufacturer (OEM) that applies discounts to the selling price of automobiles in an agent-based marketplace in order to attract customers. There exists an optimal strategy for this market assuming that agents have no history of price changes. However, when agents are allowed memory of past discounts, the original approach is no longer optimal. Here, we use variations on this optimal strategy to seed a GP population that learns a new strategy for this more complex environment. It is demonstrated that this new strategy generates over 2 million dollars in revenue and outperforms the previous optimal strategy in this new environment. Section 6 gives our conclusions.

2. The Use of Knowledge from Black and White Box Testing to Guide Design.

The goal of this section is to present an overview of both the white and black box testing methods in terms of the knowledge that can be generated by each. This knowledge will be used to guide GP design using Cultural Algorithms. This extracted knowledge will reside in the belief space and will be used to guide the heuristic search through program space. As shown in Figure 2.1, both white and black box testing approaches are employed to make contributions to the knowledge that is abstracted in the belief space. White box testing produces structural knowledge that can be related to functional knowledge using black box testing. Taken together these two processes produce a system capable of automatic knowledge-based design using GP.

During the development of the requirements for a program, a specification of system behavior is produced. In the most general sense, this consists of I/O pairs as well as a set of constraints on them. It is always assumed in standard Software Engineering practice that the results of testing can be used to guide the search for a program that fits a specification. Here, we will make that connection explicit by using specific, abstracted, knowledge to guide the GP search. A Cultural Algorithm framework approach is used to assist in the program development process by allowing it to test for and learn about system constraints.

With regards to system design, there are implicit or learned constraints determined through a reasoning process within a belief space. Once a design has reached a certain level of stability relative to the learned constraints, it is given to the black-box system to identify missing constraints that require further development work. In traditional software testing, black-box methods are often applied first to a developed program, and then white-box methods are applied (Sommerville, 1996). The black-box methods test the system against the expected I/O behavior. If there is a problem, it generally means that some constraint on the computation has been violated.

Traditionally, in using these tests to evaluate software quality, black-box testing is done to identify faults and white-box testing follows in order to identify the reasons for the fault. When these processes are used in design, the order is reversed. White-box testing is used here to develop a design from the bottom up in response to the program specification. Black-box testing then identifies the need for new or modified structures. As shown in Figure 2.1, the white-box approach is applied in order to produce generalizations to be used to derive a more successful program structure. Next, black-box techniques are applied to test the current program designs by applying abstract information in order to assist in determining deficiencies in program design.

In this section, we first define traditional white box testing along with specific techniques that will exploit the program structure. Next, traditional black-box techniques are briefly discussed, along with specific examples of its use in evolutionary design. Then, an overview of both methods applied specifically to the evolution of GP software is presented.

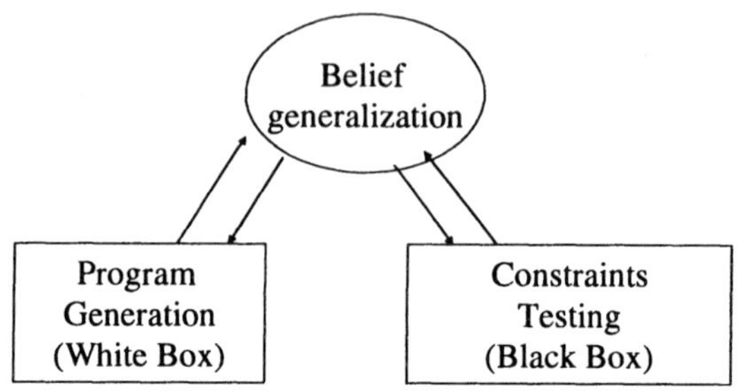

Figure 2.1 White and Black Box Testing activities interfaced with a Belief Space.

2.1 White Box Testing Methods

The goal of this section is to describe the nature of the knowledge that white-box testing can extract from a program's structural knowledge. This is the knowledge that can be used by the belief space in Cultural Algorithms to guide the design process. White-box testing methods utilize knowledge about the control and data structures of the design in order to derive test cases. This is considered testing the program's structure relative to constraints on its execution.

One popular way of assessing the effectiveness of a white-box strategy is program coverage (Kaner, 1988). The objectives of program coverage can be: 1) To ideally execute each possible path in the program relative to a given set of data: path coverage; 2) To execute each statement in a program. Since the control flow for a program can be represented as a control flow graph, white-box testing can be considered analogous to the traveling salesman problem. As such, it is within the category of NP-hard problems in the general sense (Garey, 1979). Thus, the number of paths that need to be explored to completely test even the smallest function with a cycle (loop) can be very large. This can make exhaustive techniques inadequate due to the number of test cases potentially required. Thus, it is more feasible to systematically identify subsets of the program structure that can be "covered" at some level of testing as opposed to producing "exhaustive" coverage. Examples of program coverage goals are as follows:

- Guarantee that all independent paths within a module have been exercised at least once.

- Exercise all logical decisions on their true and false sides.
- Execute all loops at their boundaries and within their operational bounds.
- Exercise internal data structures to assure their validity (Pressman 1987).

The overall goal of path testing is to execute every independent path in a program at least once for a connected graph. The starting point for path testing is a program flow graph. This is a graph model of all paths through the program. A flow graph consists of nodes representing actions and edges showing flow of control. By simplifying compound expressions into simpler conditions, one can construct the flow graph and determining what test cases are needed. The purpose of the test cases is to force the execution of each path.

2.1.1 The Use of White Box Knowledge from an Evolutionary Perspective

From a knowledge-based point of view, we can store schemata that represent a portion of a programs structure (GP-program schema) or a portion of the programs execution (GP-program slice). Shortest path techniques can be used to generate program slices (Weiser, 1984). Some work has been done using evolutionary techniques to support white-box testing. The approach to determining a minimal program slice through either static or dynamic approaches can be viewed as a minimal path program. GAs have been successfully applied to such problems. Jones, Eyres, and StHammer (Jones, Eyre, StHammer, 1996) applied Genetic Algorithms in order to automate branch testing through the automatic generation of test data. The Genetic Algorithm generated data in order to demonstrate branch coverage that would be applied to a series of problems including the quadratic equation, remainder, linear, and binary search procedures. An automatic test generation system that applied these techniques was able to detect program errors in 99% of all cases. They found the Genetic Algorithm approach able to detect certain situations where the predicate in a branch is modified, providing a higher probability of revealing errors at sub-domain boundaries.

Zannoni and Reynolds (Zannoni, 1994) have used evolutionary techniques, Cultural Algorithms, to extract GP-program schema associated with high performing individuals. Through the application of heuristic search techniques, they were able to derive subprograms that guided the development of GP programs. Here the program emerged as the result of a simulated evolution and was derived as the result of a bottom-up search process. This approach parallels that of a human novice programmer, whose search is primarily guided by the performance function, which compared the systems I/O behavior to the expected specification. (Zannoni, 1997)

2.2 Black Box Testing.

In this section, we focus on the knowledge that can be produced by black-box testing. This will be used to motivate its application on our system. Black-box testing concentrates on the functionality of software. In the black-box testing process, the software engineer tries to derive sets of input conditions that when taken together will fully execute all functional requirements of a program. The focus of black-box techniques is to derive scenarios that produce program faults without as much concern to why the fault was produced. It is distinguished from white-box testing in that the program is regarded as a black-box in this phase. Here, the tester inputs data into the system, without concern as to exactly how the program works. The test designer looks for input data and conditions that might lead to erroneous outputs. The focus of the test data search is to find "interesting" input data that is likely to expose an error in a program. Thus, the knowledge stored in the belief space produced from black-box testing can relate to intervals (classes) for variables that are likely to generate errors when variables take values in those ranges.

Black-box testing is equivalent to viewing a program and simulating a mathematical function that is specified only in terms of inputs and outputs. Since black-box testing is not concerned with the inner workings of the software, the focus is on whether the program supports the I/O specification for a problem or not. Complete functional testing, like white-box testing, is an NP-hard problem. This is due to the need for subjecting a program to all possible input sequences. Considering a binary representation of any problem, a simple 10-character input would have 2 to the 80th possible input streams of corresponding output. An exhaustive set of tests designed to exercise all of the input would take an excessive amount of time and resources. Such problems can be more feasibly approached through the establishment of equivalence partitions within the test data sets and the selection of representative test cases from each equivalence class. This is similar in principle to selecting representative paths to test in the white-box approach, as opposed to exhaustive testing

For a given program, many different equivalences classes can be identified. Biezer (1983) suggested that deriving equivalence classes based upon input and output for the program was of fundamental importance. An input equivalence class partition is a set of data, such that all of the sets members are processed in an equivalent way by the program. Likewise, the output equivalence classes partition program output into classes that have common characteristics in terms of their output. The overall objective is to generate input classes that will maintain an overlap with the output class in order to identify program faults. After a set of partitions has been identified, particular test cases from each of these partitions are chosen.

Equivalence class generation can be performed by categorizing the input test data within a set number of classes in terms of their ability to produce particular output states within the program. Even with the use of equivalence partitions, the black-box testing can become very complex as well as resource intensive.

2.2.1 Use of Equivalence Class Knowledge in Evolutionary Search

The first step in a successful black-box testing strategy is the development of the test data set. Rychtyckyj and Reynolds have used Cultural Algorithms to learn equivalent classes associated with the behavior of an expert system represented in terms of a semantic network for process scheduling (Rychtyckyj, 2000). The system was applied to utilize the attributes that are stored within the nodes to learn a defining set of attributes that in turn be used to re-engineer various portions of the network. They were able to demonstrate that in most cases Cultural Algorithms are a much more efficient tool than manual inspection to be able to identify the defining attributes and provide comparable results in a shorter period of time. This approach was used to demonstrate significant improvements as the complexity of the network increased.

Evolutionary-based approaches using equivalence partitions have also been used by Schultz, Grefenstette and DeJong (Schultz, Grefenstette, and DeJong, 1995) in evaluating intelligent controllers for an autonomous vehicle through the application of Genetic Algorithms. In this work, they discuss the representation of this knowledge in terms of the application of fault rules. Each rule would be composed of two parts: a trigger mode and a fault mode. The triggers would make up the rule conditions that must be met for the fault to occur. When the conditions specified by the triggers are met, the fault mode is obtained. Through the application of an equivalence class concept, one is able to obtain a fault scenario in order to be able to debug a program.

Chen and Rhine (1984) use a functional approach in the training of particular classes of reusable controller software components. The purpose of such an approach is to constrain the relationship function between a component's abstract and concrete levels, where the abstract level is represented by a set of fuzzy adaptive logic rules (FL). The components' concrete level is represented by means of adaptive fuzzy membership functions (MBFS). Here, they observe that faults at the concrete level of the component complements those at the abstract level. Two main functional requirements needed to reach this goal were to systematically select a fault scenario and to then identify possible faults. The Genetic Algorithms were used to search for a fault scenario by rewarding those fault scenarios that cause the highest performance degradation.

Eiben and Van Der Hauw (Eiben and Van Der Hauw, 1997) found similar success with the application of constant weights to specific formulas as a heuristic approach to guide a GA mechanism. In this approach, they were able to handle constraints in the context of a GA by defining penalties for constraint violations and use the total penalty for a chromosome as a component of its fitness.

2.3 The Potential Synergy of White and Black Box Testing

While both black-box and white-box knowledge have been used to direct Genetic search, they have not been explicitly integrated into GP design. It is suggested that while each type of knowledge may expedite search in some way, we feel that there is much potential benefit to be gained by the explicit interaction of these knowledge sources during the program design process. Knowledge from white-box testing can be used to update Genetic Program structure while black-box testing can then be used to identify flaws in these updated structures. The results from black-box testing can be used to generate new test cases for the white-box phase in order to adjust the current structure to these new demands. In the next section, we describe Cultural Algorithms and show how they can be used as a framework in which to integrate these two complimentary knowledge sources together.

3. The CAGP System: Introduction

Cultural Algorithms have been used to implement Software Engineering testing methods in order to assist in the software design process. In this section, previous uses of Cultural Algorithms to guide the generation of software systems are first detailed. Our approach differs from these earlier efforts in that we explicitly incorporate the software engineering testing phases into the design process, and tailor the approach to the design of intelligent agents. Then, a general model describing our approach as a chained Cultural Algorithm is presented.

A Cultural Algorithm is defined by Reynolds (Reynolds, 1994) as a dual inheritance system that provides for the interaction and mutual cooperation of two distinct levels of evolution: a population level and a symbolic level (the belief space). In Figure 3.1 the pseudocode for a Cultural Algorithm is presented. Initially, the population is initialized along with the belief space and communication structures. Then the population is evaluated through the application of a performance function for each individual. Next, the acceptance function selects a set of individuals and their problem-solving experiences that must be applied to adjust the belief space. The adjustment of the belief space is performed through applying a set of update procedures to modify, combine and prune the set of current beliefs. Next, the newly updated beliefs are used to control the modification of the population via the influence function. The new population is then evaluated and the process repeated until a termination condition is met.

```
Initialize population Pop(0)
Initialize belief network Blf(0)
Initialize communication channel Chl(0);
Evaluate pop(0);
T = 0;
Repeat
 accept (Pop(t),Blf(t));
 adjust (Blf(t));
 influence( Blf(t), Pop(t));
 t = t + 1
 Pop(t) = select(Pop(t-1));
    evolve(Pop(t));
    evaluate(Pop(t));
 until (termination_condition());
```

Figure 3.1 Cultural Algorithm Pseudocode.

3.1 The Chained Cultural Algorithm with Genetic Programming

Ostrowski and Reynolds proposed embedding black-box and white-box testing strategies within Cultural Algorithms in order to automate the test generation process (Ostrowski, 1999). This system is referred to as the Dual

Cultural Algorithm with Genetic Programming system or DCAGP and is shown in Figure 3.2. The first phase involves the acquisition of structural information in the belief space about components of successful programs through the use of white-box testing. This enables the second phase to become more effective as a more evolved space is being tested. The black-box phase takes a subset of the programs generated in the white-box phase and uses them as the performance function. The population in this phase is a set of input values. Each member of the population is applied as input to the evolved programs. The ones that produce more faults relative to the specified output get a higher score. A subset is then selected to update the equivalence class information in the belief space. This information is used to generate more test sets until no improvement in the scores are produced.

This equivalence class information is then fed back to the black box mechanism. There are many ways in which this can be done. For example, we can generate a new set of I/O pairs that can be added to the test set for the white-box configuration. Alternatively, if we keep information about the sub-expressions whose performance are adversely affected in the black-box testing, we can then adjust the belief statistics in the white-box phase relative to those sub-expressions that are most likely to be associated with faulty performance. The latter approach is especially useful when applying the system to more complex tasks such as the design of intelligent agents.

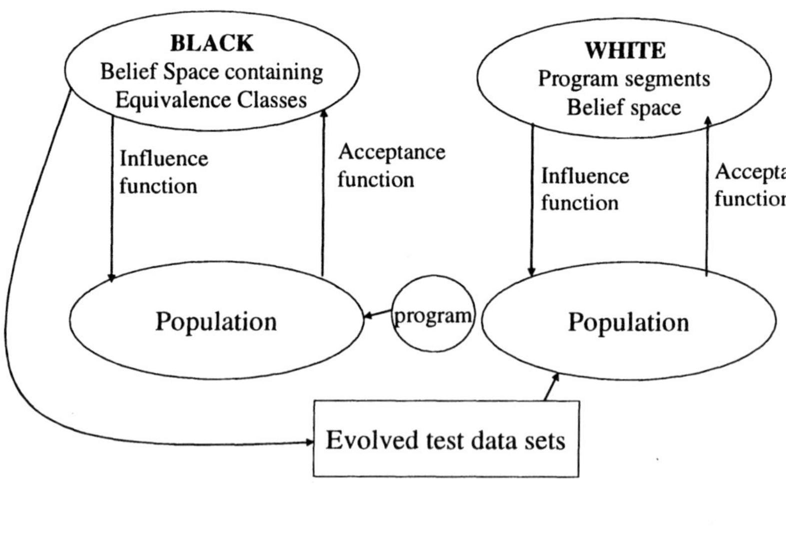

Figure 3.2 Diagram illustrating the Chained Cultural Algorithm applied to program development

4. Evolving an OEM Pricing Strategy in a Complex Agent-Based Market Place Using DCAGP

The DCAGP system described above supports the ability of an agent to autonomously develop a design within a potentially complex performance environment. In this section, we demonstrate this concept by using the system as a framework within which to evolve pricing strategies for vehicle sales by an OEM within the framework of a multi-agent market system. This application models the complex dynamics that are involved between the consumer and the manufacturer during the course of the decision process that a consumer goes through while deciding if they want to purchase a new vehicle. The overcapacity in the vehicle manufacturing arena coupled with the large number of vehicle choices in North America has led to a very dynamic and complex selling scenario where rebates, special financing offers, low interest rates and consumer confidence in the economy all interact within the marketplace in often unpredictable ways.

Ostrowski (2002) developed a multi-agent system that simulates a real-world consumer market. An optimal market equilibrium Original Equipment Manufacturer (OEM) pricing model was applied to the multi-agent system. When the linear model is run with more complex consumer behaviors, the pricing strategy appears non-optimal as evidenced by a decrease in long-term profits. Here, we use the DCAGP system to evolve a near optimal pricing strategy to maximize profits for an OEM within this more complex consumer market.

The Java-based multi-agent framework called marketScape (Ostrowski, 2002) allows us to effectively model economic scenarios in the context of an agent-based environment. There exist four substantial objects in the marketplace model here: consumers, vehicles, manufacturers, and the dealers (used vehicle). Consumer preferences are given by a homogeneous linear utility function U(a) , where a is the age of the vehicle. The utility of each agent is scaled by a heterogeneous parameter α_i uniformly distributed (0,1). Agents have a memory about the price history of new vehicles, and use their memory to make inferences about future prices. The length of the memory of an agent is heterogeneous within the population and is uniformly distributed over 1 to 5 periods.

For every period, a consumer makes an evaluation with regards to the purchase of a new or used vehicle. Consumers make their decision to maximize utility according to the following formula:

Max: $\{U(a_n) + P_c - d - P_n,\ U(a_c)\}$ where

U(a_n) represents the utility derived from a vehicle considered for purchase,

P_c represents the price of the currently owned vehicle,

d represents the trade discount,

P_n represents the price of the vehicle considered for purchase, and

U(a_c) represents the utility derived from the currently owned vehicle. If the computed utility is found to be greater than the utility associated with the current vehicle, then the newer vehicle is chosen.

The agent-based framework allowed for the demonstration of dynamic events in a market that are not easily examined using the assumptions of standard economic theory. An optimal linear pricing strategy was developed in vehicle purchase scenarios where consumers did not have memory of price histories (Porter, 1999). When consumers use memory-based strategies, firm profits declined with the OEM strategy above because of the non-linearity introduced into the process by the use of memory.

These memory-based strategies had a marked effect on firm profits when the OEM made short-term price reductions. Following these price reductions firms found that it was very difficult to increase prices again without drastically affecting sales. One example memory-based strategy employed is termed the "postpone" strategy. The postpone strategy is related to consumer expectations that vehicle prices will fall again in the future. With the postpone strategy, an agent checks their *n* period memory of prices and compares them to the current price. If there is a discounted price in their memory lower than the current price, then a purchase decision is postponed until the next period.

Figure 3: Near-Optimal Pricing Strategy with a 2-step Incentive and Postpone Strategy.

To demonstrate the impact of the "postpone strategy," we employed 200 consumer agents in a 60-period simulation in the marketScape environment as shown in figure 3. Here, we apply a 2-step price reduction (to simulate a brief incentive) incorporated with the original optimal pricing strategy with and without the postpone memory strategy. When examining the total vehicle sales, the period after incentive drops off for approximately 6 periods of no sales generating a scenario where there is a reduction in profits overall as compared to not placing the incentive on the vehicle.

Can the DCAGP system developed here be used to adjust terms in the linear model used by the OEM so as to address the non-linearity introduced by allowing agents to use price history in their strategies. There are two basic ways that the equation can be adjusted, by changing the coefficients on the existing terms, or by adding new terms and replacing old ones. Since we focus only on the situation where the agents use the postpone strategy we just address the situation where the coefficients for existing terms are modified.

Our GP population is initialized with the optimal linear OEM no-memory equation, initialized with randomized coefficients between .9 and 1.1 applied to each of the terms of the equation. The idea is then to evolve a new equation that produced better performance in the memory modified marketplace. While only the coefficients are changed in these runs, the stage is set so that the entire equation can be modified as various memory-based consumer strategies are introduced into the model.

In our results, the marketScape model was run over 60 periods with a 2-step price reduction at period 30 and 31. The results were averaged over 10 runs and examined at the 50th and 100th generations. The base Genetic Programming approach generated on average 2.124M profits in comparison to the 2.041M profits generated by the initial pricing equation. A single Cultural Algorithm using only white box knowledge performed similarly to GP. However, using both white and black box knowledge the Dual Cultural Algorithm with Genetic Programming (DCAGP) approach presented here was able to generate the highest average profits both at 50 and 100 generations. In this approach the total average profits obtained were over 2.135M by the end of 60 periods. These were also obtained with minor increases in CPU utilization. (Figure 3)

strategy	runs	cpu (secs)	max pro	var.(%
GP	50	135K	2.105M	99
CAGP	50	144K	2.111M	99
DCAGP	50	146K	2.130M	99
GP	100	271K	2.124M	99
CAGP	100	288K	2.123M	99
DCAGP	100	292K	2.135M	99

Table 6.1: Profit Comparison

The highest optimal strategy that was generated by the DCAGP was found to obtain a total profit of 2.136M. Upon examination of the coefficients produced from this near optimal solution, it was found that the highest weighted variable was the number of vehicles in the inventory. This was the only variable that was directly influenced by the price history. By applying the

DCAGP approach, we were therefore able to determine the significance of each of the parameters that are applied in a multi-agent system in response to the "postponement" strategy.

Future plans using this approach include: the addition of various other memory-based pricing strategies, the addition of multiple competing OEMS, the use of different economic situations such recession, and exploring the of globalization on pricing strategies. All of these can be handled within the DCAGP framework.

5. Conclusions

It was shown that integrating white- and black-box testing into the design process for GP using Cultural Algorithms has the potential to produce an increase in the rate of convergence as well as overall program quality. Our initial description of black and white box testing emphasized the differences in the types of knowledge produced. In fact, we can use the GP framework as a vehicle to test the utility of each knowledge source, and its overall contribution to the culture of software engineering. Combining the black- and white-box testing with a GP system produces a hybrid system that combines the best of all three approaches. The systematic search engine of the GP combined with the complementary knowledge produced by the black- and white-box testing within the Cultural Algorithm framework produced a system capable of tackling complex knowledge-based problems, such as intelligent autonomous agent design. We then demonstrated a situation in which the system used both types of knowledge to produce an innovative solution to a complex non-linear economic problem for an OEM agent.

The resultant conjecture is that GP systems, just like their human counterparts, can benefit from the knowledge provided by various software engineering techniques. Often software engineers take the tools, metrics, and concepts that comprise part of their "culture" for granted. In contrast, the GP approach allows us to reinvent these tools and truly assess their impact on the efficient production of high quality, and innovative software products. Here, we see the individual and combined impact of two of these most fundamental of concepts, black and white testing, in a way that often eludes human practitioners.

We suggest that GP systems can benefit from the evolution of it own "programming culture," a culture that reinvents and reworks existing concepts from traditional software engineering as well producing some new and novel techniques of its own along the way. That is, we envision a world where GP systems look *inward* and develop lifecycles, frameworks, and metrics to guide their own designs and *outward* towards the application of these techniques to an increasingly global production landscape.

Acknowledgements

Dr. Reynolds work on this paper was supported by a National Science Foundation Intelligent Information Systems Grant #IIS 9907257.

References

Beizer, B. (1998). Software Testing Techniques, Data Systems Analysts, Inc. Van Nostrand Rhinholdt Co.

Bilchev, George, Parmee, Ian, Darlow, Andrew. (1996). The Inductive Genetic Algorithm with Applications to the Fault Coverage Test Code Generation Problem. In *Proceedings of EUFIT '96*, Aachen, Germany.

Chen, J., Rine, D. C. (1997). Training Fuzzy Logic Based Software Components for Reuse. In *Proceedings of ISMVL*, pp. 189-194.

Eiben, A. E., van der Hauw, J. K. (1997). *Solving 3-SAT by Gas Adapting Constraint Weights.* IEEE.

Garey, Michael R., Johnson, David S. (1985). *Computers and Intractibility: A Guide to the Theory of NP-Completeness.* New York, N.Y.: W.H. Freeman and Company.

Jones, B. F., StHammer, H. H., Eyres, D. E. (1996). Automatic Structural Testing using Genetic Algorithms. *Software Engineering Journal.*

Jones, B. F., Eyres, D. E., StHammer, H. H. (1998). A Strategy for using Genetic Algorithms to Automate Branch and Fault-Based Testing. *The Computer Journal 42(2).*

Kaner, C., Falk, Jack, Nguyen, Hung Quoc, (1995). *Testing Computer Software, Second Edition.* Thompson Computer Press.

Koza, John R. (1990). Genetically Breeding Populations of Computer Programs to Solve Problems in Artificial Intelligence. In *Proceedings of the Conference on Tools for Artificial Intelligence*, pp. 819-827.

Maletic, J. (1995). The Software Service Bay: A Methodology for Knowledge-Based Software Maintenance. Phd. Thesis, Wayne State University, Detroit MI.

Miconi, Thomas. (2001). A Collective Genetic Algorithm. *Artificial Life, Adaptive Behavior* and *Agents.*

Ostrowski, D., Reynolds, R. G. (1999). Knowledge-Based Software Testing Agent Using Evolutionary Learning with Cultural Algorithms. In *Proceedings of CEC '99.*

Ostrowski, D., Tassier, T., Everson, M., and Reynolds, R. G. (2002). Using Cultural Algorithms to Evolve Strategies in Agent-Based Models. In *Proceedings of the 2002 Congress on Evolutionary Computation*, pp. 741-746.

Porter, R., Sattler, P. (1999). Patterns of Trade in the Market for Used Durables: Theory and Evidence. NBER Working Paper No. W7149.

Pressman, Roger S. (1997). *Software Engineering: A Practitioners Approach.* McGraw Hill.

Reynolds, R. G. (1994). An Introduction to the Cultural Algorithms. In *Proceedings of the 3rd annual Conference on Evolutionary Programming*, pp. 131-139. Sebalk, A.V. Fogel L.J, River Edge, NJ. World Scientific Publishing.

Rychtychkyj, N., and Reynolds, R. (2000). Assessing the Performance of Cultural Algorithms for Semantic Network Re-Engineering. In *Proceedings of the 2000 Congress on Evolutionary Computation*, pp. 1482-1491. July 16-19, La Jolla, CA, IEEE Press

Schultz, A., Grefenstette, J., and DeJong, K. (1995). "Learning to Break Things: Adaptive Testing of Intelligent Controllers." In *Handbook of Evolutionary Computation*.

Simon, Herbert A. (1986). Whether Software Engineering Needs to Be Artificially Intelligent. *IEEE Transactions on Software Engineering 7(SE-12):* 726-732.

Sommerville, I. (1996). *Software Engineering.* Addison-Wesley.

Weiser, M. (1984). Program Slicing. *IEEE Transactions on Software Engineering SE-10.*

Zannoni, Elena and Reynolds, R. G. (1994). Learning to Understand Software Using Cultural Algorithms. In *Proceedings of the third annual Conference on Evolutionary Programming.* Sebald Antony V. and Fogel, Lawrence J., editors. World Scientific Press.

Zannoni, E., and Reynolds, R. G. (1997). Learning to Control the Program Evolution Process in Genetic Programming Systems Using Cultural Algorithms. *Journal of Evolutionary Computation 5(2)*: 2:181-211

Chapter 6

CONTINUOUS HIERARCHICAL FAIR COMPETITION MODEL FOR SUSTAINABLE INNOVATION IN GENETIC PROGRAMMING

Jianjun Hu, Erik D. Goodman, Kisung Seo
Genetic Algorithm Research & Application Group (GARAGe)
Michigan State University, East Lansing, MI
hujianju, goodman, ksseo@egr.msu.edu

Abstract: Lack of sustainable search capability of genetic programming has severely constrained its application to more complex problems. A new evolutionary algorithm model named the continuous hierarchical fair competition (CHFC) model is proposed to improve the capability of sustainable innovation for single population genetic programming. It is devised by extracting the fundamental principles underlying sustainable biological and societal processes originally proposed in the multi-population HFC model. The hierarchical elitism, breeding probability distribution and individual distribution control over the whole fitness range enable CHFC to achieve sustainable evolution while enjoying flexible control of an evolutionary search process. Experimental results demonstrate its capability to do robust sustainable search and avoid the aging problem typical in genetic programming.

Key words: Genetic programming, sustainable innovation, HFC, fair competition principle

1. Introduction

As one of the most natural ways to achieve human competitive intelligence, genetic programming has shown its success in a series of application areas including electric circuit design (Koza, 1999), robot programming, molecular analysis and design, etc. Unfortunately, genetic programming is also notorious for its dependence on huge population sizes, its quick convergence after only 50 generations (the aging problem) (Luke, 2001), and the severe bloating phenomenon. Bloating prevents us from achieving solutions of the necessary complexity to solve many engineering problems. The aging problem (early death of evolutionary progress) makes it hard to achieve superior solutions even when more computing power is available. These two issues are related to the scalability of genetic programming. The lack of robustness of genetic programming search makes it not reliable enough to be used as a desktop tool for ordinary design engineers. An ideal genetic programming system should have characteristics such as robustness (reliably achieving good performance), sustainability (getting better performance with more computing effort), and scalability (evolving solutions with higher complexity when needed).

The lack of sustainable search capability of traditional genetic programming comes from two fundamental sources (Hu, Goodman *et al.*, 2003). One is the convergent nature of conventional EA models. The other is the bloating phenomenon caused by the variable length representation and its operators. When compared with the natural process of evolution, which does not normally face the issues of stagnation, bloating and lack of sustainability, two significant differences can be observed. One is that, in natural evolution, an enormous population size and diversity of environment is typically involved, while we can only use much smaller population sizes in artificial evolution, and typically in a much less diverse environment. Another major difference is that in biological evolution, evolution happens at all fitness levels, from the primitive single-cell organisms to high-level mammals. This kind of simultaneous multi-level (often corresponding to fitness level) evolution in a multitude of diverse environments may contribute to the sustainability of the biological process. A similar sustainable innovation also happens in human society, in such settings as educational systems. By educating students at all academic levels simultaneously and continuously, better and better students are trained to achieve increasing success. Although the population of students at one instant of time is of finite size, the unlimited timeframe and the continual importing of new students from kindergarten provide a non-depletable source for continuing possible innovation in the school system. In contrast, conventional GA/GP effectively terminate lower-level innovation early, as the average fitness of the population increases. Thus, the probability that the evolution process gets trapped in those "body-plans" discovered in early stage increases as the average fitness of the population increases.

In education systems, since the competition among students is segregated into different academic levels, the success of high-academic-level students (i.e., those in higher grades) never threatens the survival of students in lower grades, so the students in lower grades have a chance to grow and compete in an environment that is "fair" to them. More importantly, the lower-level knowledge framework assembled in the lower grades supports the possibility of more advanced knowledge frameworks in the higher levels. In terms of genetic programming, the early evolution stage usually establishes a basic framework for later evolution, similar to the speciation process in biological evolution. It is already well established that in tree-based genetic programming, after a certain number of generations, the structure around the root of the genetic programming tree is largely fixed.

Inspired by the sustainable evolution principle in biological and social systems, we proposed the hierarchical fair competition model (HFC) to achieve sustainable evolution for genetic programming (Hu & Goodman, 2002; Hu, Goodman *et al.*, 2002). The continuing search capability of HFC is achieved by ensuring a continuous supply and incorporation of low-level building blocks and by culturing and maintaining building blocks of intermediate levels. HFC employs an assembly-line structure in which subpopulations are hierarchically organized into different fitness levels, reducing the selection pressure within each subpopulation while maintaining the global selection pressure to help ensure exploitation of good building blocks found.

In this paper, we extract the basic ideas of the fair competition principle of HFC (Hu & Goodman, 2002) and apply it to single population evolution, which makes it more widely and readily usable in many existing GP/EA packages. Briefly, we introduce three mechanisms, called a) hierarchical elitism, b) explicit control of the breeding probability distribution over the fitness range, and c) explicit control of the distribution of the individuals over the fitness range, in order to achieve sustainable evolution for single population genetic programming and other EAs. We investigate the robustness and the effectiveness of this continuous HFC and how it helps to ameliorate the classical premature convergence problem in genetic programming. It is demonstrated that, similar to the discrete version of HFC in which the fitness range is segmented into discrete levels, the continuous HFC model (CHFC),with its explicit level maintenance mechanism, can also achieve robust and sustainable search as expected, on the problems explored.

2. Summary of HFC Principle and How it is Postulated to Help GP to Avoid Classical Problems

The HFC concept postulates that an important factor in the convergent nature of conventional EAs is their over-attention to the high-fitness individuals and neglecting of the importance of recombination and mutation among low-fitness individuals, even at later stages of evolution. The HFC concept is to sustain a supply of low-fitness individuals that represent low levels of "organization" – possessing low-order building blocks that may even be created at random, but not yet (or still) be present in higher-level individuals. These individuals are "sheltered" – allowed to recombine and mutate, regardless of the fact that higher-fitness individuals are present. The goal is not to preserve the low-fitness offspring of bad matings among high-fitness individuals, so hierarchical elitism is employed to prevent that.

High-fitness individuals in conventional EAs usually have difficulty in modifying their basic frameworks without decreasing their fitnesses dramatically, especially in the case of genetic programming. As the average absolute fitness of the population goes up, it becomes increasingly difficult for the degraded offspring of high-fitness parents to survive. The individuals that do change their basic frameworks just cannot survive with their low fitnesses. So, the probability of framework changes becomes increasingly small as evolution goes on. In addition, the frameworks of current high-fitness individuals are established with very limited testing during the early stages of evolution, which does not guarantee their global superiority. After the potential of these earliest-discovered frameworks is exhausted, no more framework innovation is likely, and genetic programming begins to exhibit stagnation. The scenario here is similar to the canalization concept of developmental biology.

This loss-of-explorative-capability (LEC) hypothesis to explain premature convergence can be understood more easily in genetic programming, where the early-stage evolution usually establishes the framework (or topological structure) of the individuals. However, LEC is also applicable to many GA problems with high epistasis, where disruption of the strongly coupled components of the genome usually leads to dramatic decrease of the fitness of the offspring. It is not the fitness diversity per se that contributes to the sustainable search capability of HFC or any other successful approach to sustainable evolution. In problems amenable to HFC, fitness levels are assumed to correspond to degrees of organization, or coupling among components, from unbiased random individuals to highly coupled, high-fitness individuals.

Sharing some similarity with HFC, Hutter (2002) proposed a Fitness Uniform Selection Scheme (FUSS) to preserve genetic diversity, which, however, is not the motivation of HFC. Compared to the CHFC proposed in this paper, FUSS lacks the explicit control of the breeding probability of all fitness levels and suffers from insufficient selection pressure to exploit high-fitness individuals. FUSS also does not have an explicit mechanism for allocating individuals to different fitness levels and thus easily suffers from unbalanced distribution of individuals. In summary, FUSS is still more like traditional diversity-oriented schemes while the CHFC proposed here is aimed at achieving sustainable search through separate control of the allocation of individuals to fitness levels and selection pressure.

The HFC mechanism aimed at avoiding loss of explorative capability are readily distinguished from many other diversity-oriented techniques. Their lack of sustainability can be attributed to their inability to make large framework modifications of high-fitness individuals at a later evolution stage, due to the fact that they cannot ensure sufficient survival time for low-fitness offspring of high-fitness parents with large modifications of their frameworks. Low fitness individuals face increasing selection pressure as the result of the constant increase of the average fitness of the population. So in the long run, large modifications of the frameworks of high-fitness individuals become increasingly unlikely.

Similarity (either genotypic- or phenotypic-based) fitness sharing, crowding and related methods (Goldberg &Richardson, 1987; De Jong, 1975; Li *et al.*, 2002;) tend to spread the individuals to the most prominent peaks, while in fitness sharing, the number of individuals allocated to each peak is proportionate to the peak fitness. First, in both methods, the average fitness of the population is constantly increasing. Second, since more and more higher peaks are identified, the number of individuals allocated to the lower-fitness peaks gradually decreases to almost 0 because of the limited population size. So after a sufficiently long evolution, low-fitness offspring generated via significant modification of the parent individual(s) can't typically survive long enough to be tuned into high-fitness individuals. There are many schemes that can achieve a high degree of fitness diversity, sharing some similarity with HFC – however, maintenance of fitness diversity is not the motivation of HFC. These schemes include Boltzmann-weighted selection (Maza & Tidor, 1991), Boltzmann tournament selection (Goldberg, 1990) and probabilistic crowding (Mengshoel & Goldberg, 1999). First, with all these methods, the average fitness of the population increases constantly. For the Boltzmann schemes, while any low-fitness individuals can survive initially when the temperature is high, the fitness range of surviving individuals becomes increasingly narrow due to the decrease of temperature. The center of the Boltzmann distribution of individuals over fitness also moves constantly away from lower-fitness toward higher-fitness. As a result, after the average fitness increases to a certain degree and the temperature is reduced to some level, low-fitness individuals with significant modifications cannot survive long enough to allow framework innovation. Similar analysis also applies to probabilistic crowding. This behavior is not entirely unwelcome, of course – in many cases, there is only a known amount of exploratory effort available to solve a problem, and a method that is tuned to exploiting that amount of effort to maximum effect is highly desirable. In contrast, a fixed CHFC strategy, without adaptation of its distributions, does not behave in that manner.

HFC addresses the above loss of exploratory capability problem by explicitly enforcing simultaneous evolution at all fitness levels. Instead of depleting the potentials of the basic frameworks of individuals established in early stages of evolution, the continuing innovation at the low-fitness level provides a constant stream of new frameworks for higher-level exploitation. These bottom-up moving individuals can survive the competition they face (thus allowing sufficient assembly of building blocks) due to the maintenance of fair competition: only individuals with similar performance levels are allowed to compete. From the perspective of avoiding local optima, instead of depending on jumping out of the local basins of attraction occupied by the highly converged population, which is often very improbable, HFC deals with trapping in local basins of attraction by allowing new individuals in other basins of attraction to emerge continuously. When examined in terms of search effort allocation, the competition among individuals is not constrained among the individuals of the current population. Instead, the allocation of computing resources takes a much stronger global view by ensuring that the assembly of building blocks happens at all absolute fitness levels. So the unit of allocation of search effort is not the individuals of the population, but the fitness levels of the whole possible fitness range.

Several forms of multi-population HFC have already been introduced (Hu & Goodman, 2002; Hu, Goodman *et al.*, 2002, Hu, Seo *et al.*, 2003). In those models, the number of fitness levels is specified by the user or may be determined adaptively. The major goals of the HFC framework can be summarized as follows:

a) To ensure fair competition among individuals by segregating competition among different levels of individuals,

   In this way, high-fitness individuals won't threaten the existence of those with lower-level fitness, as in most EAs. Extensive experimental study in Burke *et al.* (2002) suggested that a phenotypic diversity measure is superior to genotypic ones in predicting the run performance. This suggests that the straightforward and thorough phenotype-based protection of "inferior individuals" by HFC may be more reliable than other techniques such as fitness sharing or the species conserving algorithm SCGA (Li, *et al.*, 2002), in the sense that it doesn't require a distance measure which is hard to find in genetic programming, and it supports the continuing search for new frameworks at low-levels.

b) To assure that exploration and exploitation occur at all fitness levels

   Allocating computing cycles to demes of all fitness levels allows innovation to occur all the way from embryonic primitive individuals to highly evolved individuals. By cascading higher-level beyond lower-level fitness demes, a building block assembly line is established. In analogy to the Cambrian Age, HFC essentially keeps the emergence of new "species" operating, thereby enhancing the capability for sustainable evolution in genetic programming. It also greatly relieves the conventional demand for large population sizes to achieve good performance, as will be shown in the following experimental section.

c) To provide hierarchical elitism (one-way elitism)

In the multi-deme HFC, when the fitness of an offspring is lower than the fitness admission threshold of the current HFC level, the offspring will be discarded. Otherwise, it continues to stay in the current fitness level or is exported to a higher level, if appropriate. This appears as a one-way migration from lower-level demes to higher-level demes. In the context of genetic programming, since higher-fitness individuals usually have higher complexity, the elitism here can effectively remove those individuals that have unnecessary complexity for their fitness level. This mechanism may help to control the bloating phenomenon in genetic programming.

d) To incorporate new genetic material continually to provide a constant influx of evolutionary potential

A random individual generator is configured at the bottom of the HFC deme hierarchy to provide inflow of unbiased genetic material to higher fitness levels. From the bottom level to the highest fitness level, this convergent evolutionary process will never deplete its evolutionary potential. Instead, it provides a mechanism to allow innovation to happen continually at all fitness levels.

Previous work has demonstrated the capability of the HFC model for robust and continuing search capability in a series of genetic programming benchmark problems and a complex real-world circuit synthesis problem (Hu &Goodman, 2002). HFC turns out to be able to get better results more quickly and more robustly for these difficult multimodal problems in which premature convergence is of great concern. By continuing innovation, HFC achieves more robust search, which is not as sensitive to the population size as is conventional genetic programming. It gets better results by its inherent capability to avoid premature convergence.

In the original formulation of the HFC model, the whole fitness range is divided into discrete segments, each with a separate deme. It has the advantage of simplicity and efficiency. But the division is somehow arbitrary. Subsequently, the adaptive HFC model (Hu, Goodman *et al.*, 2002) demonstrated that it is not necessary to find a set of exact admission thresholds to allow fair competition since the performance is not very sensitive to the admission thresholds.

An extension of this idea is to completely remove the discrete segmentation of the demes and to use a single population to do hierarchical fair competition by introducing some special mechanisms to implement the HFC principle. This makes HFC easier to incorporate into many existing EA packages. As a complementary element, the continuous single-population HFC can also be applied to each subpopulation of the original HFC model, which is a naturally parallel model, or to any other parallel evolutionary algorithm.

3. Continuous Single-Population HFC Model (CHFC) for Sustainable Evolution

The CHFC EA model is composed of a set of components as illustrated in Fig. 6-1, including an HFC archive population, a "workshop" population, a random individual generator, a parent selector, a replacement selector, a breeding probability distribution control mechanism, and a removal control mechanism with a density estimator (determining what individuals, of what fitnesses, should be removed when new individuals enter the archive population at their specified fitness levels). In the next section, we first give an overview of the algorithm, and then introduce details of each component.

The Continuous HFC (CHFC) Algorithm

(1) Determine the following parameters besides other GP parameters:

HFC archive size, N_h

workshop size, N_w

Parameters for GP mutation and crossover operators

Breeding probability distribution and its parameter (here, based on an exponential distribution)

Density quota scaling distribution and its parameter (here we based it on an exponential distribution with the parameter λ *and the maximum ratio* K_m*)*

Pair selection method and its parameters (here, a random neighbor method with $p_{neighborhood} = 0.1$*)*

(2) Initialize the HFC archive randomly, finding the maximum and minimum fitnesses. Sort the individuals in the population by ascending fitness. The workshop deme is initially empty.

(3) Loop until termination

- According to breeding probability distribution $p_b(x)$, generate a value τ in [0, 1].

– Compute the target sampling fitness $f_t = f_{\min} + \tau\left(f_{\max} - f_{\min}\right)$
according the GP operator probabilities, decide on a current operation

If current operation is mutation, select an individual with index I in the population whose fitness is nearest to the sampling fitness; if current operation is crossover, select one parent as for mutation, then use the pair selector to select the second individual. Create one (if mutation) or two (if crossover) offspring and evaluate their fitnesses. If an offspring has a fitness greater than one of the parents, save it in the workshop deme; else discard this offspring.

Check if the workshop deme is full. If full, update the HFC archive with the new offspring in the workshop deme; else continue at top of loop.

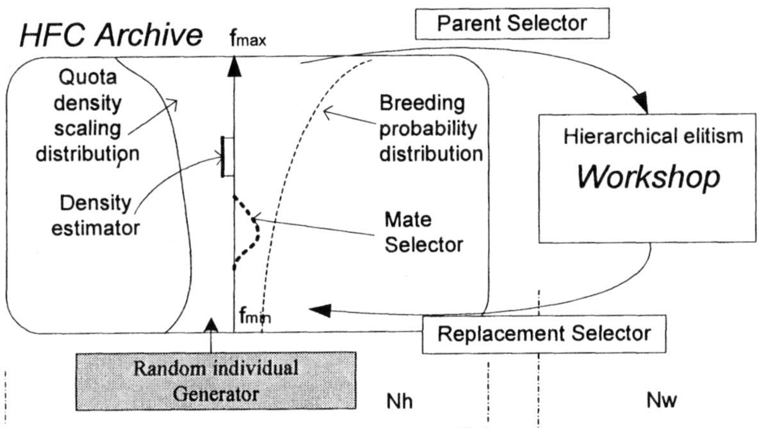

Figure 6-1. Structure of the CHFC model. Nh is the population size of the HFC archive. Nw is the Workshop size

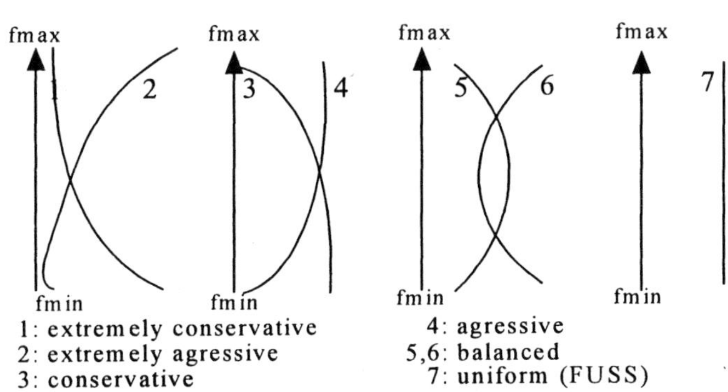

Figure 6-2. Seven types of distribution of the breeding probability of all fitness levels

HFC Archive Update Procedure

Calculate the normalized density of each individual in the HFC archive according to Eq. 3.8 below. For each individual in the workshop deme, place it in the HFC archive according to its fitness. Also, determine the individual with largest normalized density value in the archive, the "victim," and remove it. The exception is that if the victim is among the top K_e (elite group size) fitness genotypes present in the population, it is kept, and another victim is removed.

Explanation of the Continuous HFC model

The continuous HFC model is composed of the following components as shown in Fig. 6-1:

a) Two populations (only one of which is persistent) for storage of individuals. One of them is the HFC "archive" used to accommodate the breeding candidates, and works like a "normal population" in an EA. The other, "workshop," deme is used to hold temporarily the offspring generated during one generation. By varying the size of workshop, we can control the generation gap to be any desired value.
b) A random individual generator that continuously inserts some unbiased random genetic material into the HFC archive. It is the source of the fundamental genetic diversity.
c) A parent selector that is used to select breeding parents according to the breeding probability distribution for recombination or mutation. The breeding probability is introduced to ensure sufficient selection pressure to exploit early-discovered high-fitness individuals, analogous to the allocation of larger subpopulation sizes to higher HFC levels (Hu & Goodman, 2002). Rather than using the fitness uniform selection operator (Hutter, 2002), a higher probability is typically allocated to higher fitness individuals, although that is under user control. However, these higher fitness individuals do not dominate the population as in a conventional EA, because of the individual allocation control mechanism used to prevent uncontrolled growth of high-fitness individuals.

   The breeding probability distribution can be linear, exponential or any other distribution; typically, one is chosen that favors higher fitness individuals for breeding. Fig. 6-2 lists some typical distributions; Fitness Uniform Selection (Hutter, 2002) uses the last, uniform, distribution.

   In our experiments, the following (truncated, reversed exponential) distribution was used:

$$p_b(x) = \frac{\lambda e^{-\lambda(1-x)}}{1-e^{-\lambda}} \quad for \;\; x \in [0, 1] \qquad (3.1)$$

The parameter λ is a parameter to control the selection pressure or the aggressiveness of exploitation vs. exploration.

First a random value r between [0, 1] is generated according to the distribution $p_b(x)$; then the target sampling fitness is calculated as

$$f_t = f_{\min} + r*(f_{\max} - f_{\min}) \qquad (3.2)$$

After determining f_t, the individual I whose fitness is closest to f_t is determined.

The pair selection method for crossover can be implemented in the following ways:

According to the fair competition principle, the second selection is dependent on the first one and should be selected from the vicinity of the first selected individual. This is called the *Random Neighbor* approach. There are two ways to define the neighborhood of an individual (using the implicit level concept of HFC): 1) an individual (mate) with an index within *K/2* steps (offset of index number in the sorted population) away from the first individual may be randomly selected,

$$[I - \frac{K}{2} * \lambda(f_t),\ I + \frac{K}{2} * \lambda(f_t)] \qquad (3.3)$$

where K is neighborhood size, $\lambda(f_t)$ is a scaling parameter to allow different neighborhood sizes in different fitness levels. (In this paper, $\lambda(f_t)$ is set to 1 for simplicity.)

Or, 2) the entire fitness range may be divided into L levels, and the second individual (mate) may be chosen from the set of individuals with fitnesses in the range

$$[f_t - \frac{f_{\max} - f_{\min}}{2L} * \lambda(f_t),\ f_t + \frac{f_{\max} - f_{\min}}{2L} * \lambda(f_t)], \qquad (3.4)$$

using a percentage of the HFC archive size to decide the L value

$$L = p_{neighborsize} * N_h \qquad (3.5)$$

where N_h is the HFC archive size.

We have also tested the other two approaches used in FUSS (Hutter 2002) (to be reported elsewhere), the *Next Nearest* approach, where the second individual is the one whose fitness is second closest to the sampling fitness, and *Independent FUSS*, where we select the second individual independently, just as we selected the first individual.

d) A replacement selector, which decides which individuals in the HFC archive are to be replaced by the new offspring. This is based on the density estimator and the quota density scaling distribution.

e) A density estimator, which is used to estimate the density around each individual, with additional scaling by the quota density scaling distribution. By adjusting the scaling distribution, we can achieve any desired proportion of individuals at the various fitness levels. Here, the relative density concept is used. That is, we evaluate by what percentage the current density surpasses its nominal (quota) density at any point on the quota density scaling distribution. The individual with the highest percentage becomes the next individual to be replaced.

The relative density in the neighborhood of an individual can be estimated with either of two approaches:

Approach 1: Select two individuals from the sorted (ascending fitness) HFC archive with indices within K/2 steps of current individual I. The raw density is estimated as

$$D_{raw}(I) = f(I + \frac{K}{2}) - f(I - \frac{K}{2}) \qquad (3.6)$$

where $D_{raw}(I)$ is called the raw fitness density of an individual (the implicit denominator, K, is constant, so is ignored).
The raw nominal (quota) density of an individual $D_{rawq}(I)$ is computed as:

$$D_{rawq} = \frac{K}{N_h} * (f_{\max} - f_{\min}) \qquad (3.7)$$

where N_h is the population size of the HFC archive. Finally, the normalized density is computed as the relative density:

$$D_{norm} = \frac{D_{raw}(I) - D_{rawq} * r_d}{D_{rawq} * r_d} \qquad (3.8)$$

where r_d is a scaling parameter to be described below.

Approach 2: Assume the whole fitness range is divided into L levels, and count the number of individuals (M) whose fitnesses are in the range

$$[f(I) - \frac{f_{\max} - f_{\min}}{2L},\ f(I) + \frac{f_{\max} - f_{\min}}{2L}] \qquad (3.9)$$

The raw density of an individual I is $D_{raw}(I) = M$
The raw nominal quota density is:

$$D_{rawq} = \frac{N_h}{L} \qquad (3.10)$$

where N_h is population size of the HFC archive
The normalized density is then calculated as (3.8).

We use a percentage of the HFC archive size to decide the above L, where $L = p_{neighborsize} * N_h$, and N_h is the HFC archive size.

The parameter r_d in (3.8) is used to bias the allocation of individuals over different fitness levels. It is generated according to a user-specified quota density scaling distribution, which determines the relative size of high-fitness individuals and low-fitness ones. It is important to maintain sufficient high-fitness individuals for effective search. For different problems, this distribution can also have different types, as illustrated in Fig. 6-2. In our experiments, the following distribution was used.

$$r_d(x) = \frac{a + \lambda e^{-\lambda x}}{a + 1 - e^{-\lambda}} \quad where \quad a = \frac{(e^{-\lambda} - K_m)}{K_m - 1}\lambda \qquad (3.11)$$

where λ is used to control the shape of the exponential distribution. K_m is a user-specified maximum ratio of $r_d(1)/r_d(0)$. It is used to control the maximum ratio of individuals that are allowed between the highest fitness level and the lowest fitness level.

f) Hierarchical elitism is enforced during the insertion of offspring into the workshop. When the fitness of the offspring is less than that of both parents, the offspring is discarded (not used to update the HFC archive).

4. Experiments and Observations

The basic motivation of HFC is to ensure sustainable search by avoiding premature convergence. It is most useful in the case of difficult problems where local optima are of great concern. Here the Santa Fe trail artificial ant problem (Koza, 1994) is used to illustrate how CHFC can improve the performance of genetic programming, dealing with the bloating problem and exerting control over the fitness distribution of the population.

Table 6-1. Parameters Used in the Various GP Experiments on Santa Fe Trail

Common GP Parameters	Additional CHFC parameter
Function and terminal set {Forward, Right, Left, Prog2, Prog3, IfFoodAhead}	workshop size: 20
Max evaluation: 100000	Breeding probability distribution type: exponential (lambda=4.0)
Population size: 1000	Density scaling distribution: exponential (lambda=4.0, maxRatio=10)
Tournament size: 2 (for steady-state and generational GP)	Elite group size $K_e = 1$
Crossover rate: 0.9	Probability of introducing random individuals each generation: 0.2
Shrink mutation: 0.05	Percentage generated randomly in that event: 0.05
Point mutation: 0.05	
Maxtree depth: 17 mutate maxdepth: 5	
Init max depth: 6 Init min depth: 2	

The artificial ant problem involves evolving a sequence of movements in a two-dimensional grid to collect as much food (distributed on the grid) as possible. We used the Santa Fe Trail, with a size of 32*32. The maximum amount of food is 89. We defined the fitness as

$$fitness = \frac{1}{1 + \frac{\text{max food} - \text{collected food}}{\text{maxfood}}} \quad (3.12)$$

We compared the performance of CHFC with both steady-state GP and generational GP. The parameters are listed in Table 6-1.

Twenty experiments were run for each algorithm. The experimental results are summarized in Table 6-2.

Table 6-2. Peformance comparison of CHFC and conventional GP in Santa Fe Trail artificial ant problem

Algorithm	Best mean fitness	Std. dev.
CHFC-GP	0.92	0.05
Steady state GP	0.76	0.073
Generational GP	0.81	0.089

To gain insight into the evolutionary process, we drew four histograms showing the progression of the fitness distribution of the population during the run. (Since steady-state GP has similar behavior, only generational GP is compared to CHFC-GP.) Fig. 6-3 and 6-4 show the fitness distributions of generational-GP and CHFC-GP for the ant problem. It is clear that conventional genetic programming suffers a lot from the domination of early-discovered high-fitness individuals while CHFC, through its controlling mechanism based on fair competition, can always maintain a balanced distribution of both high-fitness individuals for exploitation and low-fitness individuals for "Cambrian innovation"—the innovation of the basic framework.

Fig. 6-5 plots the fitness growth process of the three algorithms. It is impressive that CHFC, by avoiding premature convergence, continues to generate innovative (and superior) solutions when the steady-state and generational GP have stagnated.

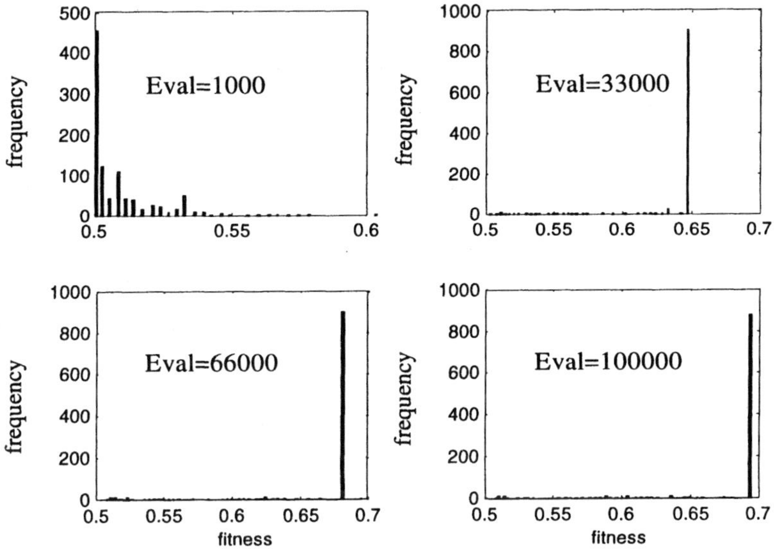

Figure 6-3. The convergent nature of conventional generational GP. The high-fitness individuals rapidly dominate the population of this typical run. The subplots were prepared by sampling at evaluation numbers 1000, 33000, 66000, and 100000.

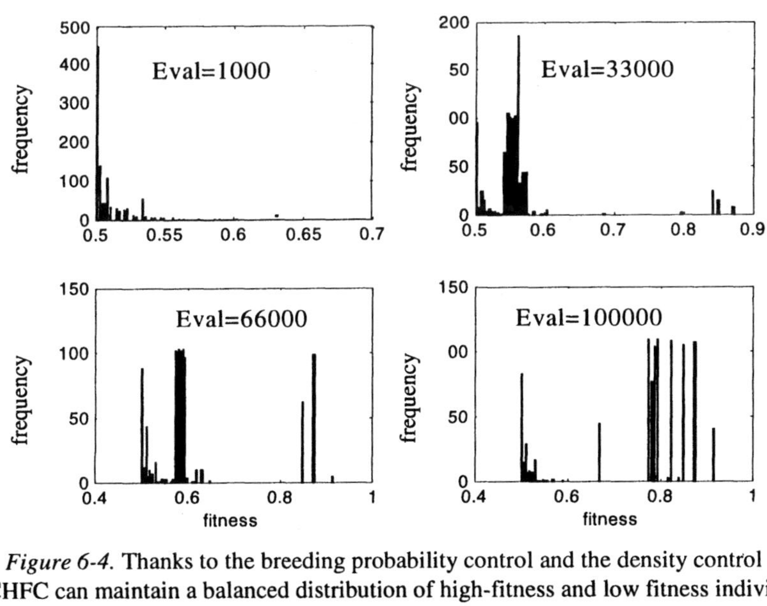

Figure 6-4. Thanks to the breeding probability control and the density control mechanism, CHFC can maintain a balanced distribution of high-fitness and low fitness individuals without any risk of domination in this typical run, evidently without sacrificing its exploration of the high-fitness range. Subplots are prepared at evaluation numbers 1000, 33000, 66000, 100000

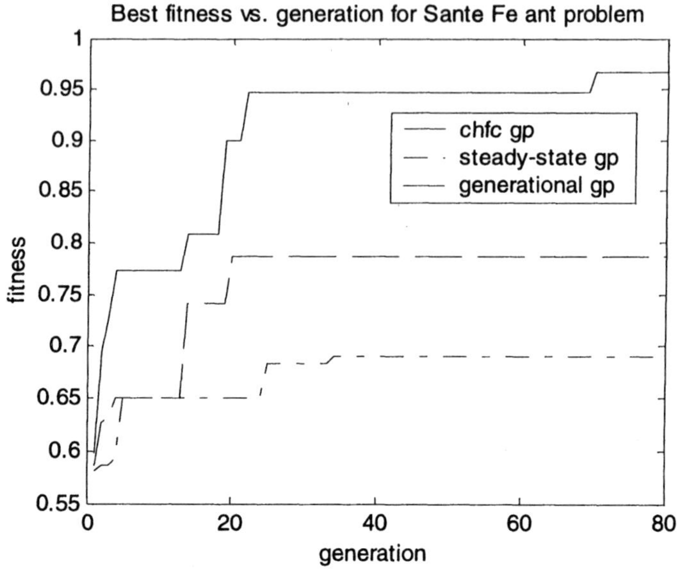

Figure 6-5. Comparison of CHFC with steady-state and generational GP. Having avoided the local optima that trapped the other runs, CHFC makes consistent progress and discovers improved solutions.

5. Conclusions and Future Work

By extracting the key ideas of the HFC model and combining some ideas from the fitness uniform selection (FUSS) scheme, this paper proposes the CHFC model for sustainable evolution with single population EAs. CHFC is characterized by its hierarchical elitism to avoid degradation of individuals, explicit control of the distribution of individuals over the fitness range to ensure a continuing supply of building blocks of different orders, and explicit control of breeding probability distribution over the fitness range to enforce strong selection pressure for exploiting early-discovered high-fitness individuals, while lessening the risk of domination and premature convergence. Indeed, CHFC is not a convergent method at all – it seeks to find high-fitness individuals without ever "pulling up the ladder" of building blocks that descends all the way to low-fitness, randomly-generated individuals. CHFC's strength appears to be its providing of separate control of selection pressure from the specification of a desired distribution of individuals across the fitness landscape.

The well-known artificial ant genetic programming benchmark problem is used to investigate the robustness and continuing search capability of CHFC-GP. CHFC provides a simple approach to ensure sustainable and robust search in this genetic programming problem. It obtained better results more robustly and more quickly while avoiding convergence, providing sustainable search. Along with the original HFC model, for which sustainability of evolution has been demonstrated on several more difficult real-world synthesis problems, CHFC adds further evidence that the HFC EA model is able to transform conventional EAs from a convergent framework into a sustainable one.

Our loss-of-explorative-capability (LEC) hypothesis provides an interesting perspective on the aging phenomenon in genetic programming. "50 generations of innovation" seems to come from the inherent convergent nature of the conventional EA framework coupled with a common property in genetic programming: large structural modification by crossover or mutation usually leads to low-fitness individuals that rarely survive in later stages of evolution. HFC algorithms devised on the basis of this theoretical understanding have allowed us to use much smaller population sizes to achieve desired solution qualities on a variety of problems, while enjoying much stronger robustness (Hu, Goodman *et al.*, 2003). The separate control of selection pressure and the individual distribution over fitness range proposed in this CHFC model may provide practitioners with much more flexible control over evolutionary search: one can be aggressive without risk of premature convergence.

Experiments have shown that there are cases in which the fitness distribution is hard to control with the current methods. A better density control mechanism needs to be devised. It will also be interesting to investigate the theoretical properties of CHFC and to seek to eliminate or routinize the setting of many of its parameters. In addition, hybridizing CHFC with the original multi-population HFC may provide a good model for large-scale parallel GP implementation. A topic for future consideration is how fitness sharing can be applied to each fitness level and whether there is any kind of synergy that can thereby be exploited.

Acknowledgement

This work is developed based on the Open beagle GP C++ framework. The authors sincerely thank its author Christian Gagne for lasting and patient help. We thank David E. Goldberg, Joseph A. Driscoll, Bill Worzel, and Duncan MacLean for their careful and insightful suggestions in their reviews. This work is sponsored by the National Science Foundation under contract DMI 0084934.

References

Burke, E., Gustafson, S. and Kendall, G. (2002). A Survey and Analysis of Diversity Measures in Genetic Programming. In *Proceedings of the Genetic and Evolutionary Computation Conference.* New York, July.

De Jong K. A. (1975). An Analysis of the Behavior of a Class of Genetic Adaptive Systems Doctoral Dissertation, University of Michigan. *Dissertation Abstracts Int.* 36 5140B (University Microfilms 76-9381).

Goldberg, D. E. (1990). A Note on Boltzmann Tournament Selection for Genetic Algorithms and Population-oriented Simulated Annealing. *Complex Systems 4*: 445-460

Goldberg, D. E. and Richardson, J. (1987). Genetic Algorithms with Sharing for Multimodal Function Optimization. In *Proceedings of the 2nd Int. Conf. on Genetic Algorithms.* Cambridge, MA, J. J. Grefenstette.

Hu, J. and Goodman, E. D. (2002). Hierarchical Fair Competition Model for Parallel Evolutionary Algorithms. In *Proceedings of the Congress on Evolutionary Computation, CEC 2002, IEEE World Congress on Computational Intelligence*, Honolulu, Hawaii.

Hu, J., Goodman, E. D., Seo, K. and Pei, M. (2002). Adaptive Hierarchical Fair Competition (AHFC) Model for Parallel Evolutionary Algorithms. In *Proceedings of the Genetic and Evolutionary Computation Conference*, pp. 772-779. New York.

Hu, J., Goodman, E. D., Seo, K., Fan, Z. and Rosenberg, R. C. (2003). HFC: a Continuing EA Framework for Scalable Evolutionary Synthesis. In *Proceedings of the AAAI-2003 Spring Symposium on Computational Synthesis*. Stanford, CA. AAAI Press.

Hu, J., Seo, K., Fan, Z., Rosenberg, R. C., and Goodman, E. D. (2003). HEMO: A Sustainable Multi-Oobjective Evolutionary Optimization Framework. To appear in *Proceedings of the Genetic and Evolutionary Computation Conference.* Chicago.

Hutter, M. (2002). Fitness Uniform Selection to Preserve Genetic Diversity. In *Proceedings of the 2002 Congress on Evolutionary Computation*, 783-788. Hawaii.

Koza, J. R. (1994). *Genetic Programming II: Automatic Discovery of Reusable Programs.* Cambridge Massachusetts: MIT Press.

Koza, J., Bennett, F. H, Andre, D., and Keane, M. (1999). A. *Genetic Programming III.* San Francisco, CA: Morgan Kauffmann.

Langdon, W. B. (1998). *Genetic Programming and Data Structures.* Dordrecht: Kluwer.

Li, J. P., Balazs, M., Parks, G. T., and Clarkson, P. J. (2002). A Species Conserving Genetic Algorithm for Multimodal Function Optimization. *Evolutionary Computation* 10(3): 207-234

Luke, S. (2001). When short runs beat long runs. In GECCO-2001: In *Proceedings of the Genetic and Evolutionary Computation Conference*, pp. 74-80. Lee Spector *et al*, eds. Morgan Kaufmann.

Maza, M. de la and Tidor, B. (1991). Boltzmann Weighted Selection Improves Performance of Genetic Algorithms. A.I. Memo 1345. Artificial Intelligence Laboratory, MIT, Cambridge.

Mengshoel, O. J. and Goldberg, D. E. (1999). Probabilistic Crowding: Deterministic Crowding with Probabilisitic Replacement. In *Proceedings of the Genetic and Evolutionary Computation Conference,* pp. 409-416. Orlando, Florida, USA.

Waddington, C. H. (1942). Canalization of Development and the Inheritance of Acquired Characters. *Nature 150*: 563-565.

Chapter 7

WHAT MAKES A PROBLEM GP-HARD?

A Look at How Structure Affects Content

Jason M. Daida

The University of Michigan, Center for the Study of Complex Systems and Space Physics Research Laboratory, 2455 Hayward Avenue, Ann Arbor, Michigan USA 48109-2143, daida@umich.edu

Abstract: This chapter summarizes theoretical work at The University of Michigan concerning the question: "What makes a problem difficult for genetic programming to solve?" It specifically describes linkages between content, tree structures, and problem difficulty in genetic programming. It focuses on the significance of structure in influencing problem difficulty.

Key words: GP theory, tree structures, problem difficulty, GP-hard, test problems

1. Introduction

Most of what I have learned about doing large, significant problems with GP has come from working on real-world applications. In particular, if I wanted to do a significant (and presumably difficult) problem nowadays, I would look to Koza (Koza et al., 1999). A short list of what I get from that work would be the following:

- Use large populations
- Use structure-altering operations
- Specify GP solutions as meta-programs and choose functions and terminals accordingly.

My list might differ from anyone else's, if only because my take on Koza's work has been influenced by a *structuralist* view of complex systems. I define a *structuralist* as one who examines the relationship of parts to a thing. I would subsequently be interested in the structure and topology of interconnections between nodes in GP. With this view, my group has focused on the theoretical aspects of why some GP problems are more difficult than others. Still, it is ironic that I would turn to applications for understanding on how to *really* solve problems.

While GP theory has come a long way in the past decade, there is still much that theory has not explained. Even simple questions have yet to be answered. For example, although we have hypotheses, we still do not know why some problems are more difficult than others for GP. We also do not have a bound for what can or cannot be solved with even the most basic of GP systems. We have no theory that would tell us what to do if GP were to get stuck on a problem.

On the other hand, there are many success stories, anecdotes, and *ad hoc* methods that have been proven to work—if only for a particular instance. Do *X*, because we have shown it to work for problem *Y*. For a while, this approach resulted in one paper per application; it was justifiable, too. Every new application (seemingly) resulted in a new operator or a major adjustment to GP. I should know... guilty as charged.

I would suggest that the situation between theory and application in GP is like navigating the eastern seaboard of the United States. In the late 1800s, travel there was treacherous and difficult. Good maps (theory) were absent. A person relied on the experience (application) of navigators just to remain alive. An observer wrote in 1888, "...so little was known of the dangers attending navigation along our extensive seaboard, that those who engaged in commercial ventures were constrained to rely upon local knowledge and the reports of the hardy navigators who might carry their venture to success (Ogden, 1888)." I would say that the success of GP lies now with the practitioners, the "hardy" that make a living with GP. The "mapmaker" theoreticians are playing catch-up...

The purpose of this chapter is to highlight what our theory says in relationship to the question "What makes a problem GP-difficult?" It is not meant to be a rigorous treatment of theory. Instead, it summarizes what my group has learned. Much of what we know supports what practitioners have implicitly known for years. This is not to say that there is no room for surprises. In this exercise of making implicit knowledge explicit, surprises can and still do happen. Counterintuitive findings can occur by logically extending explicit knowledge. (My college professors liked to point this out when I took physics as an undergraduate.) Explicit knowledge also defines what we do not know—but would be good to know to improve the practice of GP.

This chapter specifically discusses three claims. The first is "content drives difficulty" (Section 2). The second is "structure matters" (Section 3). The third is "structure determines limits" (Section 4). Each claim is followed by an overview, followed by a summary of what our group knows, followed by a summary of what our group does not know. Section 5 discusses all of these claims in the broader context of GP practice.

2. "Content Drives Difficulty"

2.1 Overview

A response to the question of "What makes a problem GP-hard?" starts with the substance (or programmatic content) of a problem. Substance is "distilled" from the nuances of our world into code by using Koza's preparatory steps (see Koza et al., 1999, p. 33). This includes a determination of a problem's function and terminal sets. This also includes matters of determining a fitness measure, the parameters for controlling a run, and termination criteria.

Substance can be classified. As it turns out, many problems share similar function and terminal sets. Consequently, such problems can be grouped together. Such a grouping roughly describes the *domain* of a problem. In GP, a *domain* would refer to the broad category of problems that can share from a common pool of functions and terminals. So for example, there is the domain of Boolean problems of which the *6-input multiplexer* problem is an instance.

A common view in the GP community is that the domain of a problem determines, in broad measure, that problem's difficulty. The reasoning for this view would be something along these lines:

- A problem's domain determines the functions and terminals to be used in GP
- These functions and terminals determine the solution outcomes for a problem, since a solution would consist of some combinatorial permutation of these functions and terminals
- The likelihood of finding a solution among all combinatorial permutations of functions and terminals determines that problem's difficulty.

Consequently, a typical approach to testing GP systems or theory has been through a test suite. Although there is no formal standard of what is in a test suite, the idea is simple. Pick problems from a few representative domains. So a typical test suite might consist of *sextic* (a data-modeling domain), *multiplexer* (a Boolean domain), and *ant* (an agent domain). The use of at least two problems from two different domains has been ingrained in many papers that feature a new method or an empirical approach to theory.

2.2 What We Know

We know that the substance of a problem can significantly affect its difficulty. This understanding comes from our work with a specific instance in data modeling—the *binomial-3* problem[1] (Daida et al., 1999; Daida et al., 2001).

The *binomial-3* represents one of several tunably difficult problems in GP. It does so not by changing the number of terminals that is used to solve the problem, but by changing the content within those terminals. Since a portion of that content is presumed to be samples from the set of real numbers, one simply changes from where those samples are taken to effect tuning. For the *binomial-3* problem, the tuning parameter that changes the content within terminal nodes is α.

A simple metric that measures problem difficulty is a count of success. Since GP is stochastic, we can count the number of times GP solves a problem and divide that number by the total number of times GP was run. A very easy problem would then result in GP solving the problem correctly, say 1,000 times out of 1,000 trials. A very difficult problem would likely result in GP not solving the problem at all for the same number of trials.

The *binomial-3* exhibits a wide range of tunability. Figure 7-1 illustrates tunability for three different settings. (See Daida et al., 2001 for GP and *binomial-3* setup.) Each histogram summarizes 600 trials. The best attainable score is 50. More trials scoring 50 indicate an easier setting. Consequently, the leftmost plot shows the results for a "moderately difficult" setting (219 successful trials out of 600 trials total). The center plot shows the results for a "difficult" setting (9 out of 600). The rightmost plot shows the results for a "very difficult" setting (3 out of 600 trials, with a cluster of trials that failed with scores of less than 10).

[1] The *binomial-3* requires GP to solve for the equation that fits a set of data points. For test purposes, the model underlying these data points is known and is generated by the equation $f(x) = (1 + x)^3$. The function set typically consists of addition, subtraction, multiplication, and protected division. The terminal set is x and a set of random real number constants that have a range of $-\alpha$ to α, where α is some real number value. Tuning is done by changing α.

2.3 What We Don't Know

With the *binomial-3*, all we have done was to identify a single contributor for making a problem difficult—interactions between functions and terminals. What we don't know is still abundant. In particular:

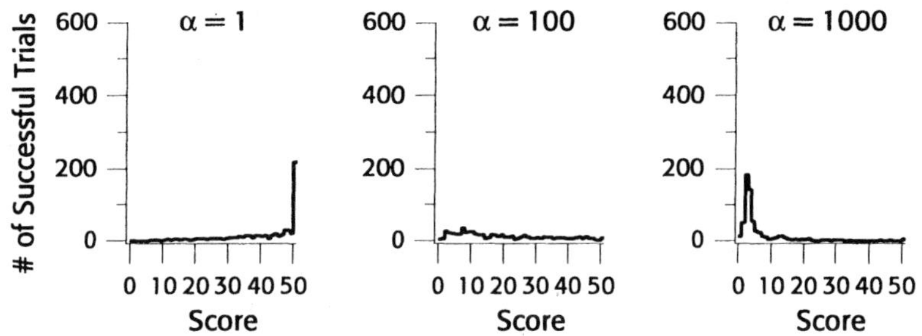

Figure 7-1. Tunable difficulty. The *binomial-3* problem can be made more or less difficult by changing α. Easier versions of the problem have many more trials that score higher. A score of 50 is perfect.

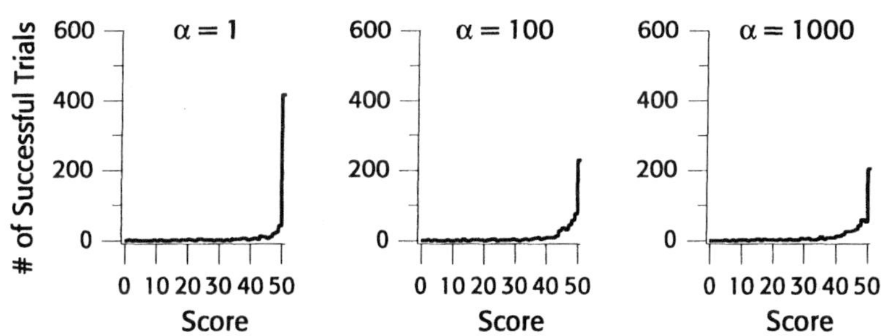

Figure 7-2. Consequence of using tournament selection. All of the parameter values were the same as those used to generate the data shown in Figure 7-1. The exception was that this data was generated using tournament, instead of fitness-proportionate selection.

- We don't have a metric that can tell us something useful about interactions between functions and terminals.
- We don't know how much, how little, or even what kind of interaction is needed to make a problem tractable.
- We don't have a way to measure the effectiveness of a set of components for a given problem.
- We don't know how *binomial-3* findings apply to other domains.

To illustrate just how little we know, Figure 7-2 summarizes results for the same three settings as shown in Figure 7-1. The difference is that instead of fitness-proportionate selection—which was used to generate the results in Figure 7-1—tournament selection was used. Simply changing the method of selection resulted in significant improvement. The smallest improvement was 90%. The largest improvement was over 6,600%. What was "very difficult" under fitness-proportionate selection becomes "moderately difficult" under tournament selection. Why?

Theoretical work by Bickle and Thiele showed that tournament selection in GP would result in a loss of diversity (Bickle and Thiele, 1995). Experimental work like (McPhee and Hopper, 1999) supports this claim. That a rapid loss in diversity is linked with greater problem solving ability is counterintuitive in GA (genetic algorithms). In GA, premature convergence has generally been associated with a consequence of losing diversity (Mitchell, 1996). However, the practice of losing diversity was adopted early on in GP, beginning with Koza's use of greedy over-selection (Koza, 1992).

The linkage between function / terminal interactions, selection methods, diversity, and problem difficulty has not been formally described. In some manner, tournament selection has altered the way that components interact. What has happened, though, is unclear. Regardless, the GP community has favored tournament selection over fitness-proportionate selection. The reasons for choosing tournament selection have had more to do with computational ease than with problem solvability (Banzhaf et al., 1998).

3. "Structure Matters"

3.1 Overview

By *structure*, I refer to the links that connect functions and terminals to each other. Its meaning is broader than *syntax*, which requires knowing something about the content that components have. Structure implies a graph, which consists of nodes (vertices) and links (edges). The classic GP solution is expressed as a tree structure.

Some of the earliest attempts to improve GP modified the structure of potential solutions. For example, Koza described automatically defined functions (ADFs) in structures that enable code reuse (Koza, 1994). The idea is that with reuse, GP obtains smaller, denser code. Nowadays, to achieve their human-competitive results, Koza et al. have developed a contemporary method similar to ADFs, i.e., architecture-altering operations (Koza et al., 1999).

Meanwhile, theoreticians have tried to understand the role of structure in the simplest of GP systems. Rosca and Ballard suggest that solution outcomes are determined, in part, by rooted tree schemata (Rosca, 1997; Rosca and Ballard, 1999). They argue that during a run, GP identifies these structures first and subsequently builds upon these structures to form solutions.

Other researchers have independently supported Rosca and Ballard's theory, which highlight the importance of root structures. This support has come largely in the form of empirical studies that feature selected benchmark problems. These would include (McPhee and Hopper, 1999; Gathercole and Ross, 1996). Of these two, McPhee and Hopper's work makes the strongest empirical case that the root structure plays a key role in determining solution outcomes for GP in general.

Structure, of course, plays a significant role in other studies in GP. For example, structure was a key determinant of fitness in one of the earliest adaptations of the Royal Road problem (Mitchell et al., 1992) (in GA) to GP (Punch et al., 1996). Work by Goldberg and O'Reilly resulted in the development of the *ORDER* and *MAJORITY* fitness problems in which the structures of solution outcomes were driven by the problem at hand (Goldberg and O'Reilly, 1998; O'Reilly and Goldberg, 1998). Structure has played a key role in the analysis by Soule, Foster et al., who have looked at the evolution of the shape in GP outcomes (Soule et al., 1996; Soule and Foster, 1997; Soule and Foster, 1998). Soule and Foster's work on the evolution of tree shapes has since been carried forward and extended by Langdon (Langdon et al., 1999; Langdon, 2000; Langdon and Poli, 2002).

Structure plays a key role, too, in schema theory for GP. Poli's work (and later Langdon and Poli) presumes schemata that are structurally based (i.e., have a certain size and depth) (Langdon and Poli, 2002; Poli and Langdon, 1997; Poli and Langdon, 1998; Poli, 2000). Bloating has also been associated with structure; it has been Langdon and Poli's contention that bloating occurs partly as a result of a random walk over a landscape of tree-shape distributions (e.g., Langdon and Poli, 1997; Langdon, 2000).

3.2 What We Know

Until recently, we have only been able to speculate about the role of structure in GP. We have had to infer what those structures were like based on indirect observations. Direct observations were impractical, if only because the visualization of trees does become unwieldy after 60 or so nodes (Knuth, 1997). However, we recently published a method for visualizing large trees in (Daida, 2002; Daida et al., 2003a).

Much of the difficulty of visualizing large trees stems from unused space. As trees get larger, more space gets devoted to their leaves. This means that the more unused space there is, the closer one is to the root.

We can use an analogy of looking at real trees to get a sense of our method for visualizing large tree structures. The conventional method for visualizing a tree is like a side view of a real tree. Our method is like a top view. The root node is placed at the center. Branches radiate outwards. Figure 7-3 shows what a full tree would look like using our method.

We can similarly visualize arbitrary tree structures. However, instead of fanning branches out in an arbitrary manner, we branch them out as if they

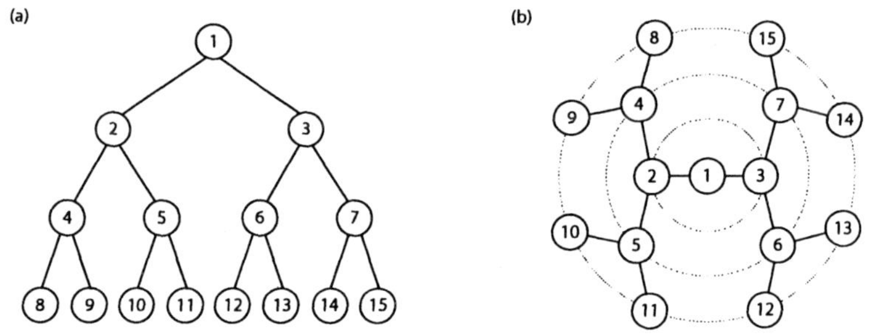

Figure 7-3. (a) Conventional view of a tree. (b) A view of the same tree using our method.

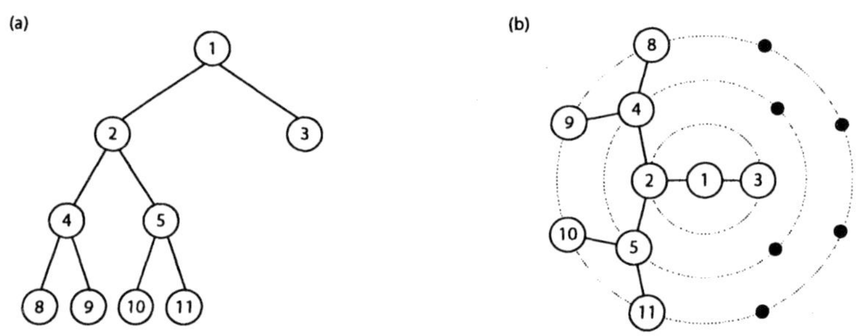

Figure 7-4. (a) Conventional view of an arbitrary tree. (b) A view of the same tree using our method. The dots are for reference only.

were part of a full tree. Figure 7-4 shows what an arbitrary tree would look like. Note that the labeling scheme remains consistent between full and arbitrary trees.

We can use this method to visualize large trees. Figure 7-5 shows two larger trees. One is a full tree of depth 10. The other is a GP tree of depth 26. The gray circle around the tree on the right represents depth 26, which is shown for reference.

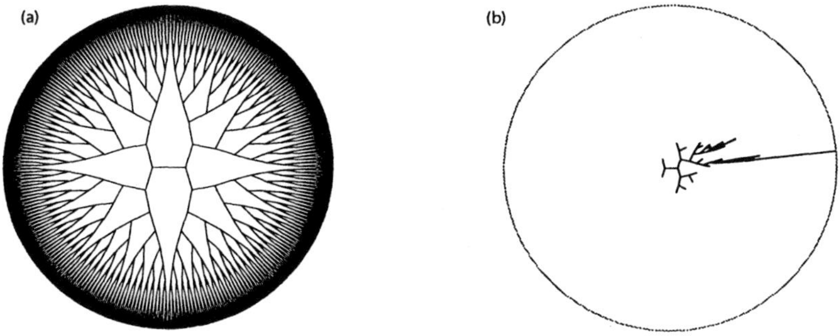

Figure 7-5. (a) Full tree of depth 10, 2047 nodes. (b) GP tree of depth 26, 461 nodes.

The graphics in Figure 7-5 can be made smaller and still allow one to pick out visual cues about a tree's structure. We can use this property to see the structures of many trees. Of interest would be an examination of what trees look like for a representative sample of a population.

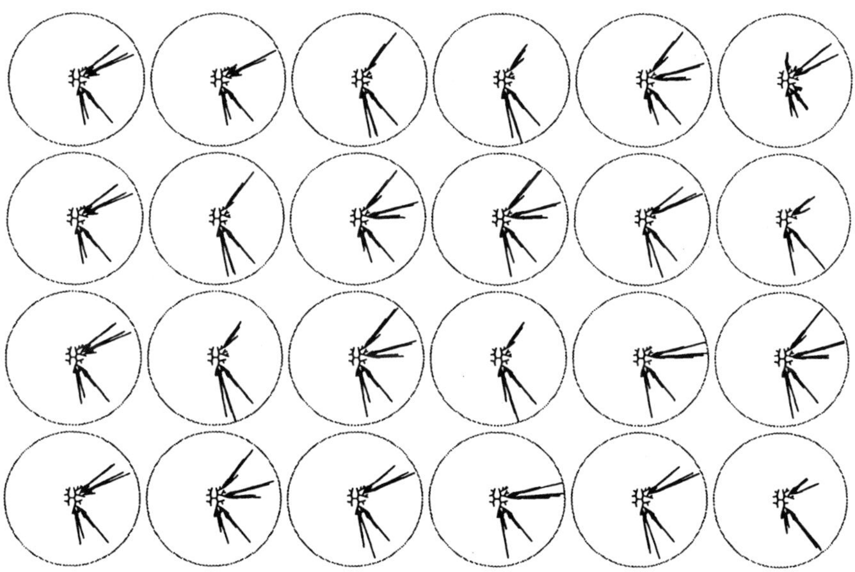

Figure 7-6. A sample from a GP population (fitness-proportionate selection).

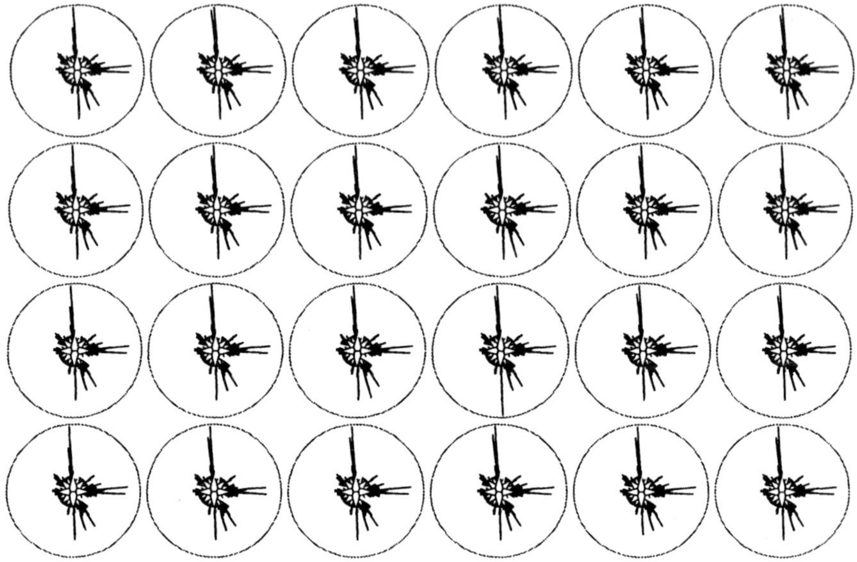

Figure 7-7. A sample from a GP population (tournament selection).

Figures 7-6 and 7-7 each show a sample of 24 individuals out of a population of 500. Both populations were from *binomial-3* trials. Both trials were conducted under the "very difficult" setting, which was discussed in Section 2. Figure 7-6 corresponds to fitness-proportionate selection. Figure 7-7 corresponds to tournament selection. Keep in mind that tournament selection outperformed fitness-proportionate selection by over 6,600%.

Both of these figures indicate that regardless of selection method, the resulting GP trees are generally sparse. Closer inspection reveals some differences, though. One significant difference may be that the samples obtained under fitness-proportionate selection seem to be structurally more diverse than those obtained under tournament selection. A discussion of this difference is given in (Daida et al., 2003a).

3.3 What We Don't Know

Although the visualizations show more tree structures than have been previously possible, at best they provide only a clue. A number of questions remain, including: What drives the shape of these structures? Are these figures representative of GP as a whole? Are there limits to the kinds of structures that can be generated in GP? How do structures evolve over time? Why is there a lack of structural diversity? Did the type of selection used influence the kind of structures that were generated? How does content correlate with structure? Which metrics best characterize these structural features? How do structures vary between trials? Is there a connection between GP trees and the difficulty of a problem? Clearly, there is a degree of structural similarity in these small samples. Unfortunately, there are more questions that are raised by a suggestion of structural pattern than can be adequately answered in this chapter.

4. "Structure Determines Limits"

4.1 Overview

In the GP community, it is thought that structure matters. However, it is also thought that if structure matters, then content matters *more*. Surely that must be the case when it comes to problem difficulty, or is it?

Conventional wisdom dictates that content is a necessary condition for the determination of problem difficulty. Selection occurs on content, not structure. Several studies have suggested that structure is an *emergent* property of selection. For example, Goldberg and O'Reilly have shown that selection drives the shape of solutions (Goldberg and O'Reilly, 1998; O'Reilly and Goldberg, 1998). Langdon and others (e.g., (Langdon, 2002; Banzhaf and Langdon, 2002)) have shown that structural aberrations like bloat are also (emergent) outcomes of selection (Langdon and Poli, 1997). However, does this mean that content is a *necessary* condition for determining problem difficulty?

There have been clues that content is *not* a necessary condition. Punch et al. described a tunably difficult problem that they called the *Royal Tree* (Punch, 1996). The *Royal Tree* problem was one of the earliest instances of borrowing from the *Royal Road* function, which has been used in the GA community for some time. Solutions to the *Royal Tree* problem were based on structural specifications—there was no content, per se, involved in evaluation. In particular, the *Royal Tree* problem required a solution of a full tree, which had to be built up by successive depth levels.

Surprisingly, it did not take much to turn the *Royal Tree* problem from a "hard" problem into an "impossible" one, in spite of the significant computational resources that were used by Punch et al.. For the most part, however, the community has apparently dismissed the problem as anomalous: "real" coding solutions do not appear like *Royal Tree* programs; it's "obvious" that the *Royal Tree* problem is hard because the solutions need to be full trees; and so on.

Punch et al.'s work demonstrated conclusively that content is a *sufficient* condition for determining problem difficulty. They have demonstrated that for one particular structure, structure alone is a determinant for problem difficulty. The story might not yet end there, either. It may turn out that it is not just full tree structures that GP has difficulty in deriving. It may be most of them...

4.2 What We Know

As of this writing, what we know is a hypothesis that is in the process of being validated.

We have hypothesized that structure imposes significant limits on solutions (Daida, 2002; Daida and Hilss, 2003). This would happen because the iterative growth of trees generates fractal-like structures. Such structures are constrained to the types of shapes that are allowable. Consequently, only certain ranges of structural shapes would normally occur during a typical GP run.

The particular analogy that can be used to describe tree growth in GP is diffusion-limited aggregation. Diffusion-limited aggregation is a physical process that has been discovered and modeled by Witten and Sander (Witten and Sander, 1981; Witten and Sander, 1983). A common way of describing this process has been a drunkard's walk.

A drunk staggers around until either he finds something to lean on or he leaves the area. Say he finds a lamppost. He stops and leans on it. A second drunk stumbles and either he finds something to lean on or he leaves the area. By chance, the second drunk leans onto the first. The process is repeated. The resulting pattern of drunks is one that looks like Figure 7-8.

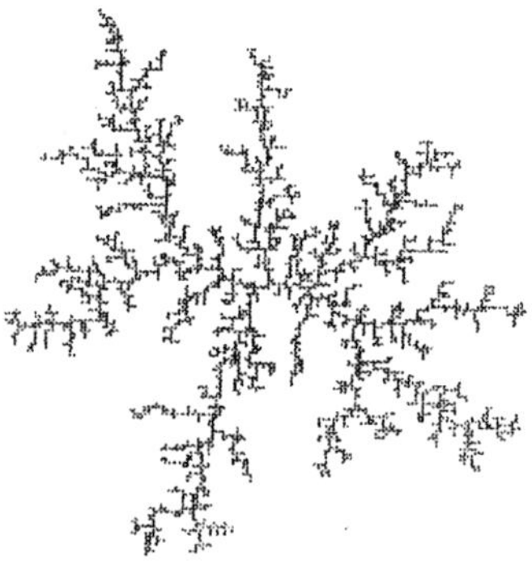

Figure 7-8. Diffusion-limited aggregation (Witten and Sander, 1983). (From T.A. Witten and L.M. Sander, "Diffusion-Limited Aggregation," *Physics Review B*, vol 27, (1983) pp. 5686-5697, reproduced by permission of The American Physical Society and L. Sander.)

It is not by accident that the pattern of drunks is sparsely packed. The likelihood of a stumbling drunk bumping into a leaning one is greater at the periphery of the pattern than at its center. There comes a time when new growth at the center of a pattern becomes next to impossible. Nearly all growth eventually occurs at the tips of the resulting dendrites. This pattern of dendrites is typical of diffusion-limited aggregation. Physical objects that have been created by diffusion-limited aggregation have limits as to how tightly packed they can be. As the object grows, it can become next to impossible to fill the voids that are left behind during growth. As it turns out, such patterns exhibit a limited range of associated sizes and shapes.

For GP, instead of "drunks," substitute "subtrees." Although operators like crossover do not involve a subtree that "leans" or "sticks" to an existing tree, there are structurally equivalent operations that would make it seem so. Figure 7-9 illustrates an example. Figures 7-9 (a) and (b) depict two parent trees with arrows indicating the crossover point. Figure 7-9 (c) depicts the resulting child tree. Figure 7-9 (d) shows the child's structural equivalent, with bold indicating the added structure relative to a parent. It is as if a three-node subtree has been "stuck" to one of the original parents.

We can model this process of growing a tree by randomly "sticking" subtrees to a "seed." Figure 7-10 shows a sequence of growth for a simulated example. The figure shows six "snapshots" during growth. The outer gray ring is for reference only and corresponds to depth 15. Details of this process, called the lattice-aggregate model, can be found in (Daida, 2002; Daida and Hilss, 2003).

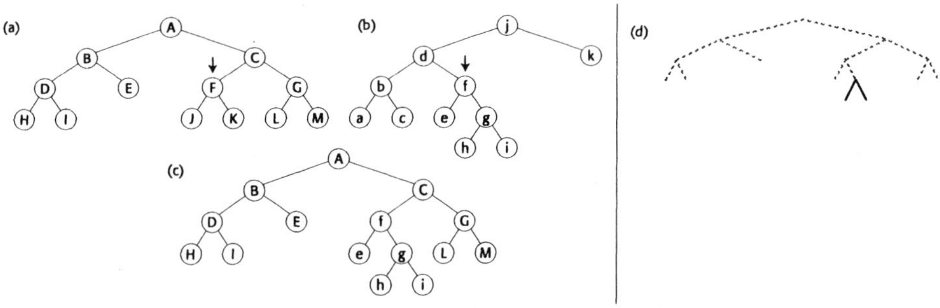

Figure 7-9. Crossover and a structurally equivalent operator. In this example, it is as if a three-node structure attached to one of the parents.

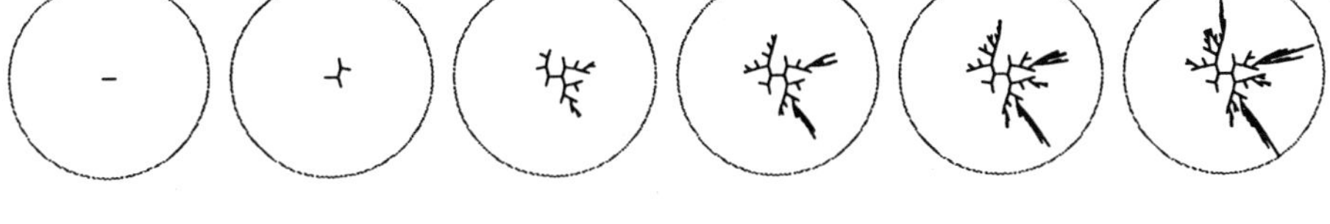

Figure 7-10. Simulated growth of a tree (Daida et al., 2003a). Trees that are grown in a random, iterative fashion exhibit some of the characteristics of the dendritic pattern of diffusion-limited aggregation.

Tree structures that were created with our analog of diffusion-limited aggregation also display a certain sparseness. In other words, a simulation of tree growth would indicate that some structures (i.e., sparse ones) are more likely than others (i.e., dense ones).

One can map the location of likely structures by measuring the trees that result after many trials of simulated tree growth. When we did this with our model, we found four regions. Figure 7-11 shows this map.

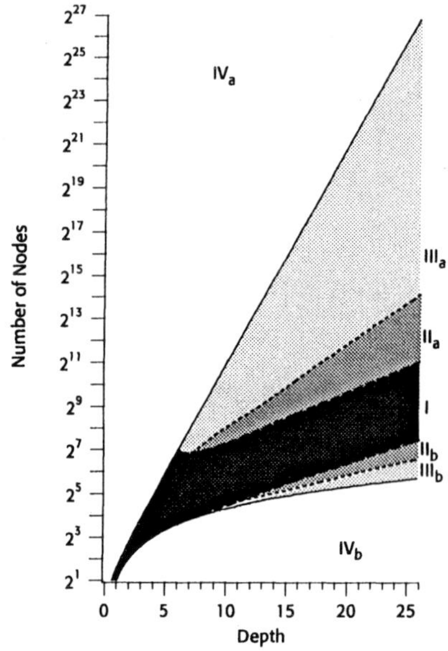

Figure 7-11. Predicted regions. A map of likely tree structures can be derived based on a process that is analogous to diffusion-limited aggregation.

The map of likely tree structures in GP can be thought of as a bottle. Region I is where most GP trees would occur—this would be the inside of the "bottle." Region II is where increasingly fewer trees would be found—this would be the walls of the "bottle." Region III is where few trees would typically occur—this would be the outside of the "bottle." (Region IV represents impossible configurations of trees.)

The bottle occupies only a small fraction of the total allowable search space. The bottle only seems large because it is depicted on a logarithmic scale. For the particular map shown in Figure 7-11, the interior of the bottle represents less than 0.01% of the entire allowable search space in size and depth.

We can compare this map with empirical results from GP. Figure 7-12 shows a comparison with data from the *binomial-3*. This data corresponds to the histograms that were shown in Sections 2 and 3. Each plot shows 600 points—one point for each trial that was run. Each point represents the highest-scoring ("best") solution out of a population of 500 from that trial. In spite of the wide range of problem difficulty, nearly all of the solutions were found within the "bottle."

We can repeat this comparison with results from other test problems as well. Figure 7-13 shows this for *sextic*, *6-input multiplexer*, and *11-input multiplexer*. The setup and implementation for *sextic* is based on (Koza, 1994), while *6-* and *11-input multiplexer* are based on (Langdon, 2000b). All

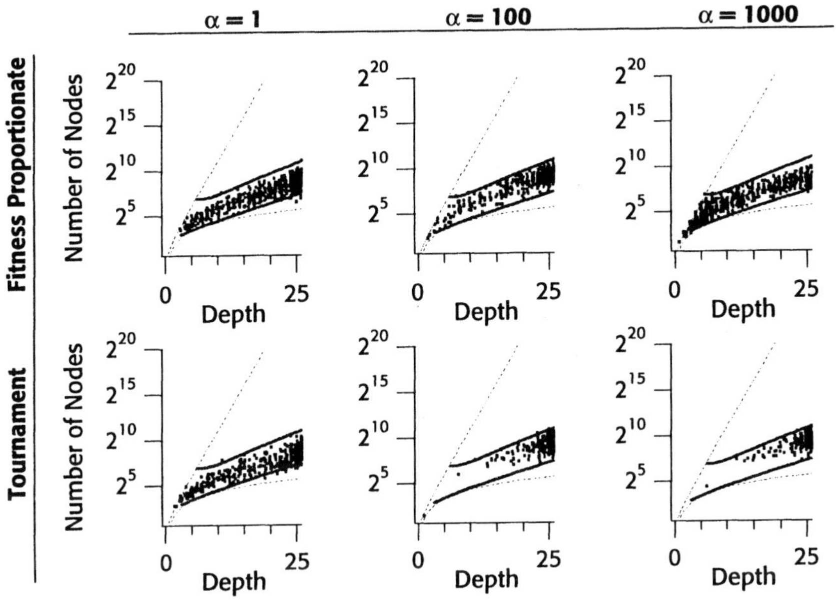

Figure 7-12. Empirical comparison of the *binomial-3* with the "bottle." The boundaries of Region I are superimposed.

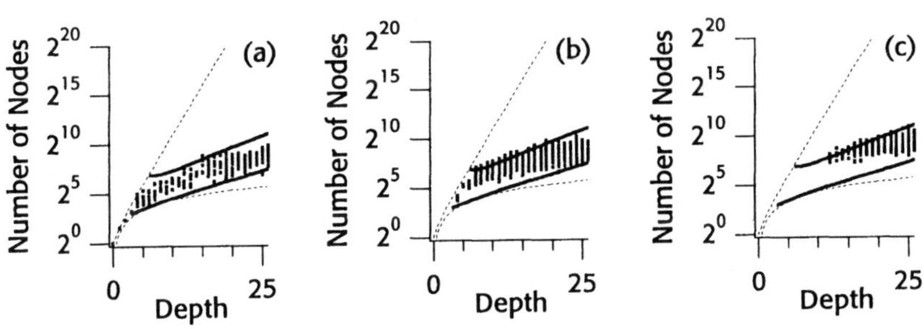

Figure 7-13. Other empirical results. (a) Sextic, (b) 6-input multiplexer. (c) 11-input multiplexer.

three of these problems featured populations of 4,000 and used tournament selection. Each plot summarizes the results for depths 0 – 26 for 1,000 trials. Again, for where the boundaries of the regions have been derived, nearly all of the solutions were found within the "bottle."

The existence of the "bottle" implies that most tree structures would be difficult for GP to derive. It anticipates the possibility that other tunably difficult problems can be based on structural specification alone. Easy variations would be located in Region I; difficult variations, Region III. In fact, this has been done and is described in (Daida et al., 2003b).

4.3 What We Don't Know

There is much that is not known since we are only beginning to validate our hypothesis. For example, suppose that results from a validation experiment are found to exist outside the "bottle." Would it be because the theory is wrong or would it be because something else is involved?

We know that the map is an approximation. For depths greater than 26, better approximations would be needed. Although we have an idea, exactly how those approximations are to be determined is unknown.

We also do not know what the value is of the "bottle" towards problem solving. The reason is that the "bottle" could represent a trade-off that is necessary for GP to balance between exploration and exploitation. For example, would it be desirable to break out of the "bottle" or would it be better to remain in it? If remaining in the "bottle" is necessary, how does one increase the complexity of programs for GP to create? If breaking out of the "bottle" is necessary, how would GP keep from getting lost in a search space? What is the role of the "bottle" in shaping GP outcomes?

Finally, we do not understand the role of the "bottle" in what makes a problem GP-hard. On one hand, it seems to have little bearing. For example, the solutions remained within the "bottle" for the *binomial-3*, regardless of problem difficulty. On the other hand, one could ask what would be the effect on problem difficulty if a desired solution lay outside the "bottle"? Would GP try to compensate by coming up with an alternative solution that remains in the "bottle"? Is this yet another reason for code bloat?

5. Discussion

This chapter has examined three claims: "content drives difficulty," "structure matters," and "structure determines limits." To some degree, each has been known to contribute to problem difficulty. Together, however, these claims suggest a view of problem difficulty that differs from the prevailing view. A short sidebar of the prevailing view is in order.

The prevailing view for what makes a problem difficult in GP is the prevailing view for what makes any problem difficult in the field of genetic and evolutionary computation. Under this view, what makes any problem difficult can be described by the metaphor of an adaptive landscape in evolutionary biology (e.g., (Langdon and Poli, 2002; Fogel, 1997; Deb et al., 1997)). The adaptive landscape, as posed in (Wright, 1932), has suggested to practitioners an optimization of fitness in a multi-dimensional, multi-modal search space. It is within this environment that genetic and evolutionary algorithms "walk." This interpretation of the metaphor of an adaptive landscape is not without precedent within evolutionary biology. After all, Wright's illustration of adaptive landscapes looked like topographic maps (which Simpson noted in his seminal work (Simpson, 1944, p. 49). Dobzhansky took Wright's figure of speech a step further and mapped Wright's abstraction of "hills" and "valleys" to *real* mountains and valleys (i.e., the San Bernadino Mountains, California, USA) (Dobzhansky, 1941). Under this prevailing view, then, hard problems would map to "tortuous terrain" in multi-dimensional, multi-modal search space.

There are difficulties in applying the landscape metaphor to GP (Daida et al., 2001; Langdon and Poli, 2002). In spite of these difficulties, we can use this metaphor to illustrate the differences between the prevailing view and ours. Figure 7-14 shows two different "landscapes" for two problems in GP. Under the prevailing view, GP should and can take a path anywhere through the landscape. The map given in Figure 7-11 suggests otherwise. Structure limits most of the allowable solutions to Region I. This happens in spite of wide variations in problem difficulty. There apparently is only one exceedingly narrow path through a landscape. Moreover, that path does not change location, regardless of the landscape upon which it is superimposed.

It may well turn out that this path is the only reasonable one for standard GP to take. To take other paths may mean that too much time and resources be spent on exploration instead of exploitation. Still, there may be sound reasons to look off of the path. One suspects that is where solutions to the real-world problems lie.

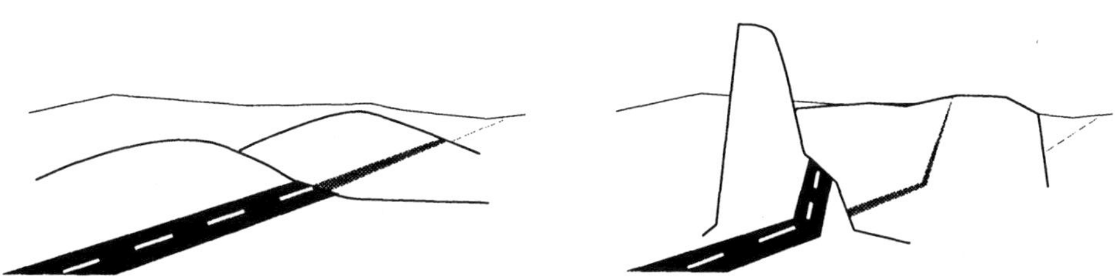

Figure 7-14. In spite of the differences in problem "landscapes," GP tends to cut a narrow path in a search space. Furthermore, this path does *not* change location, regardless of problem.

There are two practices that Koza uses that suggest a way off of this path (Koza et al., 1999). The first practice is his use of architecture-altering operations. Architecture-altering operations are structure-altering ones. They have the potential to change the overall structure of a solution in a way that random iterative growth cannot easily attain. Of the structure-altering operations that Koza mentions, the ones involving subroutine creation, duplication, and deletion strike me as those that might significantly increase the density of a tree structure.

The second practice is what I would think is the more potent one. In designing analog circuits, Koza uses GP to create a meta-program as a solution. A meta-program is executed to fashion a labeled cyclic graph. Of interest is that, while the meta-program is subject to the constraints of a tree structure, the resultant cyclic graph is not. It may be possible to sidestep the issue of structural constraints through the use of meta-programs. While this is a matter of speculation, I would note that most of the human-competitive results have been derived through the use of GP meta-programs.

6. Concluding Remarks

This chapter highlights what my group has learned in response to the question "What makes a problem GP-difficult?" We have made three claims: that content drives difficulty, that structure matters, and that structure determines limits. In view of these statements, we would say that although content determines problem difficulty, structure might, in turn, determine content.

I can cast my short list of what I would need to do to solve a difficult GP problem in light of these claims. I would use structure-altering operations because structure alone may impose too hard a limit on what my solution can be. I would also design GP to generate meta-programs, programs that generate other programs. Meta-programs themselves might be structure limited. However, the solutions that these meta-programs produce would likely not be limited in the same way. Finally, I would use large populations, since structural limits might prevent useful combinations of content from occurring.

In the process of highlighting what has been learned, this chapter makes a point of describing what is not known. If anything, it shows that GP theory still has a long way to go.

Acknowledgments

I gratefully acknowledge the help and assistance from the following individuals and organizations: I. Kristo, S. Daida, L.M. Sander, CSCS U-M / Santa Fe Institute Fall Workshops, W. Worzel, D. MacLean, A. Hilss, D. Ward, R. Tang, S. Long, M. Samples, P. Chiusano, B. Hart, and R. O'Grady. I thank the reviewers W. Langdon, E. Goodman, and J. Hu.

References

Banzhaf, W., *et al.*, (1998). *GP: An Introduction.* San Francisco: Morgan Kaufmann.

Banzhaf, W. and Langdon, W. B. (2002). Some Considerations on the Reason for Bloat. *Genetic Programming and Evolvable Machines 3:* 81 – 91

Bickle T. and Thiele L. (1995). A Mathematical Analysis of Tournament Selection. In *Proceedings of ICGA '95*, San Francisco: Morgan Kaufmann; 9–16

Daida, J. M. (2002). Limits to Expression in GP: Lattice-Aggregate Modeling. In *Proceedings of ICEC 2002,* Piscataway: IEEE; 273–278

Daida J. M., *et al.*, (1999). Analysis of Single-Node (Building) Blocks in GP. *Advances in GP 3*, pp. 217–241. Cambridge: MIT Press

Daida J. M., *et al.*, (2001). What Makes a Problem GP-Hard? Analysis of a Tunably Difficult Problem in GP. 2001; *Genetic Programming and Evolvable Machines 2*: 165–191

Daida, J. M., *et al.* (2003a). Visualizing Tree Structures in GP. In *Proceedings in GECCO 2003.* Berlin: Springer-Verlag

Daida, J. M., *et al.*, (2003b). What Makes a Problem GP-Hard? Validating a Hypothesis of Structural Causes. In *Proceedings in GECCO 2003.* Berlin: Springer-Verlag

Daida, J. M. and Hilss, A. (2003). Identifying Structural Mechanisms in Standard GP. In *Proceedings in GECCO 2003.* Berlin: Springer-Verlag

Deb, K., *et al.* (1997). Fitness Landscapes. In *Handbook of Evolutionary Computation*, Bristol: Institute of Physics Publishing; B2.7:1–B2.7:25

Dobzhansky, T. (1941). *Genetics and the Origin of the Species*, Second ed. New York: Columbia University Press

Fogel, D. B. (1997). "Principles of Evolutionary Process." In *Handbook of Evolutionary Computation*, Bristol: Institute of Physics Publishing; A2.1:1– A2.1:3.

Gathercole, C. and Ross, P. (1996). An Adverse Interaction Between Crossover and Restricted Tree Depth in GP. In *Proceedings in GP '96,* Cambridge: MIT Press; 291–296

Goldberg, D. E. and O'Reilly, U.-M. (1998). Where Does the Good Stuff Go, and Why? In *Proceedings in EuroGP '98*, Berlin: Springer-Verlag

Knuth, D. E. (1997). *The Art of Computer Programming Vol 1*, Third ed. Reading: Addison–Wesley

Koza, J. R. (1992). *Genetic Programming*. Cambridge: MIT Press

Koza, J. R. (1994). *Genetic Programming II*. Cambridge: MIT Press

Koza J. R., *et al.*, (1999). *Genetic Programming III*. San Francisco: Morgan Kaufmann

Langdon, W. B. (2000a). Size Fair and Homologous Tree Crossovers for Tree GP. *GPEM*; *1*:95–119

Langdon, W. B. (2000b). Quadratic Bloat in GP. In *Proceedings of GECCO 2000*, San Francisco: Morgan Kaufmann; 451–458

Langdon, W. B., et al. (1999). The Evolution of Size and Shape. In *Advances in GP 3*, Cambridge: MIT Press; 163–190

Langdon, W. B. and Poli, R. (1997). "Fitness Causes Bloat." In *Soft Computing in Engineering Design and Manufacturing*, London: Springer-Verlag; 23–27

Langdon, W. B. and Poli, R. (2002). *Foundations of GP*. Berlin: Springer-Verlag

McPhee, N. F. and Hopper, N. J. (1999). Analysis of Genetic Diversity through Population History. In *Proceedings of GECCO '99*, pp. 1112 – 1120. San Francisco: Morgan Kaufmann

Mitchell, M., *et al.* (1992). The Royal Road for Genetic Algorithms: Fitness Landscapes and GA Performance. In *Proceedings in ECAL '92*, pp. 245–254. Cambridge: MIT Press.

Mitchell, M. (1996). *An Introduction to GA*. Cambridge: MIT Press

Ogden H. G. (1888). The Survey of the Coast. *National Geographic 1:* 59–77

O'Reilly, U.-M. and Goldberg, D. E. (1998). How Fitness Structure Affects Subsolution Acquisition in GP. In *Proceedings in GP '98*, pp. 269–77. San Francisco: Morgan Kaufmann

Poli, R. (2000). Exact Schema Theorem and Effective Fitness for GP with One-Point Crossover. In *Proceedings in GECCO 2000*, pp. 469–476. San Francisco: Morgan Kaufmann

Poli, R. and Langdon, W. B. (1997). A New Schema Theory for GP with One-Point Crossover and Point Mutation. In *Proceedings in GP '97*, pp. 279–85. San Francisco: Morgan Kaufmann

Poli, R. and Langdon, W. B. (1998). Schema Theory for GP with One-Point Crossover and Point Mutation. *Evolutionary Computation 6:* 231–252

Punch, W., *et al.* (1996). The Royal Tree Problem, A Benchmark for Single and Multiple Population GP. In *Advances in GP 2*, pp. 299–316. Cambridge: MIT Press

Rosca, J. P. (1997). Analysis of Complexity Drift in GP. In *Proceedings in GP '97*, pp. 286–94. San Francisco: Morgan Kaufmann.

Rosca, J. P. and Ballard, D. H. (1999). Rooted-Tree Schemata in GP. In *Advances in GP 3*, pp. 243–271. Cambridge: MIT Press

Simpson, G. G. (1944). *Tempo and Mode in Evolution.* New York: Columbia University Press.

Soule, T., *et al.* (1996). Code Growth in GP. In *Proceedings in GP '96*, pp. 215–223. Cambridge: MIT Press

Soule, T. and Foster, J. A. (1997). Code Size and Depth Flows in GP. In *Proceedings of GP '97*, pp. 313–320. San Francisco: Morgan Kaufmann

Soule, T. and Foster, J. A. (1998). Removal Bias: A New Cause of Code Growth in Tree Based Evolutionary Programming. In *Proceedings in ICEC '98*, vol. 1, pp. 781–786. Piscataway: IEEE Press

Witten, T. A. and Sander, L. M. (1981). Diffusion-Limited Aggregation: A Kinetic Critical Phenomenon. *Physics Review Letters 47:* 1400–1403

Witten, T. A. and Sander, L. M. (1983). Diffusion-Limited Aggregation. *Physics Review B: 27:* 5686–5697

Wright S. (1932). The Roles of Mutation, Inbreeding, Crossbreeding and Selection in Evolution. In *Proceedings of the Sixth International Congress of Genetics, 1:* 356–366

Chapter 8

A PROBABILISTIC MODEL OF SIZE DRIFT

Justinian Rosca
Siemens Corporate Research
Princeton, NJ 08550
rosca@scr.siemens.com

Abstract

Drift towards growing size of genotypes is one outstanding and constantly disputed invariant in an overwhelming number of applications of evolutionary algorithms with variable-size structures. In contrast to previous work to reveal its fundamentals, we probabilistically analyze genotype growth by building on the idea of a "representation-less" model by Banzhaf and Langdon. Our model, called the *fitness-size model*, corresponds to a simple evolutionary algorithm using overproduction selection and mutation working on abstract objects retaining only fitness and size information.

The probabilistic analysis offers some surprises counter to present credence. The analysis predicts that average effective and noneffective lengths (and thus overall size) tend to be invariant over time. The same is true for the variance of the effective length. In contrast, the variance of the noneffective size feature increases linearly in time, and its variation shows the trademark of a *diffusion process*. Drift to increasing size manifests if search biases favor boundary conditions. We present experimental results with both the implementation of the theoretical model and a standard genetic programming algorithm. Statistical results with the two implementations are similar and fit the theoretical predictions.

Keywords: Evolutionary algorithms with variable-size structures, complexity drift, code growth, introns, rooted-tree schemata, effective and noneffective code, fitness-size model, small size effect, diffusion, fitness landscapes.

1. Introduction

Drift towards growing size of genotypes is a well-known and studied problem in evolutionary algorithms (EA) with variable-size representations. This phenomenon manifests under a wide range of representations, operators, algorithm variants, and values of control parameters (Blickle and Thiele, 1994; Nordin *et al.*., 1995; Langdon and Poli, 1997; Langdon and Poli, 1998b; Langdon, 2000; Luke, 2000a). The phenomenon shows serious disadvantages for tree structures used in genetic programming (GP) (Soule and Heckendorn, 2002), as it has been observed that the GP manipulates mostly noneffective code in later stages of evolution. By *noneffective* (also *intron* or *inviable*) code of an object, we refer to coding regions that can be modified without inducing any effect on the fitness value of the object for any fitness case. In contrast, *effective code* is code that, if changed, would affect fitness for at least one fitness case.

While there is generally consensus that code growth and noneffective code growth are correlated (Banzhaf and Langdon, 2002), hypotheses in the literature are empirical and indiscernible. Recent analysis of several prevalent hypotheses appears in (Banzhaf and Langdon, 2002; Soule and Heckendorn, 2002). Exceptions from the purely empirical explanations of code growth are the work of (Nordin *et al.*., 1995) and (Rosca and Ballard, 1996a). In spite of present explanations, the code growth phenomenon is elusive. Consequently its explanation is constantly disputed.

We pursue the goal of revealing the fundamentals of code growth from a probabilistic perspective. We start with a simple model as used in (Banzhaf and Langdon, 2002). Rather than resorting to simulation-based analysis of various hypotheses, we attempt to estimate in closed-form the expected value and the variance of the effective and noneffective code size, and their evolution over time. The model we use is called the *fitness-size model.* It corresponds to a simple evolutionary algorithm using overproduction selection and mutation working on abstract objects retaining only fitness and size information. Section 2 reviews two early formal models for the analysis of size drift. It then introduces the fitness-size model. Section 3 presents a probabilistic analysis of this model. The results derived address special conditions for size drift such as "boundary" and "small-size" conditions. Derived formulae will show that effective size varies slowly and does not disperse over time. In contrast, noneffective code is subject to a process of diffusion. Section 4 shows results of experiments designed to test the theory. Introns do not grow exponentially in size, and moreover they do not grow at all on average unless special conditions are met. We conclude with discussions of limitations of present work and ideas for future work.

2. Formal models and algorithms in portraying variable-size representations

In formalizing the effect of genetic operations in a problem independent way, we wish to abstract away syntactic and semantic features and focus on structural and fitness properties that directly affect survival and reproduction of genotypes. We briefly review two early formal approaches, *effective fitness* and *rooted-tree schemata*, as motivations for a simplified model of an evolutionary algorithm, the *fitness-size model.*

The typical mechanism of a genetic algorithm (GA) is reviewed in Figure 8-2 below. It is of interest to us when it operates on arbitrary length structures. Let $x^i(t), L^i_e(t), L^i_n(t)$ represent the fitness (real number), effective length or size (positive integer number), and noneffective size (integer number) of individual i at time t. The total size of the individual is simply $L^i(t) = L^i_e(t) + L^i_n(t)$. We drop the time argument wherever the context is clear.

At any time, a genetic operation may modify either the effective or the non-effective region of the genotype, but not both simultaneously. The latter is not possible given the definition of effective code. Variation is governed by the following simple rule:

Variation Rule 1. Given a genotype $(x^i(t), L^i_e(t), L^i_n(t))$ subject to variation, the probabilities of variation in the effective and noneffective parts of the object are $\frac{L^i_e(t)}{L^i(t)}$ and $(1 - \frac{L^i_e(t)}{L^i(t)})$ respectively.

Effective fitness and code growth

Consider crossover and reproduction operations only. The fitness proportionate selection step (2.ii in Figure 8-2) ensures that the average proportion of program i in the next generation is given by the product of the present proportion, the ratio of its fitness to average population fitness, and the fraction of copies not damaged by crossover according to Rule 1 above (assuming constructive crossovers are negligible). The latter factor is given by the ratio $\frac{L^i_e}{L^i}$. By aggregating all the factors, the fitness of i is thus "reduced" by the fraction of copies altered in the crossover process, in what is called *effective fitness* (Nordin *et al.*., 1995):

$$x^i_{\text{eff}} = x^i - p_c \cdot \frac{L^i_e}{L^i}$$

where p_c is the probability of crossover.

Assuming that the average population fitness is stationary from time t to $t+1$, x^i_{eff} appears to determine how many copies of this parent program will be chosen for reproduction next generation. Individual i can thus increase its chances of selection by increasing x^i_{eff}, i.e. L^i, or more precisely its noneffective size L^i_n. This could explain code growth (Banzhaf *et al.*, 1998), at least in a stationary regime of evolution.

The rooted-tree schema model

The rooted-tree schema concept for GP tree structures was defined in (Rosca and Ballard, 1996a). A rooted-tree schema is a contiguous tree fragment containing the tree root (see Figure 8-1).

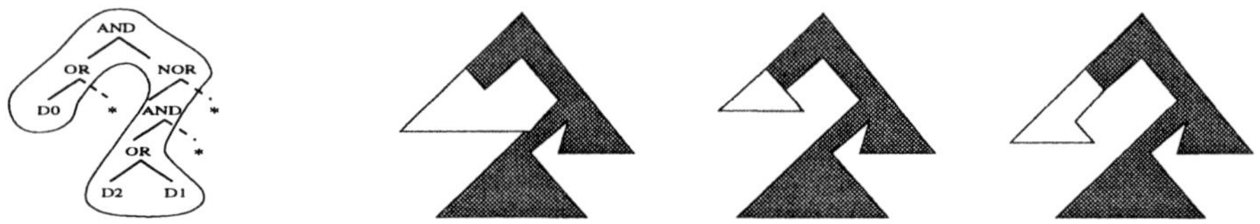

Figure 8-1. (a) Rooted-tree schema (RTS) example for parity problem; (b) Three different tree structures defined by the same RTS.

The schema analysis theory demonstrates the conditions under which a GP individual propagates the rooted-tree schema in the next generation. In summary, this will happen if either (1) the individual increases its fitness; or (2) the individual increases its size while preserving the rooted-tree schema, see (Rosca and Ballard, 1999) for details.

The formal analysis of the rooted-tree schema property shows how a simple structural property over the set of GP structures explains the role played by variable size, and further by a parsimony penalty. It also suggests how to control parsimony. Note that a rooted-tree schema does not distinguish between effective and non-effective code, which will be essential in the model presented next.

The fitness-size model

Now we review and elaborate another abstract model from (Banzhaf and Langdon, 2002), where individuals are given by objects capturing only the fitness, effective length, and noneffective length features. Denote the population at a given generation t by $\mathcal{P}_t = \mathcal{P}(t)$. Each of the M population individuals is abstracted as an object $\mathcal{P}^i(t)$, $i = 1 \cdots M$. At time t, $\mathcal{P}^i(t)$ is described by:

$$\mathcal{P}^i(t) = (x^i(t), L_e^i(t), L_n^i(t)) = (f, L_e, L_n)^i|_t \tag{8.1}$$

We also abstract the way genetic operators work. As a result of variation operations, offspring objects may have the fitness and size features changed. To simplify analysis, we assume that only one of the effective or non-effective length features can change at a time, as described in rule 1. More specifically, the following rules dictate variation:

Variation Rule 2. A variation in the noneffective part of an object produces an offspring with a varied noneffective length $L_n^{'}$ but same fitness and effective length:

$$\mathcal{P}^i(t) \overset{\text{Vary}}{\Rightarrow} \mathcal{P}^{i'}(t+1) = (x^i(t), L_e^i(t), L_n^{'}) \tag{8.2}$$

Variation Rule 3. A change in the effective part of an object produces an offspring with a varied effective length $L_e^{'}$ and new fitness x' but same noneffective length:

$$\mathcal{P}^i(t) \overset{\text{Vary}}{\Rightarrow} \mathcal{P}^{i'}(t+1) = (x^{'}, L_e^{'}, L_n^i(t)) \tag{8.3}$$

We further assume that typical genetic operators modify the abstract objects in an unbiased fashion as follows:

1 *Mutation.* Effective or noneffective length can increase or decrease with α units with equal probability for each case e.g.:

$$L_e(t+1) = L_e(t) \pm \alpha$$

$$L_n(t+1) = L_n(t) \pm \alpha$$

where for instance $\alpha \in \{1, 2\}$;

2 *Crossover.* Effective or noneffective lengths of two objects i and j participating in a crossover operation can change, complementarily, with β units, e.g.:

$$L_a^i(t+1) = L_a^i(t) \pm \beta$$
$$L_b^j(t+1) = L_b^j(t) \mp \beta$$

where for instance $\beta \in \{1, 2, 3\}$ and either of L_a and L_b can be an effective of noneffective length.

The α and β parameters are chosen to be small. The rationale for small changes suggested above has to do with typical changes in tree-based representations, as in GP. On one hand, we would like to study the effects of small variations, on the other hand, it is known that most genetic changes in GP occur towards the fringe of the tree representations (Rosca, 1997a; Rosca, 1999). For instance, the average height of crossover points in complete binary trees is approximately two, and experimentally one can trace the average height of crossover points for any shape of tree to be between two and three. In general, maximum values for α and β would depend on time.

Two classes of assumptions about the *fitness landscape* are made in this work. (1) *Uncorrelated* fitness landscapes. We assume that any fitness change resulting from a structural change in the effective part is equivalent to resampling independently from a static distribution of fitness values with p.d.f. $f_{\mathbf{x}}(x) = Prob(\mathbf{x} = x)$. Thus, the fitness of an offspring is uncorrelated to the fitness of the parent. This is the same as the uncorrelated fitness landscape assumption in (Banzhaf and Langdon, 2002). Moreover, we assume that all lengths of encodings have the same fitness distribution, $f_{\mathbf{x}}(x)$. In this case the distribution of raw fitness values, $f_{\mathbf{x}}(x)$, is taken to be Gaussian with zero mean and unit variance. (2) *Correlated* fitness landscapes. The uncorrelatedness assumption from (1) is not strictly necessary. Here we make a weak correlation assumption: offspring fitness is most often smaller than parent fitness (negative differential fitness in (Rosca, 1997a)).

Algorithms for the fitness-size model

We consider two basic evolutionary algorithms and specific selection schemes. The first one, (A), follows the standard GA/GP approach as defined in (Holland, 1992; Koza, 1992). The second one, (B), is similar in spirit to evolution strategies (ES) (Bäck and Schwefel, 1993) but simpler. These two schemes are contrasted in (Banzhaf *et al.*, 1998). Their pseudo-code description appears in Figures 8-2-8-3. In either case the population is initialized as follows:

$$\mathcal{P}_0 = \{(x, L_e, L_n)^i, i = 1, ..., M; x \sim f_{\mathbf{x}}(x); L_{e/n}^{\min} \leq L_{e/n} \leq L_{e/n}^{\max}\}$$

where $L_{e/n}^{\max}$ is the maximum value of the effective (or noneffective) size.

Algorithm A$(f_{\mathbf{x}}(x), M, p_m, p_c, p_r, T, S, L_n^{\min/\max}, L_e^{\min/\max})$

1 t=0; Randomly initialize $\mathcal{P}_0(f_{\mathbf{x}}(x), M, L_n^{\min/\max}, L_e^{\min/\max}; \mathcal{P}_0)$

2 Repeat, while $t < T$

   (a) Generate $\mathcal{P}_{t+1}$ by reproduction, crossover, mutation($\mathcal{P}_t; \mathcal{P}_{t+1}$)

      i Select genetic operation O $(p_r, p_c, p_m; O)$

      ii Select parent individual(s) $\mathcal{W}$ from $\mathcal{P}_t$ $(O, \mathcal{P}_t, S; \mathcal{W})$

      iii Generate offspring(O,$\mathcal{W}$;$\mathcal{P}_{t+1}$)

   (b) Next generation: $t = t + 1$

Figure 8-2. Skeleton of a standard GP algorithm (A) with mating selection. T is the total number of generations and p_m, p_c, p_r are the mutation, crossover and reproduction rates. Selection is performed by a tournament of size S.

One can argue that schemes A and B are roughly equivalent, as the variation and selection steps are apparently reversed. They are called "mating selection" and "overproduction selection," respectively, in (Banzhaf *et al.*, 1998). In contrast to A, which can alter individual genomes by crossover and mutation, B

Algorithm B($f_{\mathbf{x}}(x), M, p_m, p_r, T, L_n^{\min/\max}, L_e^{\min/\max}$)

1 t = 0; Randomly initialize $\mathcal{P}_0(f_{\mathbf{x}}(x), M, L_n^{\min/\max}, L_e^{\min/\max}; \mathcal{P}_0)$

2 Repeat, while $t < T$

(a) Generate $\mathcal{P}_{t+1}$ by reproduction and mutation($\mathcal{P}_t; \mathcal{P}_{t+1}$)

i Randomly select parent individual w ($\mathcal{P}_t$; w)

ii Select genetic operation ($p_r, p_m; O$)

iii If reproduction: copy parent($w, \mathcal{P}_{t+1}$)

iv If mutation:

A Generate offspring($\mathcal{P}^{i'}$)

B Select fittest between parent and offspring ($\mathcal{P}^i$,$\mathcal{P}^{i'}$;$\mathcal{P}_{t+1}$)

(b) Next generation: $t = t + 1$

Figure 8-3. Skeleton of a simple evolutionary strategy-like algorithm (B) showing "overproduction selection" but no self-adaptation of mutation parameters. T is the total number of generations.

uses mutation only. B is simpler and will be used as a basis for our probabilistic analysis of the fitness-size model, which is presented next.

3. Probabilistic analysis of algorithm B

Step 2 of algorithm B in Figure 8-3 describes an iterative operator I that is to be applied repeatedly to the population at any given time: $\mathcal{P}(t) = I(\mathcal{P}(t-1)) = I(I(...I(\mathcal{P}(0)))) = (I \circ I \circ ... \circ I)(\mathcal{P}(0))$. Here, we are interested in characterizing statistical changes in the population of objects, in particular the equilibria and evolution of the features of interest:

- What is the expected/average *effective* and *noneffective* size of evolved objects after T time steps?
- What is the variance of the effective and noneffective size?
- What is the expected fitness value in the population?

The main goal below is to determine closed-form estimation formulae and to answer the first two questions above for algorithm B. In Section 4, we will experimentally obtain statistics of the fitness-size features as they evolve over time and compare the results with the theoretical predictions. Thus we consider the *expected value* of a particular feature *over the population*, denoted by $E[.]$.

Elementary events in applying iterator *I*

Assume that mutation is applied to vary individual $\mathcal{P}^i(t)$. Then the possible ways of variation are represented in Figure 8-4, and the associated transition probabilities are given on the arcs of the tree. We have used labels p, p_v, q, and r for the following probabilities:

- p is the probability that the effective part of the individual will be modified:

$$p = \frac{L_e^i(t)}{L_e^i(t) + L_n^i(t)} \tag{8.4}$$

- p_v is the probability of choosing a variation operation. Variation operations will be equally likely and use a symmetric set of changes in order to not introduce explicit search biases (Gordon and DesJardins, 1995) in the absence of knowledge about solutions. An example of a set of mutation operations for algorithm B is: two mutation operations where the first increases and the second decreases the object's size by α, $\alpha \in \{1\}$. In this case $p_v = \frac{1}{2}$, as exemplified in Figure 8-4.

- q is the probability that the fitness of the newly mutated individual is less than the fitness of an individual j randomly chosen from $\mathcal{P}(t)$:

$$q = \Pr(x \leq x_t^j) \quad \text{where} \quad x \sim f_{\mathbf{x}}(x), \;\; x^j(t) \sim f_{\mathcal{P}(t)}(x) \tag{8.5}$$

- r is the probability that the fitness of individual i is lower than the fitness of an individual j randomly chosen from $\mathcal{P}(t)$:

$$r = \Pr(x_t^i < x_t^j) \quad \text{where} \quad \mathcal{P}^j(t) \in \mathcal{P}(t), \;\; x^j(t) \sim f_{\mathcal{P}(t)}(x) \tag{8.6}$$

  It can be easily proven that $r = \frac{1}{2}$, as $\Pr(x^i < y) = \Pr(x^i \geq y)$ when x^i and y have the same distribution. Here, by abuse of notation, we consider that the population represents a distribution of fitness (or size) values with p.d.f. $f_{\mathcal{P}(t)}$. The notation $x^j \sim f_{\mathcal{P}(t)}(x)$ indicates that x^j "samples" this distribution $f_{\mathcal{P}(t)}$.

Remark. While $p_v = \text{constant}$, p, q, and r depend on i and t. The analysis that follows is independent of the choices for α and β.[1]

[1] In practice, the magnitude of variation depends on the size of individuals (and thus implicitly on time), potentially leading to superlinear size growth. The operators used in this analysis do not allow for superlinear growth, but the same analysis applies to operators that do support such growth.

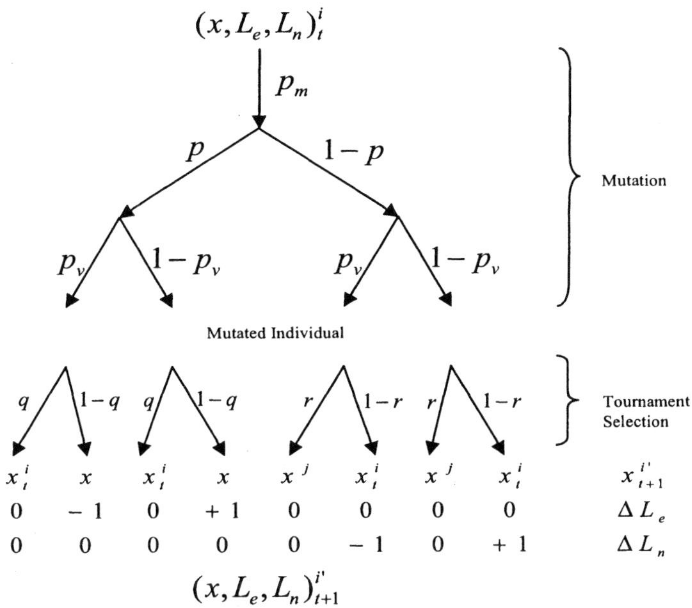

Figure 8-4. Probability tree of possible events in mutation plus overproduction selection (tournament selection of mutated individual with a second individual randomly selected from the population in this case).

Analysis of evolution

In this section we determine the average and standard deviation of the effective and noneffective sizes. We will refer to a path k in the set of all paths Γ^i of the probability tree in Figure 8-4:

$$\Gamma^i = \{(x, L_e, L_n)_t^i \overset{\text{Vary,path k}}{\Rightarrow} (x, L_e, L_n)_{t+1}^{i'}, \forall k\} \tag{8.7}$$

Effective size varies slowly and does not disperse. The average effective length at generation $t+1$ is:

$$E[L_e(t+1)] = \frac{1}{M}\sum_{i=1}^{M} L_e^i(t+1) = \frac{1}{M}\sum_{i=1}^{M}\sum_{k\in\Gamma^i} p_k \cdot L_e^i(t+1)|_k \tag{8.8}$$

After replacing the probability of each path p_k by the product of the probabilities of each arc on the path and some algebraic manipulation, we obtain:

$$E[L_e(t+1)] = E[L_e(t)] + \frac{1}{M}\sum_{i=1}^{M} pqp_v(L_e^j(t) - L_e^i(t)) \tag{8.9}$$

where the random variable $L_e^j(t)$ is drawn from the set of effective sizes of individuals from $\mathcal{P}(t)$, and p_v is constant.

Property 1. Assume that pq is relatively invariant. By averaging over multiple j's, $\frac{1}{M}\sum_y \sum_{i=1}^M pq(L_e^j(t) - L_e^i(t)) = 0$ given that $\mathcal{P}^j$ and $\mathcal{P}^i$ sample the same set given by $\mathcal{P}(t)$ and the sum order can be reversed. Thus $\frac{1}{M}\sum_{i=1}^M pq(L_e^j(t) - L_e^i(t)) = 0$ in the limit for large populations.

This and equation 8.9 give the following important result:

Lemma 1. $E[L_e(t+1)] = E[L_e(t)]$.

The standard deviation of the effective lengths is by definition

$$\text{Var}[L_e(t+1)] = E[(L_e(t+1))^2] - E[L_e(t+1)]^2, \quad \text{which becomes :}$$

$$\begin{aligned} \text{Var}[L_e(t+1)] &= E[(L_e(t))^2] - E[L_e(t+1)]^2 + \\ &+ \frac{1}{M}\sum_{i=1}^{M}\Big(\underbrace{pq((L_e^j(t))^2 - (L_e^i(t))^2)}_{\text{Bias term 1}} + \underbrace{p(1-q)}_{\text{Bias term 2}}\Big) \end{aligned} \tag{8.10}$$

One additional bias term (2) appears in the bias sum, i.e. $p(1-q)$. It can be argued, however, that $q \approx 1$ even early in the evolutionary process. For uncorrelated fitness landscapes, by definition $x \sim f_{\mathbf{x}}(x)$ while $y = x^j \sim f_{\mathcal{P}(t)}(x)$ in the probability tree of Figure 8-4. The quickly divergent dynamics of $f_{\mathcal{P}(t)}(x)$ from $f_{\mathbf{x}}(x)$ due to selection (see Figure 8-5) implies that $q = \Pr(x \leq y) \approx 1$. For correlated fitness landscapes, most of the offspring have fitness smaller than the maximum fitness of any parent (e.g. see Figure 4.2 in (Rosca, 1997a)). Again, $q \approx 1$.

Now, we can simplify equation 8.10 using Lemma 1 and an argument similar to that in Property 1 to show that the bias term given by the sum is zero when averaging over time. We can therefore state that:

Lemma 2. $\text{Var}[L_e(t+1)] \approx \text{Var}[L_e(t)]$.

In the limit, the variance of L_e is constant. Lemmas 1 and 2 show that effective length varies slowly and does not drift or diffuse over time. We will verify this experimentally in Section 4 (see Figures 8-7-8-8).

Noneffective length is subject to diffusion. The situation will be somewhat different when considering noneffective length. The average noneffective length at generation $t+1$ can be computed similarly to average effective length (see Figure 8-4):

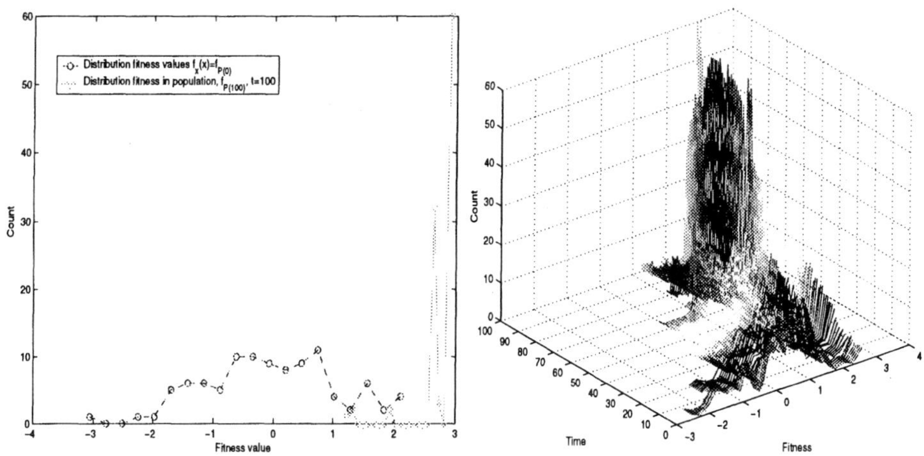

Figure 8-5. Left: Values x are sampled from static distribution of fitness values $f_{\mathbf{x}}(x)$. Values y are sampled from the "moving" distribution of fitness values given by the population. The two distributions diverge quickly over time (here $t = 100 \ll T = 10000$), as the latter one (x-dashed line) advances towards increasing fitness values; Right: Evolution of population fitness distributions for first 100 generations.

$$E[L_n(t+1)] = \frac{1}{M}\sum_{i=1}^{M} L_n^i(t+1) = \frac{1}{M}\sum_{i=1}^{M}\sum_{k\in\Gamma^i} p_k \cdot L_n^i(t+1)|_k \quad (8.11)$$

Again, after substituting the path probabilities p_k we obtain:

$$E[L_n(t+1)] = E[L_n(t)] + \frac{1}{M}\sum_{i=1}^{M}(1-p)rp_v(L_n^j(t) - L_n^i(t)) \quad (8.12)$$

where the random variable $L_n^j(t)$ samples the noneffective length of individuals from $\mathcal{P}(t)$. So far, there is no big difference compared to the case of effective length (equation 8.9). By a similar argument as before:

Property 2. Assume that the term $(1-p)r$ is invariant. By averaging over multiple j's, $\frac{1}{M}\sum_y \sum_{i=1}^{M}(1-p)r(L_n^j(t) - L_n^i(t)) = 0$ given that $\mathcal{P}^j$ and $\mathcal{P}^i$ sample the same distribution given by $\mathcal{P}(t)$. Thus $\frac{1}{M}\sum_{i=1}^{M} pq(L_n^j(t) - L_n^i(t)) = 0$ in the limit for large populations.

This and equation 8.12 imply:

Lemma 3. $E[L_n(t+1)] = E[L_n(t)]$ in the limit of large populations.

Remark. If confirmed experimentally, this would be a surprising result counter to present theories and conjectures in the GP literature (See sections 4.0 and 4.0).

The standard deviation of the noneffective lengths is:

$$\begin{aligned}\mathrm{Var}[L_n(t+1)] &= E[(L_n(t))^2] - E[L_n(t+1)]^2 + \\ &\frac{1}{M}\sum_{i=1}^{M}((1-p)r((L_n^j(t))^2 - (L_n^i(t))^2) + (1-p)(1-r))\end{aligned} \quad (8.13)$$

Again as in the case of L_e, there are two bias terms in the equation above. One can argue, as before, that the sum of weighted differences of squares is approximately zero. But this is no longer true for the second bias term:

$$\frac{1}{M}\sum_{i=1}^{M}(1-p)(1-r) \quad (8.14)$$

Assume for a moment that the bias term in equation 8.14 is a constant κ, $\forall t$. Then equation 8.13 results in the following first order recurrence relation:

$$\begin{aligned}\mathrm{Var}[L_n(t+1)] &= \mathrm{Var}[L_n(t)] + 1\cdot\kappa = \mathrm{Var}[L_n(t-1)] + 2\cdot\kappa = \\ &= \ldots = \mathrm{Var}[L_n(0)] + (t+1)\kappa, \quad \text{or}: \end{aligned} \quad (8.15)$$

Lemma 4. $\mathrm{Var}[L_n(t+1)] = \mathrm{Var}[L_n(0)] + (t+1)\kappa$

Lemma 4 is characteristic to and describes the phenomenon of *macroscopic theory of diffusion* in physics (Berg, 1993) (see left plot in Figure 8-6). It shows that noneffective length has the tendency towards diffusion, i.e. its variance increases linearly with time.

In reality, the term in equation 8.14 may not be constant. Then a more accurate interpretation is:

Property 3. **(a)** If the proportion of effective code increases (i.e. $(1-p)$ decreases) then the bias term decreases and the variance of the noneffective length narrows. This is a logical natural phenomenon after evolution discovers more effective solutions of increased size. **(b)** However, if the proportion of effective code is constant or decreases then the variance increases and the noneffective length diffuses to sharply higher and lower values.

Remark. The two regimes (3a) and (3b) that we argue for in Property 3 will generate wild variations of $\mathrm{Var}[L_n(t+1)]$ in contrast to $\mathrm{Var}[L_e(t+1)]$. This behavior should be observable in the experimental results later (see Section 4 and Figures 8-7-8-8).

Conditions for drift in population size: the "boundary" and "small-size" effects

The diffusion process related to noneffective length, in particular equation 8.12 that computes $E[L_n(t+1)]$, should be interpreted carefully. We predict that Lemma 3 will not hold true when the distribution of effective and noneffective lengths has small sizes (i.e. low tens of nodes per individual in initial population), because of (see Figure 8-6):

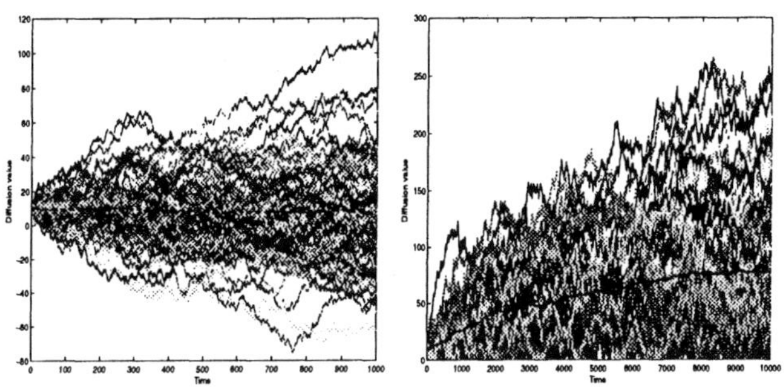

Figure 8-6. Diffusion of size in a population of 100 individuals with initial size 10, varying by ± 1 in one time step under no size constraints (left); Same process with an imposed positive lower boundary on size. Average size is represented in bold (right).

(a). Boundary conditions. Noneffective lengths can not diffuse towards zero or negative lengths and genetic operations may be invalidated. This is the *boundary effect* for small noneffective lengths.
(b). Small size effect. There is a nonuniform effect of the coefficient in the bias term of the average formula 8.12, i.e. $\lambda = (1-p) = \frac{L_n^i(t)}{L^i(t)}$, in the case of small size values of the individuals versus large size values[2]. This is the *small size bloat effect*: noneffective code will grow over time when length is small (see right plot in Figure 8-6).

4. Experimental validation

The fitness-size model was used with both algorithm B described before and algorithm A (standard GP with crossover implemented as described in Section 2.0). Now we experimentally verify the assumptions and formal results of Section 3. As will be seen, results for algorithm A are very similar to results for algorithm B, and fit the theory.

[2]The same absolute variation in L_n^i leads to very different values in the cases of small and large sizes, for instance when $L^i(t) = 10$ and $L^i(t) = 1000$. In the former case, the bias term will be positive rather than zero, and this will result in a drift in noneffective size.

Introns do not grow exponentially

Figure 8-7 presents the phenomenon of noneffective size drift when random initial effective and noneffective lengths are between 6 and 20. The results are

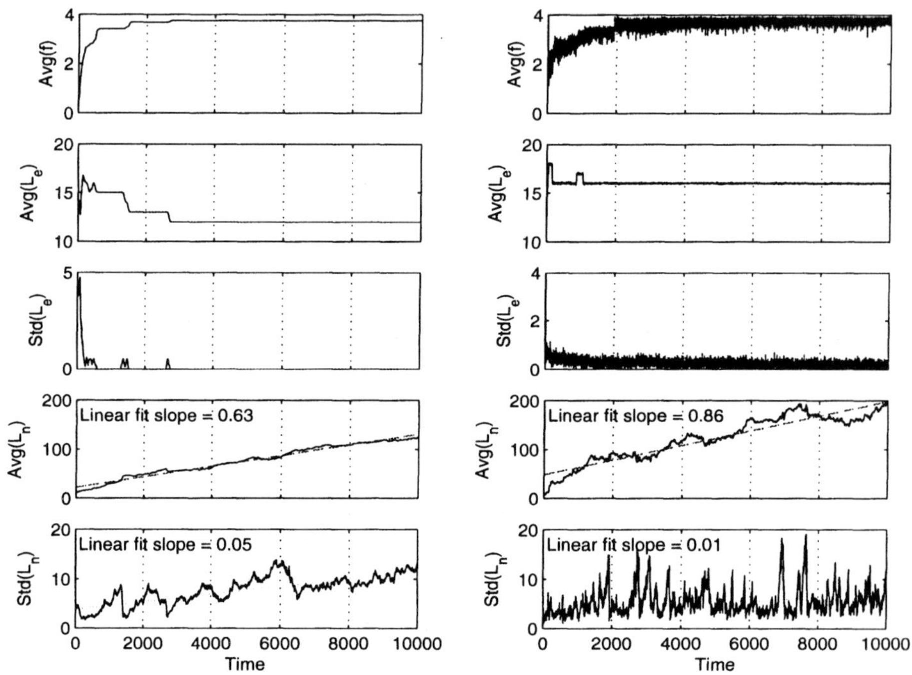

Figure 8-7. If "small size" conditions are satisfied, introns grow linearly. Algorithm B (left) $p_m = 0.1$; Algorithm A (right) $p_c = 0.1; p_r = 0.9; S = 4$. In both cases $M = 100; T = 10000$, and the initial population was generated with $6 \leq L_{e/n} \leq 20$.

in agreement with Lemma 1,2 and Property 3 (linear growth in the small size scenario). It has been stated in (Banzhaf *et al.*, 1998) and elsewhere that the rate of growth of introns is exponential. This is simply not true here. When the bias terms in equation 8.12 are non-zero, there is a linear increase in noneffective (intron) size over time. The precondition is that sizes of structures are relatively small.

Introns do not grow at all on average

Figure 8-8 presents results for random initial effective and noneffective lengths between 1000 and 1020. The linear fit for L_n has a slope approximately equal to zero for different realizations of the experiment and for a diverse range of parameters p_m, p_c, p_r and tournament size. For example Figure 8-9 (left) shows the slope of a linear fit to the variation in the average size of noneffective code L_n across a uniform range of ten values of p_m for algorithm A. A positive and increasing slope value would indicate size growth, but this does not happen.

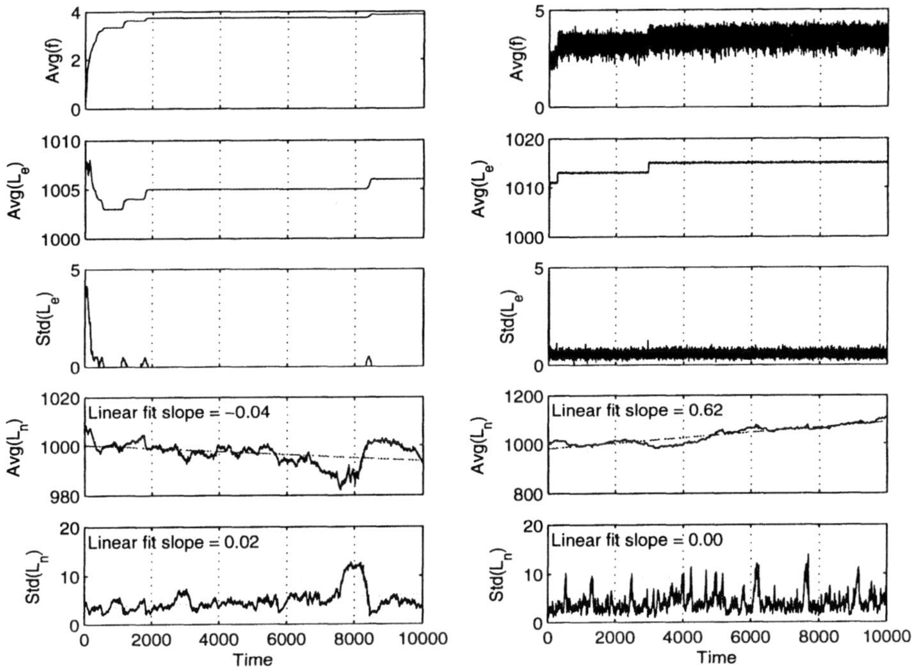

Figure 8-8. For large sizes introns no longer necessarily grow! Algorithm B (left) $p_m = 0.1$; Algorithm A (right) $p_c = 0.8; p_r = 0.2; S = 4$. Common parameters: $M = 100; T = 10000$; for both effective and noneffective lengths, the initial population was generated with $1000 \leq L_{e/n} \leq 1020$.

The no-bloat result for large effective and noneffective sizes is puzzling. In this case assumptions and results are in agreement with Lemma 3. The phenomenon has been verified for a diversity of settings of parameters (rates of genetic operations and selection pressure) for both algorithms. Theoretical results are indeed confirmed by experiments.

5. Conclusion

The fitness-size probabilistic model for EC with variable-size representations explains the bloat phenomenon in a problem and representation-independent way as a diffusion process with boundary conditions. The chapter derives closed-form formulae for the evolution of the mean and variance of effective and noneffective sizes and their interpretation under the model. When run

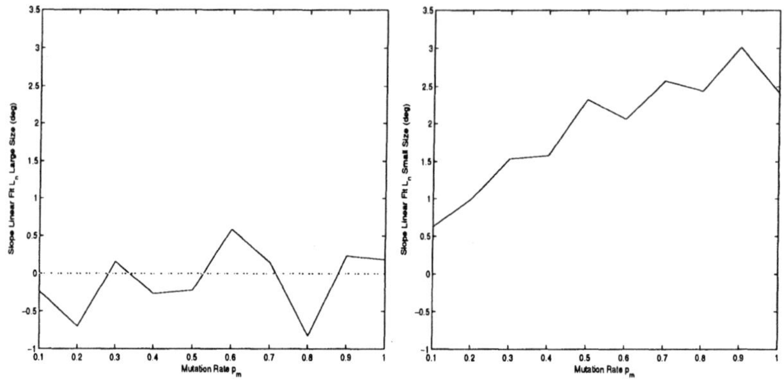

Figure 8-9. Across a wide range of mutation rates, algorithm A does not necessarily result in size growth for large sizes (left). Contrast this with positive slope of L_n increase for small sizes (right). Parameters are same as in Figure 8-8, except for $p_c = 0$, and $p_r = 1 - p_m$.

with an initial population of large individuals, the bloat effect showed in the literature is non-existent on average. However size is subject to a diffusion process, where the variance of noneffective size grows linearly in time. While on average there is no size growth effect, counter to common knowledge, for small initial sizes of objects size does grow linearly in time rather than exponentially. Theoretical results herein are confirmed both with an ES-like overspeciation mutation algorithm and with an accurate replica of the standard GP algorithm using the fitness-size representation model.

One important assumption in this study is that variation operations independently affect only one of the noneffective or effective coding regions. The probabilistic model could be extended to account, for instance, for the case when a change in the effective region also affects noneffective length. Particular properties of the fitness landscape could be included in the theoretical and simulation models as well. For instance, one should consider various types of correlation between the offspring and parent fitness. Results should be validated on large population sizes as well. We expect that a comprehensive model will have an important role in defining and analyzing practical implementations.

Acknowledgments

The author wishes to thank David Fogel for bold and useful comments on a first draft of the chapter, to Bill Langdon and Wolfgang Banzhaf for their feedback on short notice, and to Matt Streeter for discussions after the Ann Arbor workshop. This paper would not have been possible without the constant encouragement to beat the deadline from Rick Riolo.

References

Bäck, T. and Schwefel, H.-P. (1993). An overview of evolutionary algorithms for parameter optimization. *Evolutionary Computation 1(1):* 1–23.

Banzhaf, W., Nordin, P., Keller, R. and Francone, F. (1998). *Genetic Programming - An Introduction.* Morgan Kaufmann, San Francisco, CA.

Banzhaf, W. and Langdon, W. B. (2002). Some considerations on the reason for bloat. *Genetic Programming and Evolvable Machines 3(1):* 81–91.

Berg, H. C. (1993). *Random walks in Biology.* Princeton University Press.

Blickle, T. and Thiele, L. (1994). Genetic programming and redundancy. In *Genetic Algorithms within the Framework of Evolutionary Computation (Workshop at KI-94, Saarbrücken)*, Hopf, J. (Ed.), pp. 33–38, Im Stadtwald, Building 44, D-66123 Saarbrücken, Germany. Max-Planck-Institut für Informatik (MPI-I-94-241).

Gordon, D. F. and DesJardins, M. (1995). Evaluation and selection of biases in machine learning. *Machine Learning 20:* 5–22.

Holland, J.H. (1992). *Adaptation in Natural and Artificial Systems,* second edition. The MIT Press, Cambridge, MA.

Koza, John (1992). *Genetic Programming: On the Programming of Computers by Natural Selection.* MIT Press, Cambridge, MA, USA.

Langdon, W. B. (2000). Quadratic bloat in genetic programming. In *Proceedings of the Genetic and Evolutionary Computation Conference (GECCO 2000)*, Whitley, D. *et al.* (Eds.), pp. 451–458, Las Vegas, Nevada, USA. Morgan Kaufmann.

Langdon, W. B. and Poli, R. (1997). Fitness causes bloat. In *Second On-line World Conference on Soft Computing in Engineering Design and Manufacturing.*

Langdon, W. B. and Poli, R. (1998b). Fitness causes bloat: Mutation. In *Proceedings of the First European Genetic Programming Conference (Euro GP 98)*, Lecture Notes in Computer Science (LNCS), Volume 1391, Banzhaf, W. *et al.* (Eds.). Springer.

Luke, S. (2000a). *Issues in Scaling Genetic Programming: Breeding Strategies, Tree Generation, and Code Bloat.* PhD thesis, Department of Computer Science, University of Maryland, A. V. Williams Building, University of Maryland, College Park, MD 20742 USA.

Nordin, P., Francone, F. and Banzhaf, W. (1995). Explicitly defined introns and destructive crossover in genetic programming. In *Proceeedings of the Workshop on Genetic Programming: From Theory to Real-World Applications (NRL TR 95.2, University of Rochester)*, Rosca, J. P. (Ed.), pp. 6–22.

Rosca, J. P. (1997a). *Hierarchical Learning with Procedural Abstraction Mechanisms.* PhD thesis, University of Rochester, Rochester, NY 14627.

Rosca, J. (1999). Characteristics and biases of evolution in genetic programming. GECCO tutorial *www.cs.rochester.edu/u/rosca/.*

Rosca, J. P. and Ballard, D. H. (1996a). Complexity drift in evolutionary computation with tree representations. Technical Report NRL5, University of Rochester, Computer Science Department, Rochester, NY, USA.

Rosca, J. P. and Ballard, D. H. (1999). Rooted-tree schemata in genetic programming. In *Advances in Genetic Programming 3*, L. Spector *et al.* (Eds.), pp 243–271. MIT Press, Cambridge, MA, USA.

Soule, T. and Heckendorn, R. B. (2002). An analysis of the causes of code growth in genetic programming. *Genetic Programming and Evolvable Machines 3(3):* 283–309.

Chapter 9

BUILDING-BLOCK SUPPLY IN GENETIC PROGRAMMING

Kumara Sastry[1], Una-May O'Reilly[2], David E. Goldberg[1], David Hill[3]

[1] *Illinois Genetic Algorithms Laboratory*
University of Illinois at Urbana-Champaign

[2] *Computer Science and Artificial Intelligence Laboratory, Massachusetts Institute of Technology*

[3] *Department of Civil and Environmental Engineering*
University of Illinois at Urbana-Champaign

Abstract This paper analyzes building block supply in the initial population for genetic programming. Facetwise models for the supply of a single schema as well as for the supply of all schemas in a partition are developed. An estimate for the population size, given the size (or size distribution) of trees, that ensures the presence of all raw building blocks with a given error is derived using these facetwise models. The facetwise models and the population sizing estimate are verified with empirical results.

Keywords: building blocks, population size, schemas, partition, building-block supply, expression

1. Introduction

Genetic programming (GP) practitioners are often frustrated by the lack of theory available to guide them in selecting parameters for applied problems. They also lack a foundation of knowledge to explain many of their empirical findings regarding effective population sizes. To date, fundamental reasons for why varying population size while maintaining a constant number of fitness evaluations for a run results in different evolutionary trajectories and endpoints, have not been revealed.

The purpose of this paper is to start addressing this lack of theory by providing an estimate of the population size necessary to solve a given GP problem. It is hoped that like in genetic algorithm (GA) theory (Goldberg, 2002), the availability of a population sizing equation will be a valuable tool to aid GP practitioners in their efforts to understand how GP processes information. In addition, it may indicate how to adapt GP to be more competent.

The first step towards understanding population sizing is to tackle the issue of building-block (BB) supply. We forgo a temporal approach which would assume that the recombination, mutation and other diversity-generating operators will create and maintain sufficient BB diversity on an appropriate time scale. Instead, using a spatial approach, we estimate the population size required to ensure diversity and the number of BBs present in the initial population.

The objective of this study is to develop facetwise models for supply of BBs and to estimate the population size required to guarantee the presence of all raw BBs for a given tree size (or size distribution) in the initial GP population. Though ensuring BB growth supersedes BB supply in the subsequent population, BB growth will be extremely difficult if BB supply is not ensured. Also, while decision making usually governs population sizing, it is sometimes governed by BB supply. In such cases a facetwise model of BB supply is necessary for ensuring a successful GP design. Furthermore, understanding initial supply of BBs is essential for developing a practical population-sizing model.

This paper is structured as follows. We start with a brief literature review of BB supply. Section 3 provides background and states key assumptions made in this study. Details of an expression mechanism and test problems are provided in section 9.4, followed by facetwise models for BB supply in section 5. Section 6 outlines some thoughts on handling BB supply for real GP expressions. Finally, summary and conclusions are presented.

2. Brief Literature Review

The GP community is interested in identifying strategies to size populations, in order to estimate the computational effort required to solve particular problems with GP; however, few studies have addressed this topic, thus far. One approach has been suggested by (Langdon and Poli, 2002), but it has not been fully developed. This approach employs the methodology used by (Poli, 2000) for the sizing of populations in GA. In this method, Poli used Stephen and Waelbroeck's (Stephens and Waelbroeck, 1999) concept of transmission probability to develop a recursive conditional schema theory that allows for the prediction of the probability of reaching a solution to a problem in a fixed number of generations. An expression for the transmission probability for standard GP was developed by (Langdon and Poli, 2002). However, the expression is very difficult to evaluate.

The methodological and analytic foundation for our approach to deciphering selectorecombinative GA (Goldberg, 2002) and GP (Goldberg and O'Reilly, 1998; O'Reilly and Goldberg, 1998) has been stated before. Put succinctly, our approach is to analyze and understand GP's simple mechanisms before its complex ones. We predict that lessons learned from experimentation and theory on a simple case will lead to insight, and possibly, carry over to more complex cases. Therefore, we start by analyzing building-block supply in GP's initial population, before the activity of crossover and selection.

While building-block supply has been largely ignored in GP literature, many researchers have studied the BB supply in GAs. (Holland, 1975) estimated the number of BBs that receive at least a specified number of trials using Poisson distribution. A later study (Goldberg, 1989b) calculated the same quantity more exactly using binomial distribution and studied their effects on population sizing in serial and parallel computation. (Reeves, 1993) proposed a population sizing model for supply of alphabets with fixed cardinality. Recently, (Goldberg *et al.*, 2001) developed facetwise models for ensuring BB supply in the initial population for genetic algorithms. They considered a population of fixed-length strings consisting alphabets of arbitrary cardinality χ. They predicted that the population size required to ensure the presence of all competing building blocks with a tolerance of $\epsilon = 1/m$ is given by $n = \chi^k (k \log \chi + \log m)$, where χ is the alphabet cardinality, k is BB size, and m is the number of BBs.

This paper follows a similar methodology along the lines of (Goldberg *et al.*, 2001) and develops facetwise models for predicting the probability of the presence of a single schema as well as all schemas in a given partition. Before developing the models, we present some background and state assumptions used in the modeling procedure.

3. Preliminaries

In this section, we present definitions and concepts that underpin our analysis of BB supply in GP.

GP Tree Composition

Most GP implementations reported in the literature use parse trees to represent candidate programs in the population (Langdon and Poli, 2002). We have assumed this representation in our analysis. To simplify the analysis further, we consider the following:

1 A primitive set of the GP tree is $\mathcal{F} \cup \mathcal{T}$, where $\mathcal{F}$ denotes the set of functions (interior nodes to a GP parse tree), and $\mathcal{T}$ denotes the set of terminals (leaf nodes in a GP parse tree).

2 The cardinality of $\mathcal{F} = \chi_f$ and the cardinality of $\mathcal{T} = \chi_t$.

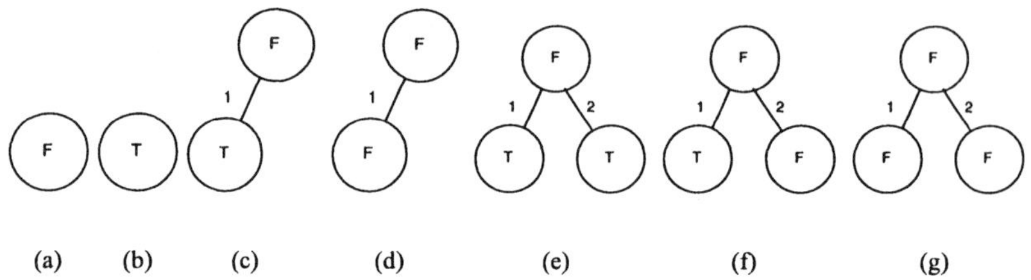

Figure 9-1. The smallest tree fragments in GP. Fragments (c) and (d) have mirrors where the child is 2nd parameter of the function. Likewise, fragment (f) has mirror where 1st and 2nd parameters of the function are reversed.

3 The arity of all functions in the primitive set is two: All functions are binary and thus parse trees generated from the primitive set are binary.

We believe that our analysis could be extended to primitive sets containing functions with arity greater than two (non-binary trees). We also note that our assumption closely matches a common GP benchmark, symbolic regression, which frequently has arithmetic functions of arity two.

Translating GA Schemas to GP Schemas Isn't Straightforward

Schemas are similarity templates that describe sets of solutions that share a common feature. The GP literature contains several alternative definitions of schemas (Koza, 1992; Altenberg, 1994; O'Reilly and Oppacher, 1995; Whigham, 1995; Rosca, 1997b; Langdon and Poli, 2002). Per (O'Reilly and Oppacher, 1995), a GP schema is a multiset of subtrees and tree fragments with nodes denoted as functions, terminals or "don't care" symbols. Tree fragments are trees with at least one leaf that is a "don't care" symbol which can be matched by any subtree (including subtrees with only one node).

Tree Fragments. While in general tree fragments refer to a multiset of tree patterns or tree templates, we restrict ourselves to a single tree pattern. A tree fragment pattern has each of its nodes labeled with the function symbol, $\mathcal{F}$, or terminal symbol, $\mathcal{T}$. However, it does not have an absolute position or positional anchor. Figure 9-1 shows the fragments our analysis focuses on. Along the edge between a function and its child node, a numeral denotes what parameter of the function the child node is (i.e. the first or second argument in the case of a binary function). A tree fragment has a length or size; that is, its number of nodes, $k = N_t + N_f$, where N_t and N_f are the number of terminal and functional nodes in the tree fragment, respectively. Furthermore, the total number of possible instances of a tree fragment is given by

$$\kappa = \chi_f^{N_f} * \chi_t^{N_t} \quad (9.1)$$

For example, for the tree fragment P_b (fragment with only terminal), $N_f = 0$, and $N_t = 1$, and therefore, the total number of instances of P_b is χ_t.

Since a tree fragment is not anchored to a position of a tree, there can be none or more than one instance of a fragment in a single tree. Yet, the smallest fragments P_a and P_b appear at least once or twice in a tree respectively. Assuming a single tree of size s[1] and the tree properties listed in section 9.3, Table 9-1 provides estimates (derived by probability of frequency) of the average quantity of tree-fragment instances, ϕ. In other words, ϕ counts the expected number of tree-fragment instances, given the tree size (or size distribution), in the population.

P	Description	ϕ	κ
P_a	function	$\frac{1}{2}(s-1)$	χ_f
P_b	terminal	$\frac{1}{2}(s+1)$	χ_t
P_c	one terminal that is the first parameter of a binary function	$\frac{1}{4}(s+1)$	$\chi_f \cdot \chi_t$
P_d	a function at the root and a function as its first parameter	$\frac{1}{4}(s-3)$	χ_f^2
P_e	a function at the root and 2 terminals as its parameters	$\frac{1}{8}\frac{(s+1)^2}{(s-1)}$	$\chi_f \cdot \chi_t^2$
P_f	a function at the root and 1 terminal as the first parameter and one function as it second parameter.	$\frac{1}{8}\frac{(s+1)(s+3)}{(s-1)}$	χ_f^3
P_g	a binary function at the root and 2 functions as its parameters	$\frac{1}{8}\frac{(s-3)^2}{(s-1)}$	$\chi_f^2 \cdot \chi_t$

Table 9-1. Designations, P_i, and descriptions of tree fragments considered in the BB supply models, the quantity of fragments, ϕ, and the number of competing schemas in the fragments, for a binary tree of size s. See also Figure 9-1.

The Tree Fragments are not Enough: How are They Expressed? While tree fragments are the parts of a physical tree, and counting number of instances of tree fragments can itself be important, what is more important are those tree fragments that get *expressed.* The expression mechanism dictates what the building blocks of a problem are and therefore affects the BB supply. Specifically, we are interested in expression of small tree fragments into partially correct subfunctions. Let us consider, for example, symbolic regression of $1+x+x^2+x^3$. Early on in the GP run, it is important to get the constant and the linear part of the symbolic equation right. Therefore, all the tree fragments

[1] It should be noted here that the average tree size of a population can be calculated for popular initialization schemes (Koza, 1992), or initialization schemes such as PCT1 or PCT2 (Luke, 2000b) can be used to generate a population which conform to an expected tree size.

that contribute to the correct constant and linear subfunctions are important and their supply is critical in the initial population.

We illustrate the methodology to incorporate expression mechanism in BB supply models by using a simple expression mechanism, called ORDER, which is explained in the next section. We choose ORDER because while it models some of the GP behavior (Goldberg and O'Reilly, 1998; O'Reilly and Goldberg, 1998), the expression mechanism can be analyzed in a straightforward manner.

4. ORDER Expression Mechanism

ORDER is a simple, yet intuitive expression mechanism which makes it amenable to analysis and modeling (Goldberg and O'Reilly, 1998; O'Reilly and Goldberg, 1998). The primitive set of ORDER consists of the primitive JOIN of arity two and complimentary primitive pairs $(X_i, \bar{X}_i)$, $i = 0, 1, \cdots, \ell$ of arity one. A candidate solution of the ORDER problem is a binary tree with JOIN primitive at the internal nodes and either X_i's or $\bar{X}_i$'s at its leaves. The candidate solution's expression is determined by parsing the program tree inorder (from left to right). The program expresses the value X_i if, during the inorder parse, a X_i leaf is encountered before its complement $\bar{X}_i$. Furthermore, only unique primitives are expressed in ORDER during the inorder parse.

Building blocks in ORDER are the sets of primitives that are part of the subfunctions that reduce error (alternatively improve fitness). In this study, we consider two test problems that use ORDER expression mechanism: 1. UNITATION: where each primitive X_i is a BB, and 2. DECEPTION: where k primitives form a BB. The following sections describe these two test problems.

UNITATION

In UNITATION, for each X_i (or $\bar{X}_i$) that is expressed, an equal unit of fitness value is accredited. That is,

$$f_1(x_i) = \begin{cases} 1 & \text{if } x_i \in \{X_1, X_2, \cdots, X_\ell\} \\ 0 & \mathit{otherwise} \end{cases} . \tag{9.2}$$

The fitness function for ORDER is then defined as

$$F(\mathbf{x}) = \sum_{i=1}^{\ell} f_1(x_i), \tag{9.3}$$

where $\mathbf{x}$ is the set of primitives expressed by the tree. The output for optimal solution of a ℓ-primitive UNITATION problem is $\{X_1, X_2, \cdots, X_\ell\}$, and its fitness value is ℓ.

DECEPTION

In DECEPTION, the primitives are divided into m subgroups, each subgroup consisting of k primitives. The fitness of each subgroup is computed using the following trap function (Goldberg, 1987; Deb and Goldberg, 1993):

$$f_k\left(u(x_1, x_2, \cdots, x_k)\right) = \begin{cases} 1.0 & u = k \\ (1.0 - \delta)\left(1 - \frac{u}{k-1}\right) & u < k \end{cases}, \tag{9.4}$$

where u is the unitation, or the number of primitives, X_i, in a subgroup:

$$u\left(x_1, x_2, \cdots, x_k\right) = \sum_{i=1}^{k} f_1(x_i), \tag{9.5}$$

where x_i is the ith primitive, δ is the difference in the functional value between the correct BB and its deceptive attractor. The fitness function of a candidate solution (tree) is then given by

$$\begin{aligned} F\left(\mathbf{x}\right) &= f_k\left(u(x_1, x_2, \cdots, x_k)\right) + f_k\left(u(x_{k+1}, \cdots, x_{2k})\right) + \\ & \cdots + f_k\left(u(x_{(m-1)k+1}, \cdots, x_{mk})\right), \end{aligned} \tag{9.6}$$

where, F is the fitness function, $\mathbf{x}$ is the expressed primitives, m is the number of BBs, and $\ell = mk$.

5. Facetwise Models of Building-Block Supply

In this section, we develop facetwise models for building-block supply for ORDER expression. First we start with addressing the supply of a single BB in a given partition. Then we extend the model to ensure the supply of all schemas in a partition. We then use the facetwise models to derive a population-sizing model dictated by BB supply. The models developed in this section are verified with empirical results for UNITATION and DECEPTION along the way.

Supply of a Single Building Block

Assuming trees of size, s, and that the expression mechanism used is ORDER, the probability that a primitive expressed by a tree is given by

$$\begin{aligned} p_{X_i^{\text{exp}}} &= p\left(\#of X_i \geq 1, \#of \bar{X}_i = 0\right) + p\left(X_i \text{ appears before } \bar{X}_i\right), \\ &= \sum_{j=1}^{n_l} \binom{n_l}{n_l - j} 2^{j-1} \left(\frac{\ell-2}{\ell}\right)^{n_l - j} \left(\frac{1}{\ell}\right)^j, \\ &= \frac{1}{2\ell^{n_l}}\left[\ell^{n_l} - (\ell-2)^{n_l}\right], \\ &= \frac{1}{2}\left[1 - \left(1 - \frac{2}{\ell}\right)^{n_l}\right] \end{aligned} \tag{9.7}$$

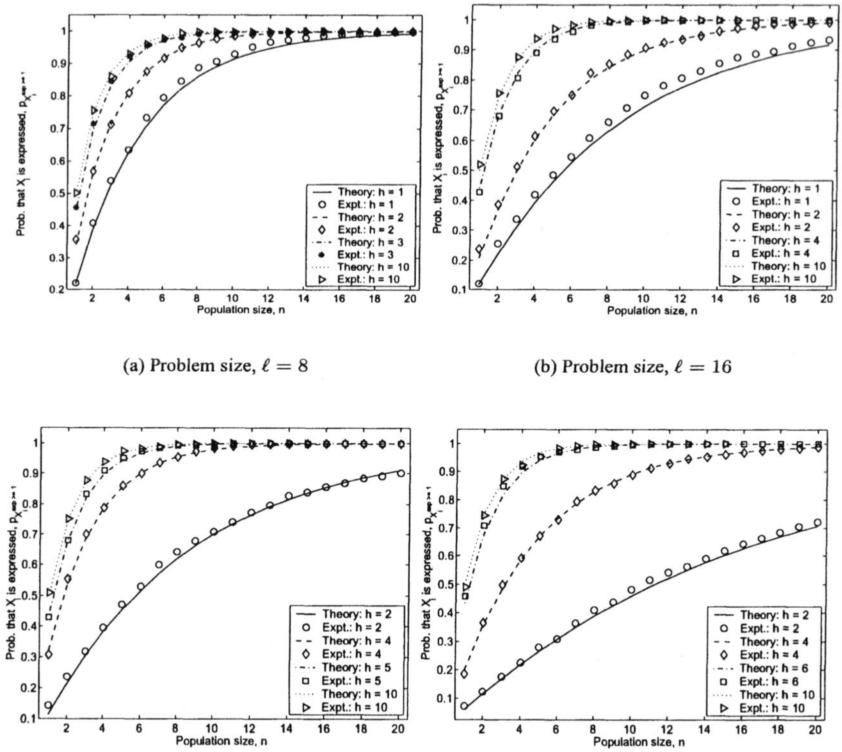

(a) Problem size, $\ell = 8$

(b) Problem size, $\ell = 16$

(c) Problem size, $\ell = 32$

(d) Problem size, $\ell = 64$

Figure 9-2. Verification of the facetwise model for a single BB supply (Equation 9.11) with empirical results for the `UNITATION` problem for different tree heights, h, as a function of population size, n. The empirical results depict the proportion of runs having at least one copy of a particular schema out of 1000 trials.

where $n_l = (s+1)/2$, is the number of leaf nodes in the tree, and $\ell = \chi_t$.

Assuming that primitives are expressed independent of each other, the probability that a order k BB (without loss of generality, we will consider $X_1 X_2 \cdots X_k$) is expressed by a tree is given by

$$p_{X^{\text{exp}}_{1\cdots k}} = p\left(X_i \text{ is expressed}\right)^k, \\ = \left[\frac{1}{2}\left\{1 - \left(1 - \frac{2}{\ell}\right)^{n_l}\right\}\right]^k. \quad (9.8)$$

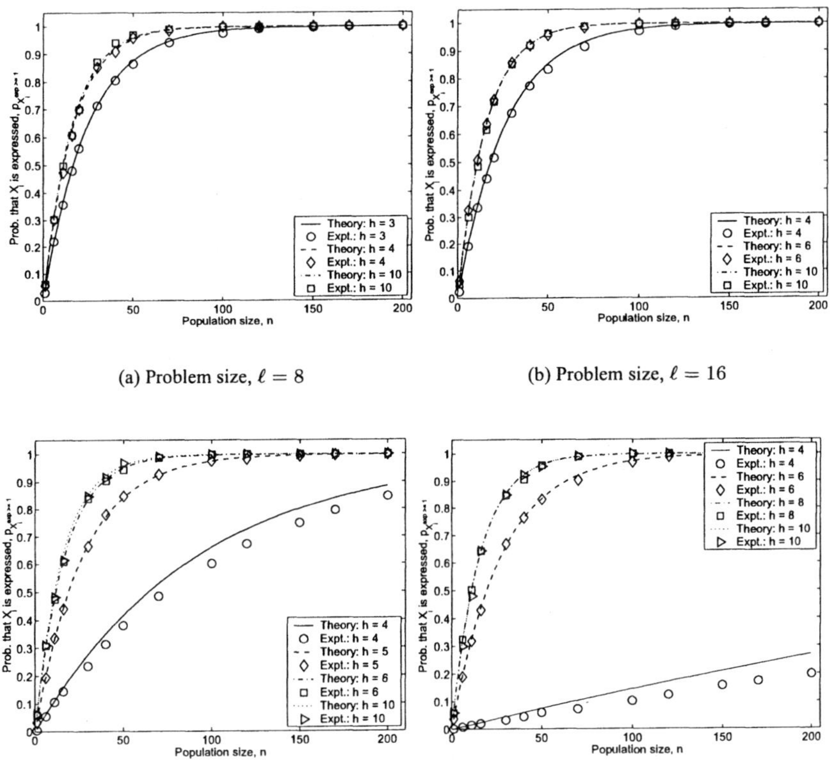

(a) Problem size, $\ell = 8$

(b) Problem size, $\ell = 16$

(c) Problem size, $\ell = 32$

(d) Problem size, $\ell = 64$

Figure 9-3. Verification of the facetwise model for a single BB supply (Equation 9.11) with empirical results for the DECEPTION problem for different tree heights, h, as a function of population size, n. The empirical results depict the proportion of runs having at least one copy of a particular schema out of 1000 trials.

The probability that the BB is not expressed by a tree is then given by

$$\begin{aligned} p_{X_{1\cdots k}^{\text{not exp}}} &= 1 - p_{X_{1\cdots k}^{\text{exp}}}, \\ &= 1 - \left[\frac{1}{2}\left\{1 - \left(1 - \frac{2}{\ell}\right)^{n_l}\right\}\right]^k \end{aligned} \tag{9.9}$$

The probability that a BB is not expressed by any of the n individuals in the population is given by

$$p_{X_{1\cdots k}^{\text{exp}=0}} = \left(p_{X_{i\cdots k}^{\text{not exp}}}\right)^n,$$

$$= \left[1 - \left[\frac{1}{2}\left\{1 - \left(1 - \frac{2}{\ell}\right)^{n_l}\right\}\right]^k\right]^n \tag{9.10}$$

Therefore the probability that a order-k BB is expressed by at least one individual in the population is given by

$$\begin{aligned} p_{X_{1\cdots k}^{\text{exp}\geq 1}} &= 1 - p_{X_{1\cdots k}^{\text{exp}=0}}, \\ &= 1 - \left[1 - \left[\frac{1}{2}\left\{1 - \left(1 - \frac{2}{\ell}\right)^{n_l}\right\}\right]^k\right]^n \end{aligned} \tag{9.11}$$

The model for single BB success given by Equation 9.11 is compared to empirical results for the UNITATION problem ($k = 1$) in Figure 9-2, and for the DECEPTION problem ($k = 4$) in Figure 9-3. The empirical results are for full trees, therefore, $n_l = 2^h$. The results show that the empirical results agree with the models.

Using the approximation, $(1 - r/s)^s \approx e^{-r}$, and recognizing that this approximation is sufficiently accurate even for modest values of s, we can simplify Equation 9.11 as follows:

$$p_{X_{1\cdots k}^{\text{exp}\geq 1}} \approx 1 - \exp\left[-n2^{-k}\exp\left\{-k\exp\left(-\frac{2n_l}{\ell}\right)\right\}\right]. \tag{9.12}$$

When $n_l \gg \ell$, $p_{X_i^{\text{exp}\geq 1}} \approx 1 - \exp\left(-n2^{-k}\right)$. In other words, the probability of a BB being expressed by at least one individual, given a population size, increases with the tree size and saturates as $2^h > \ell$, as shown in Figure 9-4 for UNITATION problem.

Supply for Partition Success

When solving real-world problems, one does not have prior knowledge about a particular schema being superior to others in a partition. Hence it is necessary to ensure that all competing schemas in a partition are present. The decision process would then be able to consider all the relevant alternative schemas. Therefore, in this section we extend the model developed in the previous section to ensure the presence of at least one copy of all the competing schemas (both the primitive and its complement) in a partition.

For ORDER, we are interested in the probability that all the 2^k possible schemas are present in the population. Assuming that individual schema success values are independent, the probability for partition success is given by

$$\begin{aligned} p_s &= \left(p_{X_{1\cdots k}^{\text{exp}\geq 1}}\right)^{2^k}, \\ &= \left[1 - \left[1 - \left[\frac{1}{2}\left\{1 + \left(1 - \frac{2}{\ell}\right)^{n_l}\right\}\right]^k\right]^n\right]^{2^k}. \end{aligned} \tag{9.13}$$

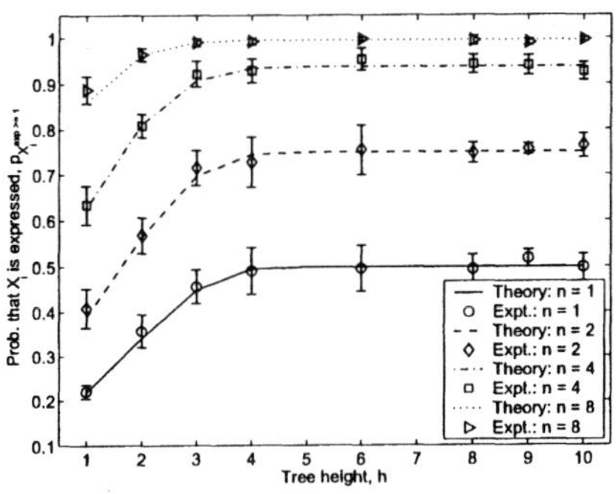

(a) Problem size, $\ell = 8$

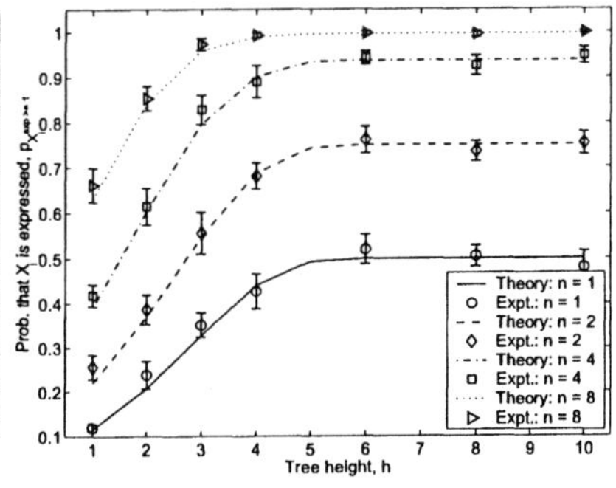

(b) Problem size, $\ell = 16$

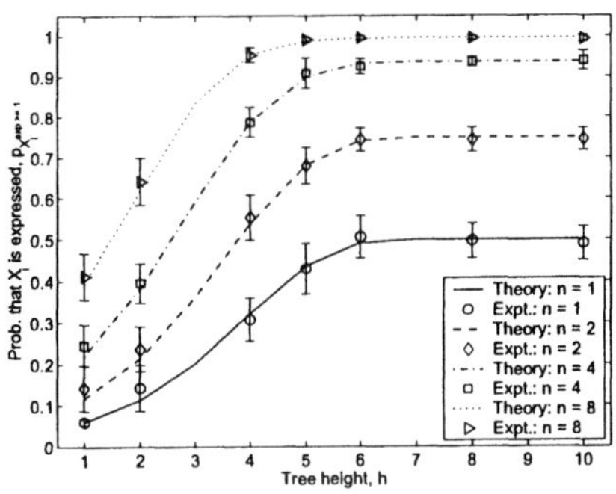

(c) Problem size, $\ell = 32$

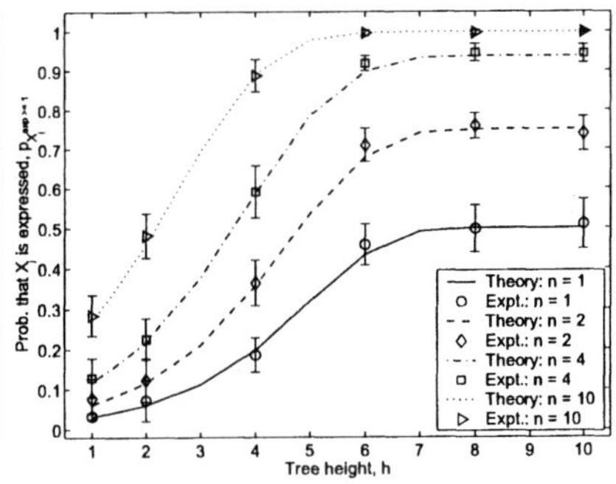

(d) Problem size, $\ell = 64$

Figure 9-4. Verification of the facetwise model for a single BB supply (Equation 9.11) with empirical results for the UNITATION problem for different population size, n, as a function of tree height, h. The empirical results depict the proportion of runs having at least one copy of a particular schema out of 1000 trials.

Using the approximation $(1 - r/s)^s \approx e^{-r}$, the above equation can be further approximated as

$$p_s \approx \exp\left[-2^k \exp\left[-n2^{-k} \exp\left\{-k \exp\left(-\frac{2^{h+1}}{\ell}\right)\right\}\right]\right]. \tag{9.14}$$

It should be noted that the independence assumption of individual schema success is an approximation and a more exact model can be derived which is

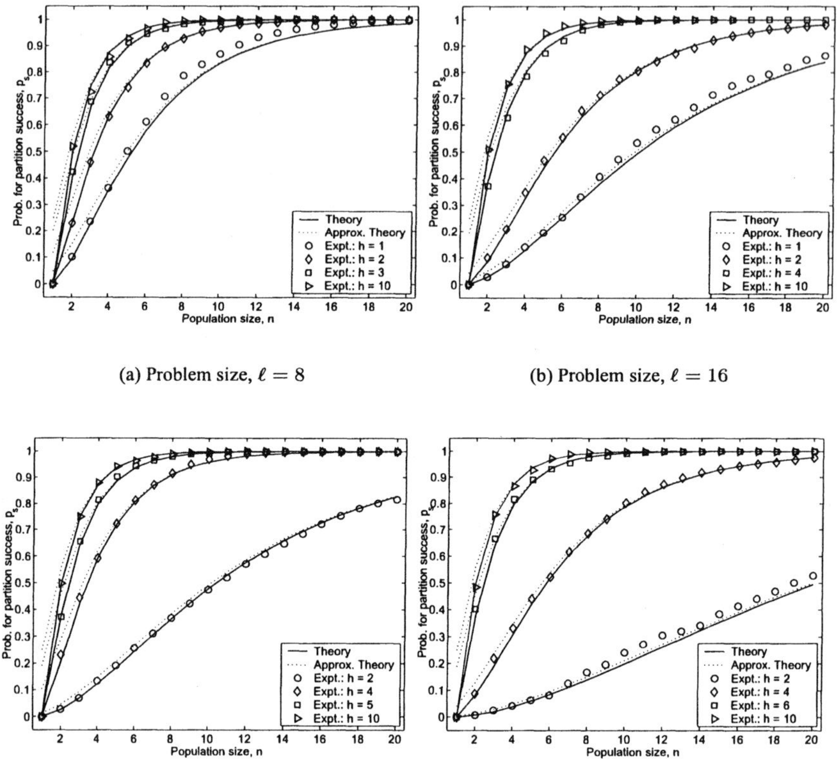

(a) Problem size, $\ell = 8$

(b) Problem size, $\ell = 16$

(c) Problem size, $\ell = 32$

(d) Problem size, $\ell = 64$

Figure 9-5. Verification of the models for BB partition success (Equations 9.16, and 9.14) with empirical results for UNITATION problem for different tree heights, h, and problem sizes, ℓ, as a function of population size, n. The empirical results depict the proportion of runs having at least one copy of a primitive and its complement in the population out of 1000 trials.

illustrated for BB of unit size ($k = 1$).

$$p_s = \sum_{i=2}^{n} \left(2^i - 2\right) \left(\begin{array}{c} n \\ n-i \end{array} \right) \left(p_{X_i^{\exp}}\right)^i \left(1 - 2p_{X_i^{\exp}}\right)^{n-i}, \tag{9.15}$$

The above equation can be rearranged as follows:

$$p_s \;=\; \sum_{j=0}^{n-2} \left(\begin{array}{c} n \\ j \end{array} \right) \left(2p_{X_i^{\exp}}\right)^{n-j} \left(1 - 2p_{X_i^{\exp}}\right)^{j}$$

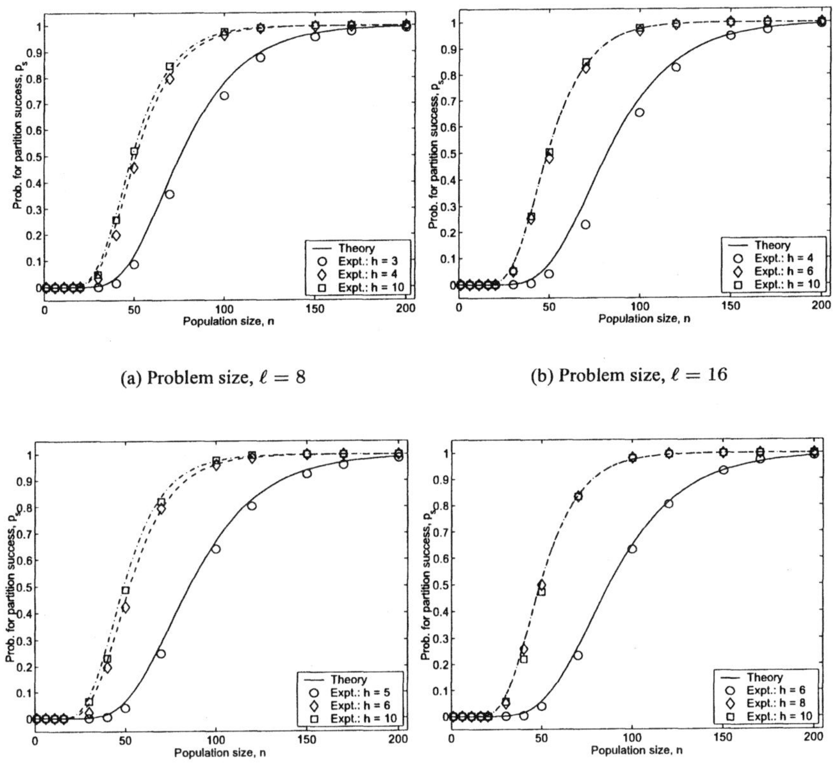

(a) Problem size, $\ell = 8$

(b) Problem size, $\ell = 16$

(c) Problem size, $\ell = 32$

(d) Problem size, $\ell = 64$

Figure 9-6. Verification of the BB partition success model (Equation 9.14) with empirical results for DECEPTION problem for different tree heights, h, and problem sizes, ℓ, as a function of population size, n. The empirical results depict the proportion of runs having at least one copy of a primitive and its complement in the population out of 1000 trials.

$$
\begin{aligned}
& -2\sum_{j=0}^{n-2}\binom{n}{j}\left(p_{X_i^{\text{exp}}}\right)^{n-j}\left(1-2p_{X_i^{\text{exp}}}\right)^{j}, \\
= \;& 1+\left(1-2p_{X_i^{\text{exp}}}\right)^{n}-2\left(1-p_{X_i^{\text{exp}}}\right)^{n}. \qquad (9.16)
\end{aligned}
$$

Equations (9.16), and (9.13) are compared with empirical results in Figure 9-5. The figures show that the approximate model (Equation 9.13) agrees with Equation 9.16 for higher population sizes and larger tree sizes. The partition success model (Equation 9.13) is compared with the empirical results for DECEPTION with $k = 4$, in Figure 9-6. Both Figures 9-5, and 9-6 clearly validate the BB supply model.

Population Sizing for Building-Block Supply

The facetwise model derived in the previous section will be rearranged in this section to estimate the population size required to ensure the presence of all BBs of a partition for ORDER, given the problem size is ℓ, and the tree height is h. Assuming that we can tolerate a probability ϵ of not having all BBs in a given partition, and setting p_s to $1 - \epsilon$, we can rewrite Equation 9.14,

$$1 - \epsilon = \exp\left[-2^k \exp\left[-n2^{-k} \exp\left\{-k \exp\left(-\frac{2n_l}{\ell}\right)\right\}\right]\right]. \tag{9.17}$$

Taking logarithms on both sides of the above equation and using the approximation, $\ln(1 - \epsilon) \approx -\epsilon$, we get

$$\epsilon = 2^k \exp\left[-n2^{-k} \exp\left\{-k \exp\left(-\frac{2^{h+1}}{\ell}\right)\right\}\right]. \tag{9.18}$$

After taking logarithms on both sides of the above equation and rearranging the resulting equation, we can write

$$n = 2^k \left(k \ln 2 - \ln \epsilon\right) \exp\left[k \exp\left(-\frac{2n_l}{\ell}\right)\right]. \tag{9.19}$$

If we assume tree size to be big enough ($n_l \gg \ell$), then the above equation can be simplified as $n \approx 2^k \left(k \ln 2 - \ln \epsilon\right)$. Furthermore, if we assume that the supply error is inversely proportional to the number of BBs, m, i.e., $\epsilon = 1/m$,

$$n \approx 2^k \left(k \ln 2 + \ln m\right). \tag{9.20}$$

It is interesting to note that the above population-sizing equation for BB supply in DECEPTION is identical to that developed by (Goldberg *et al.*, 2001) for selectorecombinative GAs.

6. Some Thoughts On Modeling Realistic GP Expressions

The last section developed BB supply models for ORDER expression mechanism and verified it for two test problems for different parameter values. This section provides a brief outline on how to develop BB supply models for realistic GP expressions. First we start by addressing the supply of raw tree fragments, or in other words, we consider that every tree fragment in the tree is *expressed*.

Tree Fragment Supply

Single BB Success. The probability that a tree does not contain a partition, P_i, is given by

$$p\left(\#of P_i = 0\right) = \left(1 - \frac{1}{\kappa}\right)^{\phi} \tag{9.21}$$

Recall that the values for κ, and ϕ for different partitions are given in Table 9-1. From the above equation, we can write the probability that the population contains at least one copy of the partition, P_i, as

$$p_k = 1 - \left[\left(1 - \frac{1}{\kappa}\right)^{\phi}\right]^n. \tag{9.22}$$

Using the approximation, $(1 - r/s)^s \approx e^{-r}$, and recognizing that this approximation is sufficiently accurate even for modest values of s, we can write

$$p_k \approx 1 - \exp\left(-\frac{n\phi}{\kappa}\right). \tag{9.23}$$

Furthermore, from table 9-1, we can see that $\phi \approx 2^{-k}s$, where $k = N_t + N_f$. Substituting this approximation for ϕ in the above equation, we get

$$p_k \approx 1 - \exp\left(-\frac{n \cdot s}{\kappa \cdot 2^k}\right). \tag{9.24}$$

It should be noted that that the approximation for ϕ is an underestimation for the tree fragments, P_b, P_c, P_e and P_f, and an overestimation for the tree fragments, P_a, P_d, and P_g.

Partition Success. As in section 9.5, we assume that the schema partition success values are independent. Then the probability of at least one success of each of the κ schemas, p_s is given by $p_s = p_k^{\kappa}$:

$$p_s = \left[1 - \exp\left(-\frac{n \cdot s}{\kappa \cdot 2^k}\right)\right]^{\kappa}, \tag{9.25}$$

$$\approx \exp\left[-\kappa \exp\left(-\frac{n \cdot s}{\kappa \cdot 2^k}\right)\right]. \tag{9.26}$$

Population Sizing for Partition Success. We now proceed to model the population size required to ensure the presence of all order-k tree fragments. Assuming that we can tolerate a probability ϵ of not having all BBs in a given partition, and setting p_s to $1 - \epsilon$, we can rewrite equation 9.26,

$$1 - \epsilon = \exp\left[-\kappa \exp\left(-\frac{n \cdot s}{\kappa \cdot 2^k}\right)\right] \tag{9.27}$$

Taking the logarithm on both sides and using the approximation $\ln(1-\epsilon) \approx -\epsilon$, for small values of ϵ, gives

$$\epsilon = \kappa \exp\left(-\frac{n \cdot s}{\kappa \cdot 2^k}\right) \tag{9.28}$$

Solving the above equation for n yields

$$n = \frac{1}{2} 2^k \kappa \left(\log \kappa - \log \epsilon\right). \tag{9.29}$$

Recall that $\kappa = \chi_f^{N_f} \chi_t^{N_t}$, and $k = N_f + N_t$. Then we can rewrite the above equation as

$$n = \frac{1}{s} \left(2\chi_f\right)^{N_f} \left(2\chi_t^{N_t}\right) \left[N_f \ln \chi_f + N_t \ln \chi_t - \ln \epsilon\right] \tag{9.30}$$

This relation can be further simplified if we assume that the supply error is inversely proportional to the number of BBs, m, i.e., $\epsilon = 1/m$. Then the equation may be rewritten as

$$n = \frac{1}{s} \left(2\chi_f\right)^{N_f} \left(2\chi_t^{N_t}\right) \left[N_f \ln \chi_f + N_t \ln \chi_t + \ln m\right] \tag{9.31}$$

Incorporating Expression

While counting the tree fragments may be useful enroute with proper expression model as in section 5, on its own it is not realistic. Therefore, we have to compute the combined probability that a tree fragment is present in the population and that it expresses a correct subfunction:

$$p(\text{BB is present}) = p(\text{fragment is present}) p(\text{expression}) \tag{9.32}$$

In the above equation we assume that the events that a tree being present in the population and it being expressed are independent. It should be noted that this assumption becomes more accurate as the population size increases. The probability of a tree fragment being present in the population, $p(\text{fragment is present}) = p_k$, and is given by equation 9.24, and the expression model is incorporated by the term $p(\text{expression})$. For example, in the symbolic regression example of $1 + x + x^2 + x^3$, the probability of expression incorporates the probability of different tree fragments expressing the linear and constant subfunctions.

7. Conclusions

In this paper, a detailed analysis of building-block supply in the initial population of GP using ORDER expression has been presented. Two facetwise models are derived, one for ensuring the supply of a single schema in a partition, and the other for ensuring the supply of all competing schemas in a partition for problems which employ ORDER expression mechanism. The latter model has been employed to estimate the population size required to ensure the presence of at least one copy of all raw BBs of a partition in the initial population. The population sizing model indicates that there is a minimum tree size dependent on the problem size. Furthermore, the models suggest that when the tree size is greater than the problem size, the population size required on BB supply grounds is $2^k\,(k \ln \chi + \ln m)$. This study also shows that the population size required to ensure the presence of all instances of tree fragments (assuming that all of them are expressed) is $\frac{1}{s}\,(2\chi_f)^{N_f}\,(2\chi_t)^{N_t}\,[N_f \ln \chi_f + N_t \ln \chi_t + \ln m]$.

Acknowledgments

We thank Martin Martin, Sean Luke, Terry Soule, and Bill Langdon. We also thank Gerulf Pedersen, Ying-Ping Chen, and Tian-Li Yu for their insightful comments and suggestions.

This work was sponsored by the Air Force Office of Scientific Research, Air Force Material Command, USAF, under grant F49620-00-0163 and F49620-03-1-0129, and the National Science Foundation under grant DMI-9908252, and CSE fellowship, UIUC. The U.S. Government is authorized to reproduce and distribute reprints for government purposes notwithstanding any copyright notation thereon. The views and conclusions contained herein are those of the authors and should not be interpreted as necessarily representing the official policies or endorsements, either expressed or implied, of the AFOSR, the NSF, or the U.S. Government.

References

Altenberg, L. (1994b). Emergent phenomena in genetic programming. In *Evolutionary Programming III — Proceedings of the Third Annual Conference*, pp. 233–241, San Diego, CA. World Scientific.

Deb, K. and Goldberg, D. E. (1993). Analyzing deception in trap functions. In *Foundations of Genetic Algorithms 2*, D. Whitley (Ed.), pp. 93–108. Morgan Kaufmann.

Goldberg, D. E. (1987). Simple genetic algorithms and the minimal, deceptive problem. In *Genetic algorithms and simulated annealing*, Davis, L. (Ed.), pp. 74–88. Los Altos, CA: Morgan Kaufmann.

Goldberg, D. E. (1989b). Sizing populations for serial and parallel genetic algorithms. In *Proceedings of the Third International Conference on Genetic Algorithms (ICGA)*, J. Schaffer (Ed.), pp. 70–79. Morgan Kaufmann.

Goldberg, D. E. (2002). *The design of innovation: Lessons from and for competent genetic algorithms*. Boston MA: Kluwer Academic Publishers.

Goldberg, D. E. and O'Reilly, U.-M. (1998). Where does the good stuff go, and why? How contextual semantics influences program structure in simple genetic programming. In *Proceedings of the First European Workshop on Genetic Programming (EuroGP '98)*, W. Banzhaf *et al.* (Eds.), pp. 16–36. Springer, Berlin.

Goldberg, D. E., Sastry, K. and Latoza, T. (2001). On the supply of building blocks. *Proceedings of the Genetic and Evolutionary Computation Conference (GECCO 2001)*, pp. 336–342. Morgan Kaufmann.

Holland, J. (1975). *Adaptation in natural and artificial systems*. Ann Arbor, MI: University of Michigan Press.

Koza, John (1992). *Genetic Programming: On the Programming of Computers by Natural Selection*. MIT Press, Cambridge, MA, USA.

Langdon, W. B. and Poli, R. (2002). *Foundations of Genetic Programming*. Springer-Verlag.

Luke, S. (2000b). Two fast tree-creation algorithms for genetic programming. *IEEE Transactions on Evolutionary Computation 4(3):* 274.

O'Reilly, U.-M. and Goldberg, D. E. (1998). How fitness structure affects subsolution acquisition in genetic programming. In *Genetic Programming 1998: Proceedings of the Third Annual Conference*, W. Banzhaf *et al.* (Eds.), pp. 269–277.

O'Reilly, U.-M. and Oppacher, F. (1995). The troubling aspects of a building block hypothesis for genetic programming. In *Foundations of Genetic Algorithms 3*, D. Whitley (Ed.), pp. 73–88. Morgan Kaufmann.

Poli, R. (2000). Recursive conditional schema theorem, convergence and population sizing in genetic algorithms. In *Foundations of Genetic Algorithms 6*, W. Martin and W. Spears (Eds.). Morgan Kaufmann.

Reeves, C. (1993). Using genetic algorithms with small populations. *Proceedings of the Fifth International Conference on Genetic Algorithms*, pp. 92–99.

Rosca, J. P. (1997b). Analysis of complexity drift in genetic programming. In *Genetic Programming 1997: Proceedings of the Second Annual Conference*, pp. 286–294.

Stephens, C. and Waelbroeck, H. (1999). Schemata evolution and building blocks. *Evolutionary Computation 7(2):* 109–124.

Whigham, P. A. (1995). A schema theorem for context-free grammars. In *Proceedings of the 1995 IEEE Conference on Evolutionary Computation 1*, pp. 178–181.

Chapter 10

MODULARIZATION BY MULTI-RUN FREQUENCY DRIVEN SUBTREE ENCAPSULATION

Daniel Howard
QinetiQ, Malvern,
Worcestershire WR14 3PS
United Kingdom
dhoward@qinetiq.com

Abstract In tree-based Genetic Programming, subtrees which represent potentially useful sub-solutions can be encapsulated in order to protect them and aid their proliferation throughout the population. This paper investigates implementing this as a multi-run method. A two-stage encapsulation scheme based on subtree survival and frequency is compared against Automatically Defined Functions in fixed and evolved architectures and standard Genetic Programming for solving a Parity problem.

Keywords: Genetic Programming, Modularization, Subtree Encapsulation, Multi-run, ADF, Subtree Database, Subtree Frequency, Parity Problem.

1. Introduction

Initial generations in Genetic Programming (GP) rapidly improve the population fitness but after further generations the improvement tails-off. A possible application of multi-run, or repeated stages of evolution, is to stop the run early, and somehow to project 'key elements' of the current solution into a new run. Enriched with these key elements, the fitness in the new run would improve rapidly to new levels. This is the motivation of this research that combines multi-run with a subtree modularization scheme. The idea is to transfer the most useful 'atomized' partial solutions from one stage of multi-run to the next stage of multi-run.

This introductory section is a general discussion on modularization and multi-run. Section 2 describes the Genetic Programming modularization by means of multi-run subtree encapsulation. Section 3 applies the scheme to a model problem and compares its performance with more popular modularization schemes. Section 4 draws conclusions.

Modularization

The benefits of the hierarchical problem solving approach for methods of automatic programming are discussed in (Koza, 1994). Modularization enables Genetic Programming to address larger problems than proof of concept ones, solving them with less overall effort by an astute division into problem components.

Another important trait of Genetic Programming discussed in (Koza *et al.*, 1999), is inventiveness. Human inventiveness has been connected with a type of mental process called 'lateral thinking' and it may be useful to consider how modularization might relate to it. (De Bono, 1970) coined the phrase 'lateral thinking' to describe the mental process that can arrive at an innovation. Its basic principle is the realization that any particular way of looking at things is only one from among many other possible ways. Hence, lateral thinking explores all of the different ways of seeing things by restructuring and rearranging the information that is available. It is more preoccupied with the usefulness of a pattern in arriving at a new pattern than with the validity of the pattern. By 're-structuring' it discovers otherwise unthinkable analogies between unrelated concepts (analogy in Mathematics surprised (Halmos, 1981)).

The connection between lateral thinking and modularization is the connection between re-structuring and diversity. Modularization is a method of generating structure and of combining structure. When modularization is directed at generating more diversity, it becomes more capable of innovation than the same system devoid of modularization. This chapter illustrates this by first using modularization to create a number of distinct terminals, which are then combined to solve the problem that standard Genetic Programming on its own cannot solve.

Modularization has other important advantages for practitioners. It facilitates the method of prospective analysis enabling the solution search within a prescribed architecture (Koza, 1994; Koza *et al.*, 1999; Howard, Roberts and Ryan, 2002). It can also be safer to use. Although a problem can be conjured up to defeat any search method (see the no free lunch theorems of (Wolpert and Macready, 1997)), schemes enriched with modularization are generally more powerful solvers of real world problems than schemes without modularization. The former find solutions in 'reasonable' effort, but the latter may never do so. However, simple schemes can outperform the modularization enabled schemes for solution of simple problems. Analysis of the solution from a modularization enabled scheme may provide insights allowing design of a simpler scheme to optimize a more focused version of the problem.

Multi-Run

The strategy of replacing a computation by a series of contiguous computational stages, or multi-run, is a feature of computational engineering that can be exploited by Genetic Programming practitioners. By analogy with hierarchical decomposition, it may be simpler to solve subproblems at each stage than to solve the overall problem at once.

In the field of numerical linear algebra, the solution of non-linear differential equations sometimes requires the practice of 'continuation' (Davidenko, 1953; Wasserstrom, 1973). Continuation follows a path characterized by a parameter of the problem. A sequence of problems for different values of the parameter is solved for because one stage cannot obtain the solution for the problem with the target value of the parameter. For example, the Rayleigh number is the continuation parameter for the 'flow in a driven cavity' problem. The solution at each stage of multi-run serves as starting point for the next solution stage, ending in solution of the problem at the target value of the parameter.

Another multi-run technique from the field of numerical linear algebra is Iterative Refinement, whereby iteration improves upon the result of an LU factorization of a matrix problem. Similarly, (Streeter et al., 2002), use a second Genetic Programming stage to approximate the solution error generated by the first Genetic Programming stage.

Multi-run can implement a data decomposition strategy. For example, in machine vision, (Howard et al., 1999; Howard and Roberts, 1999) apply Genetic Programming runs consecutively on different partitions of the imagery to discover increasingly complex image filters.

Genetic Programming with Modularization

This section describes three popular Genetic Programming methods, which bring about explicit modularization.

Automatically Defined Functions (ADF) developed by (Koza, 1992), (Koza, 1994), are parameterized subroutines shown to improve performance for a wide variety of problems. ADF architectures can be specified manually or they can be evolved (Koza, 1994) with the numbers of arguments and of ADF changing during a run with an operator that is analogous to gene duplication (Koza *et al.*, 1999).

The GP-Automata developed by (Ashlock, 1997), is a directed graph architecture which includes Genetic Programming trees at nodes of this graph. Both the graph and the Genetic Programming trees are subjected to evolution so that the states of the automata are function subprograms, which can become revisited and reused (Howard and Benson, 2003).

Subtree Encapsulation, or the idea of freezing material and naming it for its reuse in a GP run, was initially introduced by (Koza, 1992), and other schemes followed based on this principle. The Module Acquisition (MA) technique (Angeline and Pollack, 1994) created modules by randomly selecting subtrees for atomization and then truncating nodes below the subtree's root to specify formal arguments. MA employed an expansion operator to overcome problems due to a lack of genetic diversity. The Evolution Defined Function approach (Ahluwalia and Bull, 1999) is a hybrid method which uses the mutation operators of MA to govern ADF. The Adaptive Representation through Learning (ARL) algorithm (Rosca and Ballard, 1996b) overcame a dependence on *random* selection for modularization by attempting to discover *building blocks* during evolution (Altenberg, 1994a; Goldberg, 1989a). ARL did this by extending the primitive set with subroutines drawn from offspring which gave the best improvement on their parents' fitness. Although ARL was shown to perform well against standard Genetic Programming and Genetic Programming using two ADF, it requires many control parameters and relies on random selection to specify formal arguments. The ARL framework has been used to compare various heuristics for module selection (Dessi et al, 1999).

2. Subtree Encapsulation with Multi-Run

A procedure can combine modularization with multi-run. A Genetic Programming tree is factored into all of its component subtrees, which vary in size (ranging from subtrees consisting of one function and its terminals to the entire tree as a subtree).

At the end of a stage of the multi-run, all of the subtrees in the population will need to be identified ('atomized' or named). How to define a subtree class? Subtrees are the same if they compute to the same result for each of the fitness cases. Moreover, the smallest structural instance of the class or its syntactic simplification becomes a class description.

The resulting subtree taxonomy is ranked by some criteria. Candidates are subtree frequency; subtree size; subtree 'fitness'. The last one is interesting. An instance of a subtree could be assigned the fitness of the tree it belongs to, and its class 'fitness' the average over all instances. There is no known best way to rank the subtrees.

Next, a number of subtrees of high rank are added to the terminal set available to the next stage of multi-run. The increase in the number of terminals available may require a larger population size. Instead, a point mutation operator is introduced to substitute atoms during the run. Initial generations of this second stage of multi-run will involve small trees, as usual. However, should these small trees include atomized terminals they will represent truly sophisticated computations. Crossover and mutation quite rapidly move and rearrange these atoms.

This multi-run scheme using subtree encapsulation was first introduced in (Roberts et al., 2001). Module selection was based on subtree survival and frequency. This scheme was shown to out-perform standard Genetic Programming on a real world problem of target detection. Section 3 investigates its behaviour for a parity problem.

Subtree Selection

Subtree selection heuristics are difficult to specify because a subtree has no associated fitness when considered outside the context of its parent tree. A frequency-biased selection of subtrees has the following justification. Subtrees which survive numerous generations are likely to be popular building blocks, and useless subtrees would tend to be driven to extinction because they do not give their parent trees any selection advantage. However, the multi-run stage does not aim to solve the problem but instead to capture promising subtrees. Initial generations quickly drive the evolution towards certain better individuals. The idea is to stop before the change in the population fitness slows, and bloat and introns become established, see (Langdon and Poli, 1998a), or (Soule and Foster, 1998), and chapters of this book by J. Rosca and T. Soule. Subtrees are then more likely to be solution components.

The modularization scheme in this study encapsulated subtrees which appeared as multiple instances in the population after a number of generations, N_{gen}, had elapsed. The value of N_{gen} was set to coincide with an expected plateau in the best-of-generation fitness, which suggested that the evolution was converging towards a local optimum (Rosca and Ballard, 1996b).

Subtree Database

A subtree database contains the description of each subtree in the population and stores its evaluation over all fitness cases. When a new individual is produced from parts of its parent(s), recalling this database information avoids re-evaluation of large portions of the new tree (genetic operations and the shape of tree structures influence the efficiency of this reuse). However, there is a tradeoff between availability of RAM plus time of search and recall, and fitness re-evaluation time.

The only motivation for having this database is speed of computation. This depends upon the number of fitness cases, the expense of the operations, the memory available, and the Genetic Programming software implementation. Its presence is not necessary to implement the modularization by multi-run subtree encapsulation scheme in this chapter, other than for convenience. Its presence could be justified for the problem in (Roberts et al., 2001), where more details are available.

It should be stressed that the subtrees are *not* encapsulated in the population during evolution, thus crossover and mutation are free to manipulate the trees as if the database did not exist. This approach overcomes the loss of genetic diversity which was found in Module Acquisition (Angeline and Pollack, 1994).

Equivalence Grouping

After N_{gen} generations, the best parallel independent run is identified. Equivalence grouping of its resulting database, groups together subtrees that give the same output over all fitness cases. It makes it possible for operationally equivalent yet structurally distinct subtrees to be grouped together, partly because of the issue of *introns* (Luke, 2003).

Monitoring the frequency of each subtree created during the evolution (the frequency in the entire population) was the original motivation of the scheme. The frequency of a particular subtree *operation* is calculated by summing the frequency of the grouped subtrees. Subtrees which occur more than once are encapsulated by representing them as terminal nodes. Further stages of multi-run exploit the subtree evaluations (prior computation over all fitness cases) as terminal data.

Encapsulation Procedure

Figure 10-1 displays the procedure for subtree encapsulation. A first evolution stage involves multiple independent Genetic Programming runs that process an initial terminal set for N_{gen} generations. The subtree database for the best run is then analyzed to group subtrees that are operationally equivalent. The evaluations for the *surviving* subtree operations are then exported. The terminal set is augmented to include a terminal for each of N_{st} randomly selected surviving subtrees. The subtree evaluations are imported as terminal data to perform multiple Genetic Programming runs in a second evolution stage, where each run begins with an initialized population.

Encapsulating subtrees from the best of a series of runs has an advantage arising from the following issues. Firstly, fitness improvement becomes increasingly more difficult during the course of evolution. Secondly, there is a strong inverse correlation between the depth of recombination point and its effect on fitness (Igel and Chellapilla, 1999). Consequently, standard Genetic Programming tends to converge on a particular root structure in early generations (McPhee and Hopper, 1999). Furthermore, recombination near the root becomes less likely as trees *bloat* (Luke, 2003). Thus multiple runs are needed to explore different root structures and to discover modules which apply in the context of the best root structure.

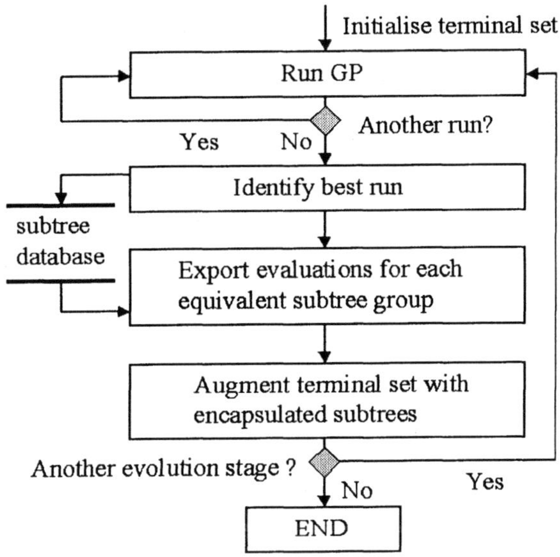

Figure 10-1. Subtree encapsulation procedure.

Figure 10-1 shows that the encapsulation procedure can be iterated for multiple evolution stages. Each stage augments the terminal set with new encapsulated subtrees. However, this iteration remains an area for further work.

3. Illustration: Parity Problem

Although it is a proof of concept problem the parity problem (Koza, 1994) clearly illustrates the advantage that ADF enjoys over standard Genetic Programming. For this reason alone, this is a good problem to discover if the modularization of subtrees scheme can also outperform standard Genetic Programming.

The experiments in this section use the two stage version of multi-run to solve the even-5-parity (a simple version of the parity problem). Fixed and evolved architecture ADF experiments are also compared. A recent review of Genetic Programming attempts for solving parity problems is presented in (Poli and Page, 2000).

The XOR problem and the more general n-order parity problems, are classes of problems which are not linearly separable (limitation of perceptron). Parity problems contain modularity, building blocks, that must be discovered for efficient solution, e.g. discovery of the XOR function is important. Methods that efficiently solve parity problems must be useful for solving real world problems with modularity.

Genetic Programming Specifications

The experimental details largely followed those for the even-5-parity problem in (Koza, 1994), the main difference being that an elitist steady-state Genetic Programming was used. The population was initialized using the ramped-half-and-half technique with a maximum initial depth of 6 levels. Each individual was trained on all of the 32 fitness cases. The raw fitness was the number of cases for which the individual was correct. The Genetic Programming parameters (same for both multi-run stages) are given in Table 10-1 where the terminal set consists of the five Boolean variables. 100 independent runs were conducted for each Genetic Programming scheme investigated.

Table 10-1. Genetic Programming parameters for the even-5-parity problem. Population size and many of the remaining parameters were not optimized but arbitrarily chosen.

parameter	setting
functions	AND, OR, NAND, NOR
terminals	D0, D1, D2, D3 and D4
population	4000
mate radius	4000
kill tournament	size 2 (elitist, never kills the best)
breed tournament	size 7
regeneration	90% crossover (single child), 10% truncation mutation
internal crossover	60%
max generations	51
max branch size	1000 nodes

The truncation mutation involved randomly selecting a node in the parent tree and replacing the subtree rooted at that node with a randomly selected terminal. This mutation operator was essential owing to the large number of terminals, O(100), which resulted from augmenting the terminal set with encapsulated subtrees. Without this operator, Genetic Programming could manipulate only the terminal contexts that were captured by chance in the initial population.

The runs using ADF were based on the fixed architecture and the evolved architecture schemes in (Koza, 1994). The evolved architecture ADF did not include the operations introduced in chapter 5 of (Koza *et al.*, 1999). Each individual consisted of a result-producing branch and a number of ADF. The fixed architecture individuals included N_F number of ADF and branch typing was used for crossover. Point typing was used for the evolved architectures.

The result-producing branch and all ADF had the function set shown in Table 10-1 augmented by references to subordinate branches. The result-producing branch could refer to any ADF and each ADF could refer to any lower numbered ADF. The result-producing branch had the terminal set shown in Table 10-1 and the ADF received only dummy variables.

Subtree Encapsulation versus Standard GP

This section compares different Genetic Programming schemes using the performance measures defined in section 4.11 of (Koza, 1994) (these are sensitive to the choice of population size). P_i is the cumulative probability of success (in %) that a particular run will solve the problem by generation i. The effort, E, estimates the number of individuals that need to be processed to solve the problem with a 99% probability. It represents the optimum trade off between the number of required runs, R_E, and the number of required generations, i_E. It was calculated using $E = R_E(i_E + 1)$ to take account of population initialization. All runs used the same population size ($M = 4000$) so E is in terms of M.

Table 10-2 summarizes the performance of Genetic Programming using N_{st} randomly selected encapsulated subtrees, where i_{min} is the minimum generation at which a solution was achieved. The first row shows the results for standard Genetic Programming (no encapsulated subtrees). It can be seen that only 2 of the 100 runs solved the problem.

Analysis of the standard Genetic Programming runs revealed that a plateau in the best-of-generation fitness tended to occur around generation 15, where the fitness did not improve for at least 5 further generations. The best run at generation 15 gave a correct result for 30 of the 32 fitness cases. The associated subtree database was analyzed to give the encapsulated subtrees for a second evolution stage. The database stored 32000 subtrees, 15000 of which occurred more than once in the population and thus were deemed to be the set of *surviving* subtrees. Analysis of these subtrees yielded 3410 groups of operationally equivalent subtrees. The most common subtree operation existed in the form of 104 structural variants, however the average number of variants in a group was four and many subtrees were operationally unique. Figure 10-2 shows the frequency of the most common subtrees after equivalence grouping, where the highest frequency was 45472.

Each run for the second evolution stage used encapsulated subtrees which were randomly selected from the 3410 possible subtrees (different runs used different terminal sets). Table 10-2 gives the cumulative probability of success at generation 35 because the encapsulated subtrees were already evolved for 15 generations. For this reason, P_{35} could be considered to relate to a total evolution of 50 generations. However, runs were terminated after 50 generations because the population was initialized for each second stage run. This is why P_{50} was also calculated.

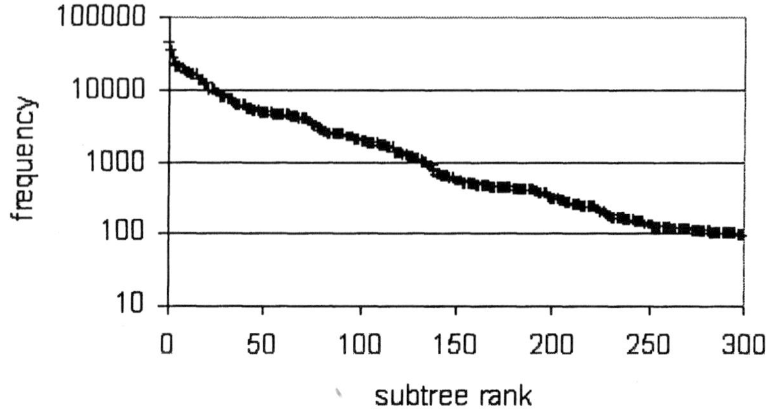

Figure 10-2. The frequency of the most common subtrees after equivalence grouping.

Table 10-2 shows that subtree encapsulation dramatically improved the overall performance, allowing solutions to be found within 6 generations. This demonstrates the importance of modularization where even frozen modules prove to have great benefit. The improved results suggest that encapsulation reduced the dimensionality of the search space. Possibly, the subtrees represented adequate partial solutions to tackle various dimensions of the problem. This is consistent with the findings in Module Acquisition (Angeline and Pollack, 1994). Furthermore, the encapsulation not only protected higher-level abstractions, but it also greatly encouraged a proliferation of these abstractions via their atomization. Evolution could explore the use of these abstractions more thoroughly than in standard Genetic Programming.

The performance improved as the number of subtrees, N_{st}, was increased, with the maximum P_{35} and a relatively low E being achieved for $N_{st} = 200$. However, Table 10-2 suggests that the search space became over-complicated when N_{st} was set to 800.

Note that the maximum program size was 1000 nodes for both standard Genetic Programming and encapsulation scheme runs. However, encapsulation allowed larger programs to be evolved because the subtrees could be expanded back into their constituent nodes. Still, the two solutions achieved by standard Genetic Programming contained about 400 nodes and so this 1000 nodes size limit did not favour encapsulation over standard Genetic Programming.

Table 10-2 shows the performance of the second evolution stage when the N_{st} most common subtrees were encapsulated. Although these results still improved

Table 10-2. Performance with randomly selected encapsulated subtrees and with the most common encapsulated subtrees. N_t is the number of subtrees. P_i is the cumulative probability of success that a particular run will solve the problem by generation i; i_{min} is the minimum generation at which a solution is achieved; the effort, E, estimates the number of individuals that need to be processed to solve the problem with a 99% probability and is the optimum trade off between the number of required runs, R_E, and the number of required generations, i_E.

	randomly selected						**most common**					
N_t	P_{35}	P_{50}	i_{min}	i_E	R_E	E	P_{35}	P_{50}	i_{min}	i_E	R_E	E
0	1	2	30	48	228	11172						
25	30	39	5	32	14	462	8	13	20	35	56	2016
50	34	42	6	17	21	378	8	17	21	28	64	1856
100	45	49	2	33	8	272	31	33	5	17	21	378
200	45	51	4	19	10	200	37	41	5	15	18	288
400	43	47	3	14	13	195	43	46	2	5	13	78
800	38	45	3	7	29	232	39	39	3	10	14	154

upon standard Genetic Programming, the performance was considerably worse than randomly selected subtrees until N_{st} was increased to 400. However, the effort required when encapsulating the 400 most common subtrees was the lowest effort of all the Genetic Programming schemes investigated.

It is not entirely clear what caused the less beneficial performance of frequency-biased subtree selection. It is possible that the most popular subtrees are less diverse amongst each other than many of the subtree sets chosen randomly. This could be investigated further. Moreover, the second evolution stage may have been hindered by a saturation of genetic material. This because the most common subtrees were necessarily the smallest possible subtrees, as a nested subtree was at least as common as its parent subtree. When the best run from the first stage was complete, 19 out of the 25 most common subtrees were represented by 3 node subtrees (before equivalence grouping). Therefore, the encapsulation of these subtrees (and their functional equivalents) promoted the proliferation of code which may have easily become widespread from normal genetic operations. Increasing N_{st} improved the performance by enforcing genetic diversity. As expected, frequency-biased subtree selection *per se* is not the optimal method of subtree selection.

Subtree Encapsulation versus ADF

Table 10-3 summarizes the performance for various ADF architectures. All ADF in the fixed architectures (with $N_F = 1$; $N_F = 2$; or $N_F = 3$) received 3 arguments, and the evolved architectures used from 0 to five ADF each taking from 1 to 5 arguments.

It was thought that the branch-size proportionate branch selection used in (Koza, 1994) introduced an undesirable dependence on bloat. An alternative branch selection scheme was employed where the result-producing branch was assigned a selection weight w_R. Thus, the result-producing branch had a selection probability of $w_R/(w_R + N_F)$ and each ADF had a probability of $1/(w_R + N_F)$.

Table 10-3 shows that the results were fairly insensitive to w_R, but $w_R = 4$ tended to give the best performance whereas $w_R = 1$ tended to give the worst. Modularization provided by ADF could be thought of as the co-evolution of a solution (the result-producing branch) and the language in which it is expressed (the ADF). In this light it is intuitive that the exploitation of the language should adapt more quickly than the language itself, i.e. $w_R > 1$.

Table 10-3. Performance using ADF for various architectures and selection weight w_R. A numeric N_F denotes a fixed architecture and *evo* denotes an evolved architecture. P_{50} is the cumulative probability of success that a particular run will solve the problem by generation 50; i_{min} is the minimum generation at which a solution is achieved; the effort, E, estimates the number of individuals that need to be processed to solve the problem with a 99% probability and is the optimum trade off between the number of required runs, R_E, and the number of required generations, i_E.

	$N_F = 1$					$N_F = 2$				
w_R	P_{50} (in %)	i_{min}	i_E	R_E	E	P_{50} (in %)	i_{min}	i_E	R_E	E
1	45	4	12	16	208	45	2	7	20	160
2	47	2	12	13	169	55	2	13	11	154
4	50	2	7	21	168	62	2	12	8	104
8	50	3	8	22	198	48	2	16	10	170
16	45	1	22	11	253	54	2	13	12	168
32	55	1	12	15	195	54	2	9	19	190
	$N_F = 3$					$N_F = evo$				
w_R	P_{50} (in %)	i_{min}	i_E	R_E	E	P_{50} (in %)	i_{min}	i_E	R_E	E
1	48	1	8	15	135	37	2	11	31	372
2	52	1	8	18	162	37	1	18	13	247
4	59	3	9	11	110	44	2	8	24	216
8	47	1	10	17	187	38	2	9	25	250
16	62	2	11	12	144	40	3	8	22	198
32	54	1	9	16	160	34	1	8	31	279

Table 10-3 shows little variation in performance as N_F was increased for the fixed architectures, where the best result of $P_{50} = 62$ and $E = 104$ was achieved using two ADF and the worst result of $P_{50} = 45$ and $E = 253$ was found using a single ADF. The fixed architectures consistently out-performed the evolved architecture. A possible explanation being that the former used branch typing whereas the latter used point typing. Consequently, the former scheme was more likely to evolve programs in which each ADF converged to distinct sub-solutions, and the convergence would tend to be quicker.

The solutions achieved by the evolved ADF architectures used between 1 and four ADF, with no obvious preference for the number of arguments. Note that in an additional experiment, 100 runs were also conducted for an evolved architecture with upper limits of three ADF and 3 arguments, but the results were worse than those shown in Table 10-3.

Comparing Tables 10-2 and 10-3 reveals that all ADF architectures gave far better performance than standard Genetic Programming, and that subtree encapsulation with $N_{st} = 200$ performed slightly better than the evolved ADF architecture but generally worse than the fixed architectures. It suggests that continually evolved modules are better than frozen modules. Probably the latter become restrictive or even obsolete as the module exploitation progresses. Note that some runs with ADF found solutions on very early generations. Possibly the initial programs could be effectively larger than those for encapsulation due to hierarchical ADF calls.

The different Genetic Programming schemes are compared in Figure 10-3 which gives plots for the encapsulation with $N_{st} = 200$, and for the ADF with $w_R = 4$ using two ADF and the evolved architecture. Significance tests using the t-test showed that the null hypothesis was rejected at 1% chance for the fixed ADF architecture versus the evolved architecture or encapsulation. However, when comparing the latter two schemes the null hypothesis could only be rejected at 32%.

The results in this section support Koza's finding that ADF in a fixed architecture generally gave better performance than ADF in an evolved architecture, and that both schemes perform much better than standard Genetic Programming (Koza, 1994).

Computation Time

Table 10-4 gives the computation time taken to complete 100 runs of each Genetic Programming scheme on a 450MHz Pentium II MMX PC. Each run was terminated after 50 generations or when a solution was found. standard Genetic Programming took the most time because it rarely solved the problem, whereas solutions were generally found the earliest when using ADF. Program evaluations took longer when more ADF were used because of the hierarchical ADF calls, and the evolved architecture also had an overhead due to point typing.

The subtree database had a large capacity (32000) and could store subtrees of any depth. This implementation allowed all the subtrees which survived evolution to be stored, but it introduced intensive database searches and also

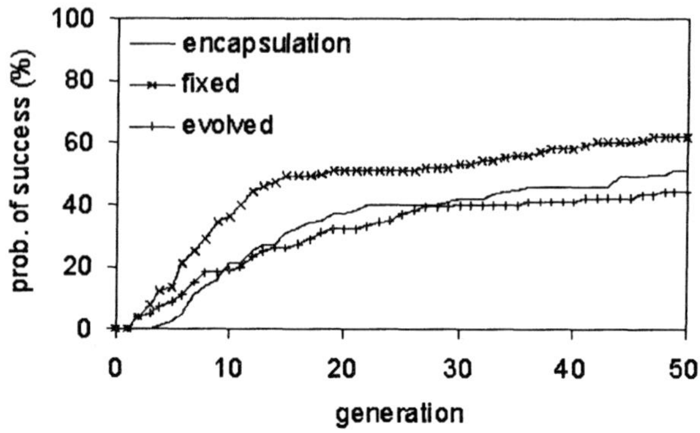

Figure 10-3. The cumulative probability of success for various modularization schemes: the encapsulation of 200 randomly selected surviving subtrees (the 2nd stage), a fixed architecture with two ADF and an evolved ADF architecture. Both ADF schemes used $w_R = 4$. Standard Genetic Programming is too poor to show.

Table 10-4. Computation time to complete 100 runs of each scheme. All ADF schemes used $w_R = 4$.

scheme	time (s)
standard Genetic Programming	82207
encapsulation $N_{st} = 200$	34312
one ADF	16881
two ADF	22834
three ADF	49497
evolved ADF architecture	46845

required trees to be alphabetized (Roberts et al., 2001). To reduce computation time, the first evolution stage was executed without the database in order to identify the best run. The best run was then repeated with the database in order to export the encapsulated subtree evaluations. The time given in Table 10-4 includes all the necessary encapsulation phases; executing 100 first stage runs for 15 generations, repeating the best run with the database and then executing 100 second stage runs for a maximum of 35 generations. It can be seen that the encapsulation was quicker than some ADF schemes.

The subtree database stores the evaluation of each subtree over all fitness cases and so each subtree is evaluated only once. A subtree's evaluations are recalled in order to evaluate a parent tree or a new subtree in which the stored subtree is nested. Consequently, the database can offer a computational saving when evaluations are more computationally demanding (when requiring floating-point arithmetic over many fitness cases). Further work is a comparative study of computation time between the database with improved search and other methods.

4. Conclusions

This chapter examined a simple modularization scheme for Genetic Programming that used subtree encapsulation and multi-run. The creation of modules was based on monitoring the survival and frequency of subtrees in the population as opposed to random selection during evolution. Multi-run was implemented in two stages. Modules were created from the best run from a *first evolution stage*, and were exploited by a *second evolution stage* to discover and combine the most beneficial modules. Structural variation (including the notion of introns) was discarded during module selection by considering operational equivalence.

The subtree encapsulation scheme greatly improved upon standard Genetic Programming for solving the even-5-parity problem because: (a) equivalence grouping produced a set of distinct terminals (different vectors of size 32) and this variety allowed the second stage to more easily search for the solution; (b) encapsulated code was protected through atomization; (c) code reuse facilitated the exploitation of regularities inherent in the problem; and (d) multi-run offered a problem decomposition through an emergent hierarchical abstraction.

Subtree encapsulation also aids phenotypic implementation, where reused components need only be processed once for each application of the phenotype. This may also aid program interpretation.

Using the same population size and other Genetic Programming parameters, the modularization by multi-run subtree encapsulation scheme kept pace with Automatically Defined Functions (Koza, 1994).

The architecture altering operations introduced in chapter 5 of (Koza *et al.*, 1999), were not implemented. These select a fit individual after some evolution has taken place, and spawn architecture altering operations by a procedure that is inspired in gene duplication. This waiting strategy can be compared with the modularization by multi-run subtree encapsulation scheme described in this chapter. A difference being that module selection by operational equivalence ensures complete module diversity at a given point in time (encapsulated terminals are all different). Further research should establish whether this offers any advantage over standard ADF practice. Perhaps an indicator of potentially useful subtrees could be obtained by analyzing which selected subtrees are most beneficial. And this might be employed to create modules *during* an evolution run, perhaps by deliberately using the best candidate subtrees to spawn ADF, or by ensuring that candidate ADF increase diversity.

Acknowledgements

The author is grateful to Simon C. Roberts for assistance with the computations, and to workshop participants for their informative comments. The idea of combining modularization and multi-run originated during John Koza's visit to our laboratory in April 2000.

References

Ahluwalia M. and Bull, L. (1999). Coevolving Functions in Genetic Programming: Classification Using K-Nearest-Neighbor, In *Proceedings of the Genetic and Evolutionary Computation Conference (GECCO 1999)*, Orlando, Florida, pp. 947-953. Morgan Kaufmann.

Altenberg, L. (1994). "The Evolution of Evolvability in Genetic Programming." In *Advances in Genetic Programming*, Kinnear, K.E. Jr. (Ed.), pp. 47-74. The MIT Press.

Angeline P. J. and Pollack J. B. (1994). Coevolving High-Level Representations. In *Artificial Life III*, C. G. Langton (Ed.), pp55-71, Addison-Wesley.

Ashlock D., 1997. GP-Automata for dividing the dollar. In *Genetic Programming: Proceedings of the Second Annual Conference*, Koza, John *et al.*. (Eds.), pp18–26. Stanford University.

Davidenko D. F. (1953). On a New Method of Numerical Solution of Systems of Nonlinear Equations. Dokl. Akad. Nauk SSSR (USSR Academy of Sciences report), 88, pp. 601-602.

De Bono, E. (1970). *Lateral Thinking: A Textbook of Creativity.* Penguin Books Ltd., England.

Dessi A., Giani A. and Starita A. (1999). An Analysis of Automatic Subroutine Discovery in Genetic Programming. In *Proceedings of the Genetic and Evolutionary Computation Conference (GECCO 1999)*, Orlando, Florida, pp. 996-1001. Morgan Kaufmann.

Goldberg, D. (1989). *Genetic Algorithms in Search, Optimization and Machine Learning*. Addison Wesley, Reading MA.

Halmos P. R., 1981. Does Mathematics Have Elements? *The Mathematical Intelligencer 3:* 147-153.

Howard D. and Roberts S. C. (1999). A Staged Genetic Programming Strategy for Image Analysis. In *Proceedings of the Genetic and Evolutionary Computation Conference (GECCO 1999)*, pp. 1047-1052, Orlando, Florida, Morgan Kaufmann.

Howard D., Roberts S. C. and Brankin R. (1999). Target Detection in SAR Imagery by Genetic Programming. *Advances in Engineering Software 30:* 303-311.

Howard D., Roberts S. C. and Ryan C. (2002). Machine Vision: Exploring Context with Genetic Programming. In *Genetic and Evolutionary Computation Conference (GECCO 2002)*, pp. 756-763. Morgan Kaufmann.

Howard D. and Benson K. (2003). Evolutionary Computation Method for Pattern Recognition of *cis*-acting Sites. *Biosystems*, special issue on Computational Intelligence and Bioinformatics (in press).

Igel C. and Chellapilla K. (1999). Investigating the Influence of Depth and Degree of Genotypic Change on Fitness. In *Proceedings of the Genetic and Evolutionary Computation Conference (GECCO 1999)*, pp 1061-1068, Orlando, Florida. Morgan Kaufmann.

Koza, John (1992). *Genetic Programming: On the Programming of Computers by Natural Selection*. MIT Press, Cambridge, MA, USA.

Koza, John R. (1994). *Genetic Programming II: Automatic Discovery of Reusable Programs*. The MIT Press, Cambridge, MA, USA.

Koza, J.R., Andre, D., Bennett, F.H. III and Keane, M. (1999). *Genetic Programming 3: Darwinian Invention and Problem Solving*. Morgan Kaufman, San Francisco, CA, USA.

Langdon W. B. and Poli R. (1998a). Fitness Causes Bloat: Mutation. In *Proceedings of the First European Genetic Programming Conference (Euro GP 98)*, Lecture Notes in Computer Science (LNCS), Volume 1391, Banzhaf, W. *et al.* (Eds.), pp 37-48. Springer.

Luke S., (2003). Modification Point Depth and Genome Growth in Genetic Programming. *Evolutionary Computation 11(1)*: 67-106.

McPhee N. F. and Hopper N. J. (1999). Analysis of Genetic Diversity through Population History. In *Proceedings of the Genetic and Evolutionary Computation Conference (GECCO 1999)*, Orlando, Florida, pp. 1112-1120. Morgan Kaufmann.

Poli R. and Page J. (2000). Solving High-Order Boolean Parity Problems with Smooth Uniform Crossover, Sub-Machine Code GP and Demes. *Genetic Programming and Evolvable Machines 1(1/2)*: 37-56.

Roberts S. C., Howard D. and Koza J. R. (2001). Evolving Modules in GP by Subtree Encapsulation. In *Proceedings of the 4th European Conference (EuroGP 2001)*, Lake Como, Italy, April 2001. Lecture Notes in Computer Science (LNCS) Volume 2038. Springer.

Rosca J. P. and Ballard D. H. (1996b). Discovery of Subroutines in Genetic Programming. In *Advances in Genetic Programming 2*, P. J. Angeline and K. E. Kinnear Jr. (Eds.). The MIT Press.

Soule T. and Foster J. A. (1998). Removal Bias: a New Cause of Code Growth in Tree Based Evolutionary Programming. In *IEEE International Conference on Evolutionary Computation*, pp. 178-186, Anchorage, Alaska, IEEE Press.

Streeter M. J., Keane M. A. and Koza J. R. (2002). Iterative Refinement of Computational Circuits using Genetic Programming. In *Proceedings of the Genetic and Evolutionary Computation Conference (GECCO 2002)*, pp. 877-884. Morgan Kaufmann Publishers.

Wasserstrom E. (1973). Numerical Solutions by the Continuation Method. *SIAM Review 15(1)*: 89-119.

Wolpert D. H. and Macready W. G. (1997). No Free Lunch Theorems for Optimization. *IEEE Transactions of Evolutionary Computation 1(1)*: 67-82.

Chapter 11

THE DISTRIBUTION OF REVERSIBLE FUNCTIONS IS NORMAL

W. B. Langdon
W.Langdon@cs.ucl.ac.uk
Computer Science, University College, London,
Gower Street, London, WC1E 6BT, UK

Abstract The distribution of reversible programs tends to a limit as their size increases. For problems with a Hamming distance fitness function the limiting distribution is binomial with an exponentially small chance (but non-zero) chance of perfect solution. Sufficiently good reversible circuits are more common. Expected RMS error is also calculated. Random unitary matrices may suggest possible extension to quantum computing. Using the genetic programming (GP) benchmark, the six multiplexor, circuits of Toffoli gates are shown to give a fitness landscape amenable to evolutionary search. Minimal CCNOT solutions to the six multiplexer are found but larger circuits are more evolvable.

Keywords: fitness landscape, evolutionary computation, genetic algorithms, genetic programming, quantum computing, CCNOT, Toffoli, low power consumption

1. Introduction

We shall show the fitness of classical reversible computing programs (Bennett and Landauer, 1985) (where fitness is given by Hamming distance from an ideal answer) is Normally distributed. If the score is normalised so that the maximum score (fitness) is 1 and the minimum is 0, then the mean is 0.5 and the standard deviation is $1/2\, m^{-1/2} 2^{-\frac{n}{2}}$. (Where n is the number of input bits and m is the number of output bits.)

Almost all genetic programming has used traditional computing instructions, such as *add*, *subtract*, *multiply*, *or*, and *and*. These instruction sets are not reversible, i.e., in general, it is impossible given a program and its output, to unambiguously reconstruct the program's input. This is because most of the primitive operations themselves are irreversible. However genetic programming can evolve reversible programs composed of reversible primitives.

A number of reversible gates have been proposed (Toffoli, 1980; Fredkin and Toffoli, 1982) which can be connected in a linear sequence to give a reversible gate array, which we will treat as a reversible computer program. At present the driving force behind the interest in reversible computing is the hope that reversible gates can be implemented as quantum gates, leading to quantum coherent circuits and quantum computing. Reversible computing has also been proposed for safety critical applications and for low power consumption or low heat dissipation.

In the absence of counter measures, most traditional computer programs degrade information, i.e., knowledge about their inputs is progressively lost as they are executed. This means, most programs produce the same output regardless of their input (Langdon, 2002b). Suppose a program has n input bits and m output bits, there are 2^{m2^n} possible functions it could implement. However a long program is almost certain to implement one of the 2^m constants. That is, the fraction of functions actually implemented is tiny as the programs get longer and worse, the fraction of interesting functions tends to zero. This is due to the inherent irreversibility of traditional computing primitives.

In the next section we describe reversible computing in more detail. In reversible computing there is also a distribution of functions which programs approach as they get longer. Instead of it being dominated by constants, every *reversible* function is equally likely (cf. Section 3). Convergence to this limit is tested in Section 5 on the Boolean 6 Multiplexor problem (described in Section 4). Section 5 shows convergence to the large program limit can be rapid. Despite the low density of solutions, evolutionary search is effective at finding them (Section 6). Initial results suggest, CCNOT and traditional computing primitives are similarly amenable to evolutionary search.

2. Reversible Computing Circuits

A reversible computer can be treated as an array of parallel wires leading from the inputs and constants to the outputs and garbage. In normal operation the garbage outputs are treated as rubbish and discarded by the end of the program.

Connected across the wires are reversible gates. Each gate has as many inputs as it has outputs. The gates are reversible, in the sense that it is possible to unambiguously identify their inputs given their outputs (see Figure 11-1). The simplest reversible gate is the identity, i.e. a direct connection from input to output. Also NOT is reversible, since given its output we know what its input must have been. Similarly a gate which swaps its inputs is also reversible (see Figure 11-2).

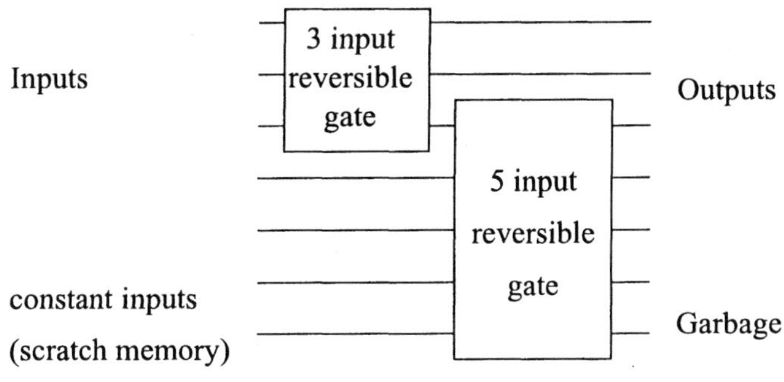

Figure 11-1. Schematic of reversible computer. In normal operation data flows from left to right. However when reversed, garbage and outputs are fed into the circuit from the right to give the original inputs to the program (plus constants).

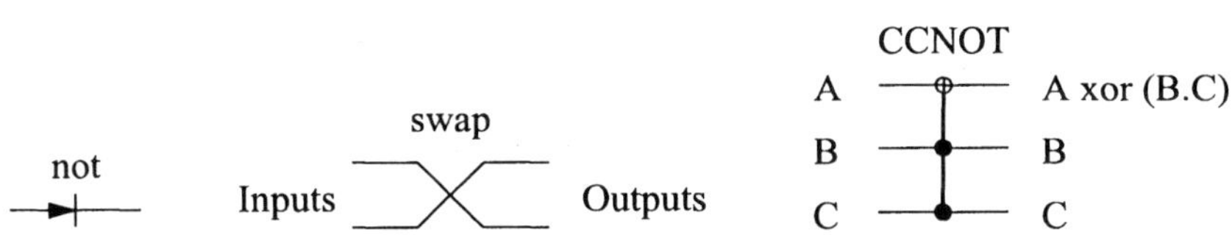

Figure 11-2. Examples of reversible gates. The CCNOT (Toffoli) gate passes all of its inputs directly to its outputs, unless all of the control lines (B and C) are true, in which case it inverts its main data line (A). (The control inputs of a CCNOT gate (B and C) can be connected to the same wire but the controlled wire (A) cannot be directly connected to either control.) CCNOT is complete, in the sense all boolean circuits can be constructed from it.

We will mainly be concerned with the controlled-controlled not (CCNOT, Toffoli) gate. Unless all of its control lines are true, the CCNOT gate passes all of its inputs directly to its outputs. However if they are all set, CCNOT inverts the controlled line. CCNOT is complete in the sense, given sufficient CCNOT gates and additional constant inputs and rubbish outputs, a reversible circuit equivalent to any Boolean function from inputs to outputs (excluding constants and rubbish) can be constructed. Since CCNOT can invert 1, the additional inputs can all be 1. Note a single CCNOT gate (plus a constant zero, e.g. provided by using another CCNOT gate to invert a one) can implement the identify function. In C code, `data[A] = (data[C] & data[B]) ^ data[A];`

While it is not necessary for the number of lines set to true, to remain constant across the circuit from left to right, a reversible computer must implement a permutation. To explain what we mean by this, consider the left hand side (of N wires) as an N bit number. There are up to 2^N possible left hand patterns. Similarly there are up to 2^N possible right hand patterns. The computer provides a mapping from left hand number to right hand number. For the mapping to be reversible, its range and domain must be the same size and a number can only appear once on the right hand side (range), i.e. the mapping must be a permutation. If all 2^N possible numbers are used, only $2^N!$ of the 2^{N2^N} possible mappings are reversible. For large N, this means only about 1 in e^{2^N} mappings are permutations and hence are reversible.

The computation remains reversible up until the garbage bits are discarded. It is at this point that information is lost. It is the deletion of information which means the computation must consume energy and release it as heat. By carefully controlling the deletion of these rubbish bits, it has been suggested that reversible computers will require less energy than irreversible computers. Present day circuits do not approach the lower bound on energy consumption suggested by their irreversibility, i.e., they require much more energy to operate gates, drive connecting wires, etc. than the theoretical bound on energy consumption due to information lost as they run.

However in the near term, energy consumption is interesting both for ultra-low power consumption, e.g. solar powered computing, and also because the energy released inside the computing circuit has to be removed as heat. The only way heat is removed at present is by making the centre of the circuit hotter and allowing heat to diffuse down the temperature gradient to the cooler boundaries of the circuit. (Active heat pumps within the circuit have been considered. Electronic refrigerators could be based upon the Peltier electro thermodynamic effect). Even today cooling circuits is a limiting factor on their operation. Increasing circuit clock speeds, despite continued reduction in circuit size, mean heat removal will be an increasing concern.

(Bishop, 1997) describes a single channel reversible system for a safety critical control application. By running the system forwards and then backwards and comparing the original inputs with those returned by traversing the system twice, he demonstrated the system was able to detect test errors injected into the system during its operation. (High reliability systems often use comparison between multiple channels to detect errors.)

3. Distribution of Large Reversible Circuits

As with a complete reversible circuit, the action of a single reversible gate across N wires can be treated as a permutation mapping one N bit number to another. Following (Langdon and Poli, 2002), (Langdon, 2002a),(Langdon, 2002b), we can treat the sequence of permutations from the start of the circuit to its end as a sequence of state transitions. The state being the current permutation. We restrict ourselves to just those permutations which can be implemented, i.e. states that can be reached. Each gate changes the current permutation (state) to the next. We can describe the action of a gate by a square matrix of zeros and ones. Each row contains exactly one one. The position of the one indicates the permutation on the output of the gate for each permutation on the input side of the gate (note the matrix is row stochastic). Now each gate is reversible, i.e., given a permutation on its output side, there can only be one permutation on its input side. This means each column of the matrix also contains exactly one one (the matrix is column as well as row stochastic, i.e., it is double stochastic). We will have multiple ways of connecting our gates or even multiple types of gate, however each matrix will be double stochastic and therefore so too will be the average matrix. Since we only consider implementable permutations, the average matrix is fully connected. If a single gate can implement the identity function, the matrix must have a non-zero diagonal element. This suppresses cycling in the limit. If we choose gates at random, the sequence of permutations is also random. Since the next permutation depends only on the current permutation and the gate, the sequence of permutations is a Markov process. The Markov transition matrix is the average of each of the gate matrices, which is fully connected, acyclic and double stochastic. This means as the number of randomly chosen gates increases each of the Markov states becomes equally likely (Feller, 1957), i.e., in the limit of large circuits each possible permutation is equally likely.

When there are many randomly connected gates and the total number of lines N is large, not only is each possible permutation equally likely but (since our reversible gates are complete) all m bit output patterns are possible. Further we will assume that we can treat each output bit as being equally likely to be on as off and almost independent of the others. If fitness f is defined by running the program on every input (i.e. running it 2^n times) and summing the number of output bits that match a target (f = Hamming distance) then f follows the Binomial distribution $2^{-m2^n} C_f^{m2^n}$. This means most programs have a fitness near the average $\frac{1}{2}m2^n$ and the chance of finding a solution is 2^{-m2^n}. While small, this is finite, whereas with irreversible gates (and no write protection of inputs) almost all programs do not solve any non-trivial problem (Langdon, 2002b).

We can also use the known uniform distribution to calculate the expected RMS error. Suppose T of the 2^n possible fitness cases are run, the expected average squared error is $\overline{\text{RMS}} = \frac{1}{k}\sum_{i=1}^{k}\sqrt{\frac{1}{T}\sum_{t=1}^{T}|P_i(\text{input}_t + 2^N - 2^n) \bmod 2^m - A_t)|^2}$. Where k is the total number of permutations, input_t is the input for the tth test, A_t = required answer and $P_i(x)$ is the ith permutation of x. As $2^m \ll k$ we can approximate the average behaviour of $P_i(t + 2^N - 2^n) \bmod 2^m$ over k cases by i over 2^m cases. So the expected average squared error is $\overline{\text{RMS}} = \frac{1}{2^m}\sum_{i=0}^{2^m-1}\sqrt{\frac{1}{T}\sum_{t=1}^{T}|i - A_t|^2}$. If only a few tests with small values are run

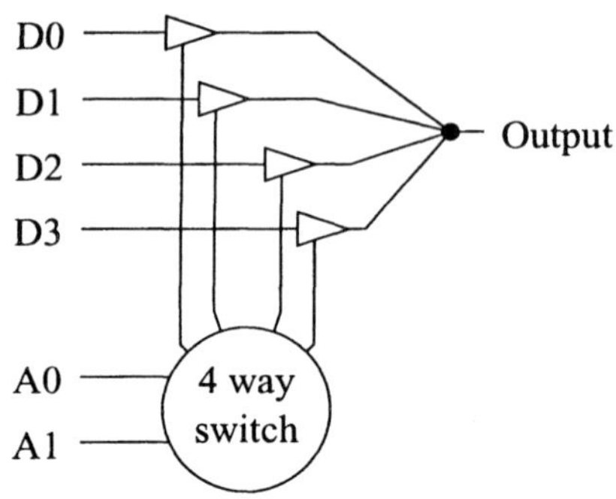

Figure 11-3. Six way multiplexor. Only one of four data lines (D0, D1, D2 and D3) is enabled; which one is determined by the two control lines (A0 and A1).

(i.e. $T \ll 2^m$ and $A_t \ll 2^m$) then the expected root mean squared error is bounded by

$$\begin{aligned}
\overline{\text{RMS}} &\approx \frac{1}{2^m}\sum_{i=0}^{2^m-1}\sqrt{\frac{1}{T}\sum_{t=1}^{T} i^2} \approx \frac{1}{2}2^m \\
\text{VAR}_{\text{RMS}} + \overline{\text{RMS}}^2 &= \frac{1}{2^m}\sum_{i=0}^{2^m-1}\frac{1}{T}\sum_{t=1}^{T}|i - A_t|^2 \approx \frac{1}{2^m}\sum_{i=0}^{2^m-1} i^2 \approx \frac{1}{3}(2^m)^2 \\
\text{VAR}_{\text{RMS}} &\approx \frac{1}{3}(2^m)^2 - \frac{1}{4}(2^m)^2 \\
\text{SD} &\approx \frac{1}{2\sqrt{3}}2^m = 0.2886751\ 2^m
\end{aligned}$$

On the other hand if exhaustive testing is carried out and the target values are uniformly spread in the range $0 \ldots\ 2^m - 1$ then $\overline{\text{RMS}} \approx \frac{7}{12\sqrt{3}}2^m$ and $\text{SD} \approx 0.23\ 2^m$.

4. 6 Multiplexor

The six multiplexor problem has often been used as a benchmark problem. Briefly the problem is to find a circuit which has two control lines (giving a total of four possible combinations) which are used to switch the output of the circuit to one of the four input lines, cf. Figure 11-3. The fitness of a circuit is the number $(0 \ldots 64)$ of times the actual output matches the output given by the truth table. Note fitness is given by number of bits in common between the actual truth table implemented by a program and a given truth table (Hamming distance).

5. Density of 6 Multiplexor Solutions

We measured[1] the distribution of fitness of randomly chosen CCNOT programs with 0, 1 and 6 additional wires at a variety of lengths, cf. Figures 11-4–11-10. These experiments confirm that there is a limiting fitness distribution and it is Binomial. Further, cf. Figure 11-9, the difference between the actual distribution and the limiting distribution falls rapidly with program length. Also we do not need many spare wires (one is sufficient) to be close to the theoretical wide circuit limit. Only without any spare lines does the limit differ from theory.

There are no CCNOT solutions to the six multiplexor problem with less than five gates. One of the smallest solutions is shown in Figure 11-11. Notice this does not use any additional wires. While the multiplexor can be solved by CCNOT without additional storage, with six lines only even fitness scores are possible. This also means, even in the limit of long programs, there is a bias towards higher fitness, increasing the mean fitness from 32 to 32.5, cf. Figures 11-4, 11-5, 11-8 and 11-10.

It is clear that CCNOT has a nice bias for solving the six multiplexor problem. On average small circuits have above average fitness and in particular (at least with six lines) the chance of solving the problem is far higher for small programs than in the limit of large programs.

6. Hill Climbing and Evolutionary Solution of the Six Multiplexor Problem

Section 3 tells us how many solutions there are but not how easy they are to find. To investigate this we carried out hill climbing and population based search on the minimal circuit and a larger circuit. Our results are summarised in Table 11-1.

Firstly we compare random search with these two more sophisticated search techniques. Figure 11-10 shows the chance of solving the six multiplexor by random search of CCNOT circuits. From the first solutions composed of 5

[1] In all the six multiplexor experiments we used speed up techniques based on those described in (Poli and Langdon, 1999).

Table 11-1. Number of runs solving the six multiplexor problem

Configuration		Hill climber	Population
6 lines	5 CCNOT	0/10	4/10
12 lines	20 CCNOT	1/10	10/10

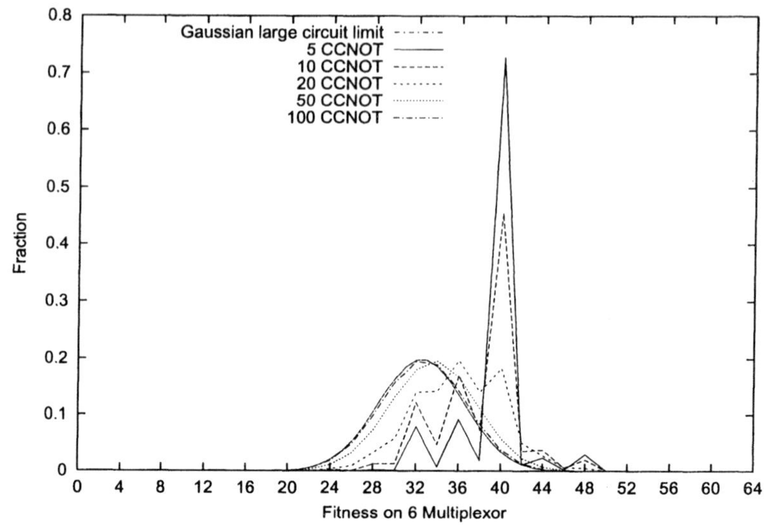

Figure 11-4. Proportion of circuits composed of controlled-controlled-NOT (CCNOT, Toffoli) gates of each fitness on the 6 multiplexor problem. Solutions have fitness of 64. (At least 100 million random circuits were tested for each length.) Since the only wires are those carrying the inputs (i.e. no additional memory) odd fitness values cannot be generated. To simplify the graph these are excluded.

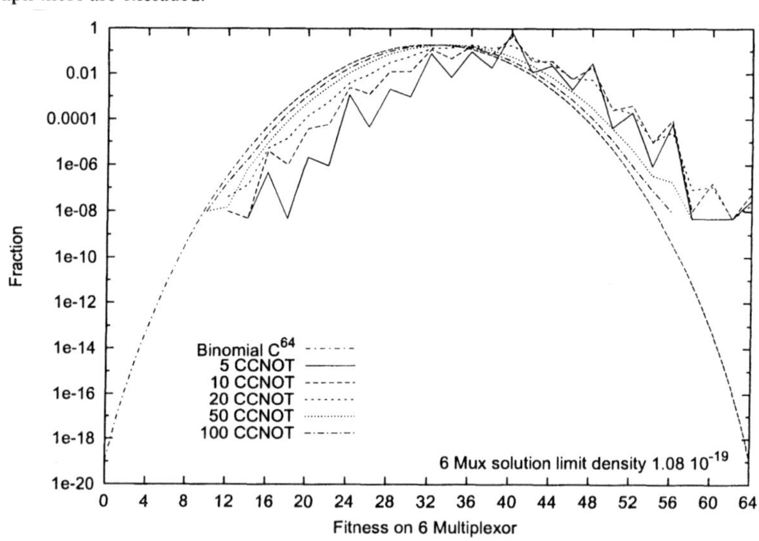

Figure 11-5. Data as Figure 11-4. In Figure 11-4 the large circuit limit is approximated by a Normal distribution with mean 32.5. Here the Binomial distribution approximates the tails near fitness 0 and 64.

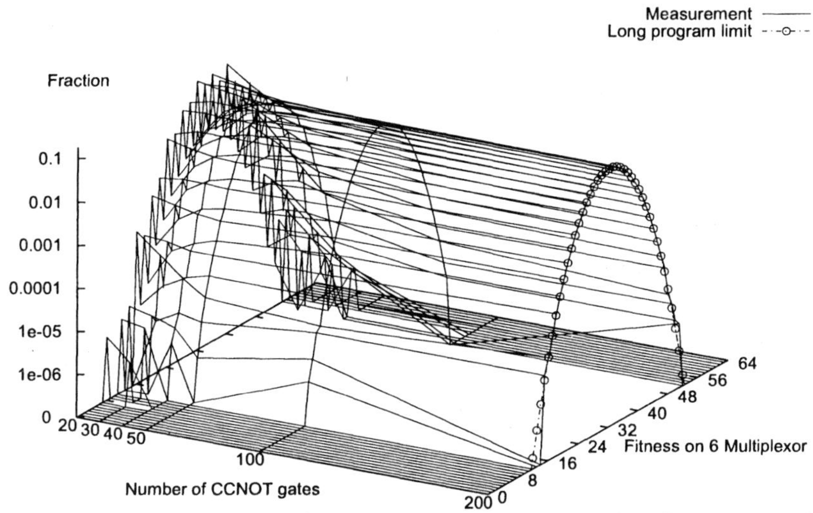

Figure 11-6. Convergence of 6 multiplexor fitness distribution as number of CCNOT gates is increased from 20 towards the large circuit limit (ringed parabola right hand end). (At least million random circuits were tested for each length.) One additional memory (garbage) line ensures all output patterns can be implemented and in the large circuit limit are equally likely, i.e., the density of solutions is 2^{-64}.

gates and no spare lines, the chance rises from about $3\ 10^{-8}$ to a peak of about $10\ 10^{-8}$ at 10 gates, and then falls towards the theoretical (non-zero) limit of $5.4\ 10^{-20}$ as the circuit size increases. (No solutions were found in more than 10,000,000 trials with either one or six spare lines.)

A local hill climber was run ten times. It mutated exactly one CCNOT gate of the five in the six wire circuit and retained the mutant only if its fitness was better. In seven runs a fitness level of 56 was reached. In two of the seven runs the hill climber reached fitness 56 significantly[2] faster than random search could be expected to. In seven of the remaining runs there is no significant difference between the hill climber and random search. (However random search always has a chance of solving the problem. While the hill climber was stuck at local optima and could never proceed.) In the remaining run, the hill climber took significantly longer to reach fitness level 48.

Our hill climber found one solution in ten runs, when the number of CCNOT gates was increased to 20 and six spare lines were added. Six of the runs attained a score of 56, two runs 48 and the last 52. None of these nine runs first reached its final fitness level significantly faster or slower than random search might be expected to. Except again random search might improve, whereas

[2]All significance tests in Section 6 use a 5% two tailed test.

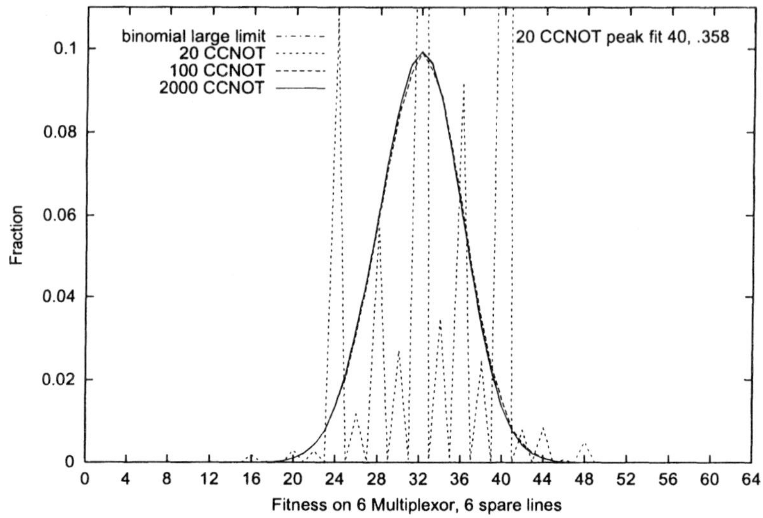

Figure 11-7. Distribution of fitness on the 6 multiplexor problem of circuits of CCNOT gates and with 12 lines.

the hill climber was stuck at local optima and could never proceed. Each 20 CCNOT 12 wire program has $20 \times (2 \times (12 - 2) + (12 - 3)) = 580$ or $20 \times 3 \times (12 - 2) = 600$ neighbours, so our hill climber will take on average no more than 4185 attempts to try them all, cf. The Coupon Collector's problem (Feller, 1957, page 284). All runs had many more trials than this.

In contrast the search space turns out to be very friendly to evolutionary search. Using the same mutation operator and a population of 500 (see also Table 11-2) in four out of ten runs minimal solutions to the six multiplexor were found, i.e., five CCNOT without spare wires. These solutions took between 20 and 100 generations. In the remaining six runs fitness level 56 was reached (in five cases significantly faster than random search). Introducing six spare wires and extending the circuits from five to 20 CCNOT makes the problem significantly easier. Ten out of ten runs (with the same population size etc.) found solutions. The solutions were found after 30–220 generations. This is significantly better than hill climbing and population search of the smaller circuit size.

The effort (Koza, 1992, page 194) required to find a non-minimal reversible solution to the six multiplexor using a population approach (87,000) is somewhat similar to that required by genetic programming to find a non-reversible one (Koza, 1992, Table 25.2). In other words, using CCNOT with spare lines has not been shown to be uncompetitive with existing approaches.

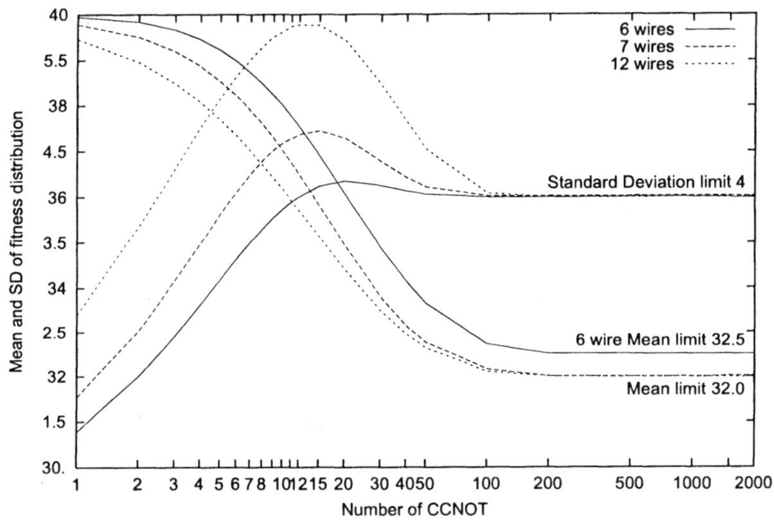

Figure 11-8. The fitness distribution of small CCNOT circuits are asymmetric with mean near 40. As number of Toffoli gates is increased, both mean and standard deviation converge to theoretical binomial limits (32 and 4), except for circuits without spare wires, in which case the mean converges to 32.5.

7. Discussion

It is important to realise the limiting distribution results hold in general for reversible computing. Not just for the six multiplexor or similar problems and for any reversible gate, not just the Toffoli (CCNOT) gate. An interesting extension would be quantum computing. Random matrices theory may give a formal bound on the size of circuits needed to approach the limiting distribution.

The benchmark can be solved with no spare wires. Indeed (Toffoli, 1980, page 636) describes a set of gates for which no more than m spare wires are needed for any finite reversible function. However allowing modest increases in the size of solution by allowing more gates and spare wires appears to make the fitness landscape more evolvable. I.e. easier for evolutionary search to find solutions. It is not clear whether additional gates or wires or both is primarily responsible. In our example at least, and we suggest perhaps to other problems, insisting upon *minimal* solutions rather than sufficient solutions, which may be bigger, makes the problem unnecessarily hard.

While NFL (Wolpert and Macready, 1997) applies to reversible computing, we expect evolutionary search also to be better than hill climbing and random search when used with other reversible gates, such as the Fredkin gate, on this and similar problems.

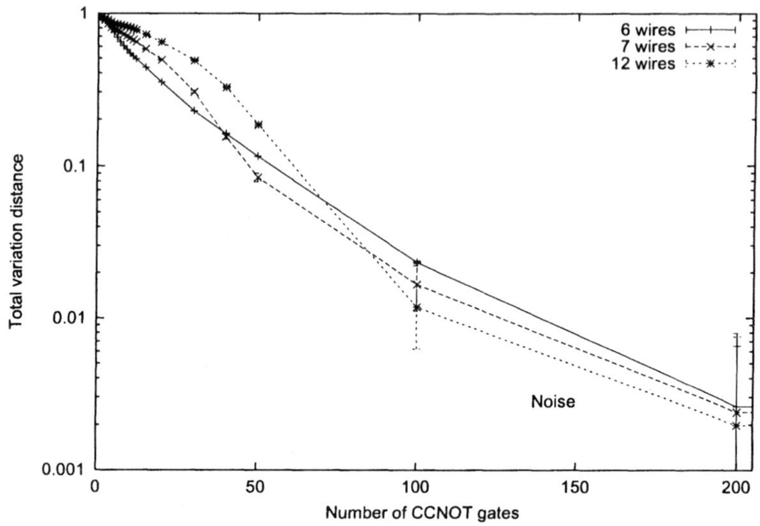

Figure 11-9. Discrepancy between measured distribution of fitness on CCNOT 6 multiplexor problem and large circuit limit (calculated as total variation distance (Rosenthal, 1995)). Rapid convergence to theoretical limit as program size increases is shown. We would expect more spare lines to mean bigger programs are needed for convergence. The plots are reminiscent of exponential decay shown for non-reversible programs (Langdon, 2002a) suggesting group theory might lead to results on the rate of convergence.

The number of programs or circuits of a given size increases exponentially with circuit size. Thus average behaviour across all programs is dominated by the behaviour of the longest programs. Almost all of these will behave as the limiting distribution suggest. Thus considering only the limiting distribution is sufficient to describe the vast majority of programs.

Where spare wires are included, for most programs, the Binomial distribution can be approximated by a Normal distribution with the same mean and variance. I.e. where fitness is given by a Hamming distance, the average fitness is $\frac{1}{2}m2^n$ and the variance is $1/2\ 1/2\ m2^n$. If we normalise fitness to the range $0 \ldots 1$, then the mean becomes 0.5 and the standard deviation is $\frac{2^{-\frac{n}{2}}}{2\sqrt{m}}$. Even with modest numbers of input and output wires, the Hamming fitness distribution becomes a needle, with almost all programs having near average fitness.

In non-trivial problems n and m rapidly become too large to allow exhaustive testing. However the limiting distribution still applies. In the limit the chance of a random program passing non-exhaustive testing is given by the number of bits which are checked. I.e. if T tests are run and only a p precision answer is needed, the chance of passing a test case is 2^{-p}. The chance of passing all the test cases is $2^{-T \times p}$. But note that, like non-reversible programs, randomly

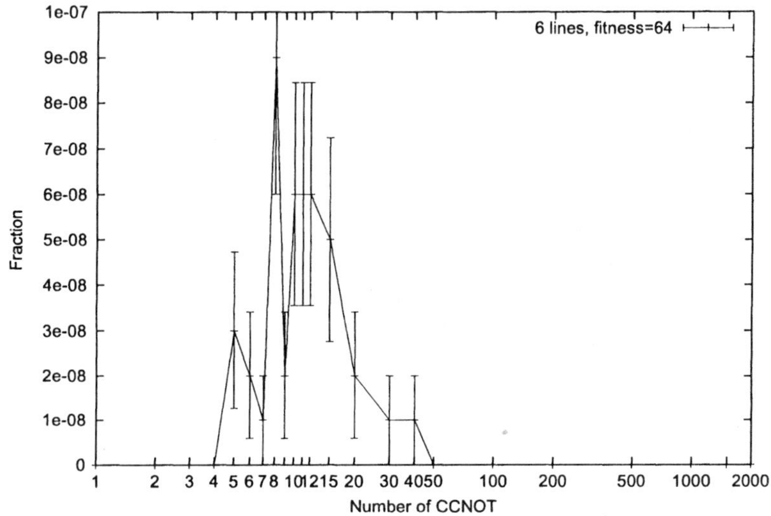

Figure 11-10. Measured density of solutions on CCNOT 6 multiplexor problem. (Based on measurements of at least 100,000,000 random circuits of each length. In one million random samples at each length, with either one or six spare lines no solutions were found.)

Table 11-2. Parameters for Multiplexor Problem

Objective:	Find a reversible function whose output is the same as the Boolean 6 multiplexor function
Inputs:	D0 D1 D2 D3 A0 A1 (plus either 0 or 6 "true" constants)
Functions set:	CCNOT
Fitness cases:	All the 2^6 combinations of the 6 Boolean arguments
Fitness:	number of correct answers
Selection:	Tournament group size of 7, non-elitist, generational
Pop size:	500
Program size:	5 or 20 CCNOT (Toffoli) reversible gates
Parameters:	100% mutation (Exactly one CCNOT gate is randomly chosen. One of its three wires is chosen at random, and replaced by a randomly chosen, different, but legal, wire.)
Termination:	fitness=64 or maximum number of generations G = 500

passing the test cases gives no confidence that the program will generalise. If an additional independent test is added, the chance of randomly passing it is only 2^{-p} (Langdon and Poli, 2002). In contrast general solutions have been evolved via limited numbers of test cases by genetic programming (Langdon, 1998) suggesting GP has a useful bias for problems of interest.

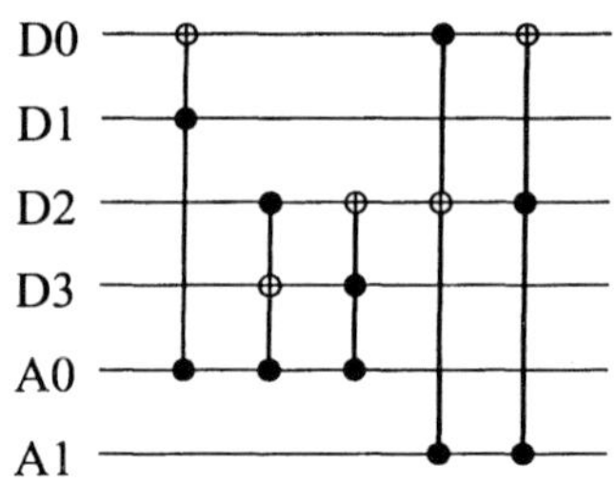

CC	CC	CC	CC	CC
0	3	2	2	0
1	2	4	5	2
4	4	3	0	5

Figure 11-11. Example evolved minimal circuit (left) of controlled-controlled-NOT (CCNOT, Toffoli) gates implementing a six way multiplexor. Genetic representation on right. Two address lines direct one of the four data inputs to the output. Circles with crosses indicate controlled wire. Note there are no additional memory (garbage) lines and only five gates are required.

Table 11-3. Distribution of Fitness of Large and Wide Reversible Circuits

Fitness function	Mean	Standard Deviation	Perfect Solutions
Hamming	$\frac{1}{2}m2^n$	$\frac{1}{2}\sqrt{m2^n}$	2^{-m2^n}
Normalised Hamming	$\frac{1}{2}$	$\frac{1}{2\sqrt{m2^n}}$	2^{-m2^n}
RMS (small T)	$\frac{1}{2}2^m$	$0.29\ 2^m$	2^{-mT}
RMS (large no. tests)	$0.34\ 2^m$	$0.23\ 2^m$	2^{-mT}

8. Conclusions

As with traditional computing, as reversible circuits get bigger the distribution of their functionality converges to a limit. Therefore their fitness distribution must also tend to a limit. Table 11-3 summarises the limit (with many wires) for fitness functions based on Hamming distance and root mean error squared. In the limit, every implementable permutation is equally likely. Note, unlike traditional computing, in the limit there is a finite chance of finding a solution.

Experiments on the six multiplexor problem have found solutions, including minimal solutions, cf. Figure 11-11. These experiments suggest the fitness landscape is amenable to evolutionary search, particularly if non-minimal solutions are allowed. Which in turn suggests the use of variable length evolution. Performance with CCNOT (Toffoli) gates is similar to that of genetic programming with non reversible programs. Simple hill climbing is liable to become trapped at sub optima, particularly if constrained to search for minimal solutions.

We suggest that the common emphasis on minimal solutions is misplaced. These examples provide additional evidence that requiring tiny solutions hurts evolvability (and other search techniques). There may only be one smallest program but there are exponentially many larger solutions.

Acknowledgments

I would like to thank Tom Westerdale, Ralph Hartley, Tina Yu, Wolfgang Banzhaf, Joseph A. Driscoll, Jason Daida, Lee Spector and Tracy Williams.

References

Bennett, Charles H. and Landauer, Rolf (1985). Fundamental physical limits of computation. *Scientific American 253 (July):* 48–56.

Bishop, Peter G. (1997). Using reversible computing to achieve fail-safety. In *Proceedings of the Eighth International Symposium On Software Reliability Engineering*, pp. 182–191, Albuquerque, NM, USA. IEEE.

Feller, William (1957). *An Introduction to Probability Theory and Its Applications*, Volume 1. John Wiley and Sons, New York (2nd edition).

Fredkin, Edward and Toffoli, Tommaso (1982). Conservative logic. *International Journal of Theoretical Physics 21(3/4):* 219–253.

Koza, John (1992). *Genetic Programming: On the Programming of Computers by Natural Selection*. MIT Press, Cambridge, MA, USA.

Langdon, William B. (1998). *Genetic Programming and Data Structures*. Kluwer.

Langdon, W. B. (2002a). Convergence rates for the distribution of program outputs. In *Proceedings of the Genetic and Evolutionary Computation Conference (GECCO 2002)*, Langdon, W. B. *et al.* (Eds.), pp 812-819. Morgan Kaufmann. Bull, Jonoska, N.,

Langdon, W. B. (2002b). How many good programs are there? How long are they? In *Foundations of Genetic Algorithms VII*, Torremolinos, Spain. Rowe, J. *et al.* (Eds.). Morgan Kaufmann.

Langdon, W. B. and Poli, R. (2002). *Foundations of Genetic Programming*. Springer-Verlag.

Poli, R., and Langdon, W.B. (1999). Sub-machine-code Genetic Programming. In *Advances in Genetic Programming 3*, Spector, L., *et al.* (Eds.), pp. 301–323. The MIT Press.

Rosenthal, Jeffrey S. (1995). Convergence rates for Markov chains. *SIAM Review 37(3):* 387–405.

Toffoli, Tommaso (1980). Reversible computing. In *Automata, Languages and Programming, 7th Colloquium*, de Bakker, J. W. and van Leeuwen, Jan (Eds.), Volume 85 of *Lecture Notes in Computer Science*, pp. 632–644, Noordweijkerhout, The Netherland. Springer-Verlag.

Wolpert D. H. and Macready W. G. (1997). No Free Lunch Theorems for Optimization. *IEEE Transactions of Evolutionary Computation 1(1):* 67-82.

Chapter 12

DOING GENETIC ALGORITHMS THE GENETIC PROGRAMMING WAY

Conor Ryan
CSIS Department
University of Limerick
Ireland
Conor.Ryan@ul.ie

Miguel Nicolau
CSIS Department
University of Limerick
Ireland
Miguel.Nicolau@ul.ie

Abstract This paper describes the GAuGE system, *Genetic Algorithms using Grammatical Evolution*, which uses Grammatical Evolution to perform as a position independent Genetic Algorithm. Gauge has already been successfully applied to domains such as bit level, sorting and regression problems, and our experience suggests that it evolves individuals with a similar dynamic to Genetic Programming. That is, there is a hierarchy of dependency within the individual, and, as evolution progresses, those parts at the top of the hierarchy become fixed across a population.

We look at the manner in which the population evolves the representation at the same time as optimising the problem, and demonstrate there is a definite emergence of representation.

1. Introduction

One of the difficulties with applying Genetic Algorithms is the identification of an appropriate representation. If crucial genes are geographically disparate, then useful building blocks are liable to be broken up. Conversely, when genes are adjacent to each other, they are less likely to be disrupted. Ideally, the most important, or salient genes, should be placed near each other.

This paper focuses on Gauge (Genetic Algorithms using Grammatical Evolution), a position independent GA that uses a GE type mapping. We discuss its relationship to GP and GE, and present an attribute grammar that can be used by any grammar based system to perform as a position independent GA, before looking at the way the representation effectively co-evolves with the population.

The work is partly motivated by the manner in which GP produces hierarchical individuals. That is, the gene at the root of a tree is often the most important, because all others are evaluated in its context. Similarly, the further one goes down a tree, the less effect the genes located there have on the tree in general.

GP differs from most other EAs in that individuals in the same population are not morphologically identical. While this is particularly true for initial and early generations, it becomes less so as the population evolves. Research on the so-called *Eve* phenomenon (McPhee and Hopper, 1999) has shown that the top few levels of most trees in a particular population converge to the same set of genes. Often, a single ancestor is responsible for contributing the top few levels.

If one could capture this dynamic for a GA, one could effectively co-evolve the representation with the individuals. Instead of a hierarchy of instructions, one would have a hierarchy of gene *positions*, each of which would require a *value* associated with that position.

There are two possible strategies to prevent under- and over- specification occurring. The first is to perform a repair on each individual after crossover to ensure that the exact number of positions are coded for, while the second is to create a distinction between the genotype, the hierarchy of positions, and the the phenotype, the actual string to be evaluated, by conducting a mapping from one to the other.

We choose the second strategy, and employ Grammatical Evolution to perform the mapping. Grammatical Evolution is a Genetic Programming system that uses a grammar to map a linear genome to a program. In this case we use an *attribute grammar* to specify which genes code for which positions. Clearly, a tree based grammar system could also be used to a similar effect, but the linear nature of GE means that it is a trivial task to ensure that the exact number of gene positions are coded for.

Our system, Gauge (Ryan *et al.*, 2002), is described using an attribute grammar. The system is then applied to a modified version of BinInt which was designed to specifically test how the system evolves the representation. We show that the more salient genes tend to move towards the start of the genome, and that the population evolves towards the ideal representation.

2. Grammatical Evolution

Grammatical Evolution (GE)(Ryan *et al.*, 1998)(O'Neill, 2001) presents a unique way of using grammars in the process of automatic programming. Variable-length binary string genomes are used, with each codon representing an integer value where codons are consecutive groups of 8 bits. The integer values are used in a mapping function to select an appropriate production rule from a grammar definition, the numbers generated always representing one of the rules that can be used at that time. GE does not suffer from the problem of having to ignore codon integer values, because it does not generate illegal values due to its property of intrinsic polymorphism (Keijzer *et al.*, 2001).

When tackling a problem with GE, a suitable grammar definition must first be decided upon. This can be either the specification of an entire language or, perhaps more usefully, a subset of a language geared towards the problem at hand. Complete grammars are freely available for languages such as C, and these can easily be plugged in to GE.

The Mapping Process

The genotype is used to map a start symbol onto *terminals*, that is, items which can appear in the actual programs, by reading codons of eight bits to generate a corresponding integer value, from which an appropriate production rule is selected by using the following mapping function:

$$Rule = (Codon\ integer\ value)$$

$$MOD$$

$$(Number\ of\ rules\ for\ the\ current\ nonterminal)$$

Consider the following rule, i.e., given the nonterminal *op* there are four production rules to select from.

```
(2) <op> :: =    +          (0)
               | -          (1)
               | /          (2)
               | *          (3)
```

If we assume the codon being read produces the integer 6, then

$$6\ MOD\ 4\ =\ 2$$

would select rule (2) (<op>→/). Each time a production rule has to be selected to map from a nonterminal, another codon is read. In this way, the system traverses the genome.

During the genotype-to-phenotype mapping process it is possible for individuals to run out of codons, and in this case we wrap the individual and reuse the codons. This is quite an unusual approach in EAs, as it is entirely possible for certain codons to be used two or more times.

In GE, each time the same codon is expressed it will always generate the same integer value, but depending on the current nonterminal to which it is being applied, it may result in the selection of a different production rule.

Crossover in Grammatical Evolution

Although crossover takes place at the linear level in GE, it can easily be visualised at the tree level, as illustrated in figure 12-1.

The grammar used to produce this individual is :

```
E ::= (+ E E) | (- E E) | (* E E) | (% E E) | x | y
```

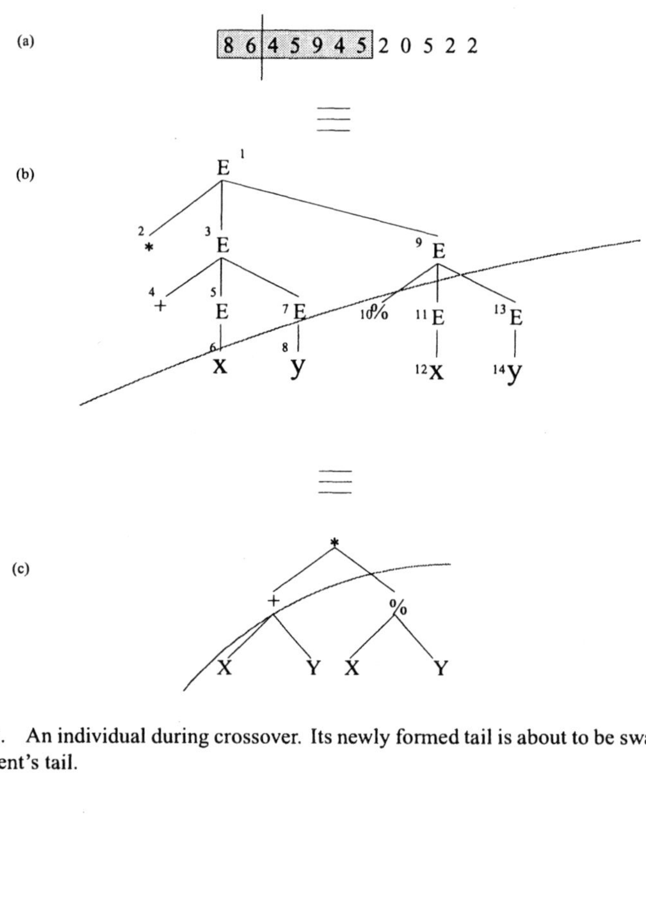

Figure 12-1. An individual during crossover. Its newly formed tail is about to be swapped with another parent's tail.

In this case, only part of the genome is required for mapping (the shaded part) and the crossover point is chosen within this region. This performs a cut across the tree, leaving gaps called "ripple sites". These sites will then be filled by using the genetic material from the other parent. However, if the material from the other parent had been used in a different context, it will change to suit the current individual. Thus, each gene is *intrinsically polymorphic*, that is, it can applied to any non-terminal in the grammar.

Once a change is made, it "ripples" through the rest of the genome. This means that the first parent, who provides the left side of the genome, passes on their genetic material unchanged, while the material passed on by the other parent *may* change, if its left side is very different from the other parent. We shall see below that this is a crucial aspect of Gauge.

3. Gauge

Gauge is a Genetic Algorithm that attempts to overcome the position dependence issue by extending GE with an attribute grammar that makes the system behave like a position-independent Genetic Algorithm. The system is far less susceptible to the disruption often associated with crossover, as it can automatically move important genes closer together as a run progresses.

The common view of Genetic Programming is that, given a particular problem statement, a program that satisfies the fitness function is to be generated. In other words, *given a set of terminals, how should they be arranged so that the fitness function is satisfied?*

The approach Gauge takes is that *all* problems, even those traditionally considered the topic of Genetic Algorithms, can be looked upon as automatic programming ones. In particular, if one considers a problem to have two parts, that is, not only *what values should the genes have?* but also *where should the genes reside?*, it is reasonable to compare this class of problem with those above.

When applying Gauge to a Genetic Algorithm problem, one uses a set of codon pairs, one for each gene position in the original problem. An individual is processed in a similar manner to GE, by moding each gene by an appropriate value to produce a useful value. Like GE, the codons are 8-bit values, also giving a degenerate encoding. For Gauge, the values required are the position that the pair will code for, and the value for that position.

To produce these values, a list of *unspecified* positions is maintained, that is, those positions that have not been given a value yet. Thus, even when evolving individuals that are binary strings, Gauge always has a distinct genotype and phenotype.

Although it may seem a somewhat perverse move to apply a mapping process to a binary string with a view to producing *another* binary string, the value added by the mapping process is position independence.

The manner in which these pairs are manipulated can be represented by the grammar below. There are a number of attributes associated with the developing string; of particular importance is the $array$ attribute, which maintains an array of free locations (positions that are as yet unspecified) in the phenotype string. When a new <P> is specified from the non-terminal <S> a desired location is given by the next codon. The desired location is copied to the non-terminal $<F>_{val}$ using the attribute, val, where it is used to determine what the final phenotypic location will actually be. To achieve this some management attributes are required, namely, v and c. v counts through the number of *vacant* positions encountered, and c counts the *total* number of positions encountered (both vacant and occupied). The three rules in <F> determine the following respectively:

1 This is the correct location AND it is vacant. Return the value of c, create a <V> production for the value and give a recursive call to <S> for the next pair (i.e., position and value).

2 This is not the correct location, but it is vacant. Increment the counter into the array AND the count of vacants.

3 This is not vacant - note that this is the ONLY check that has to be made, we don't care if it is the desired location or not. Increment the counter into the array only.

By simply generating a non-terminal <F> in the developing string the rules of this production call themselves recursively until the specified phenotypic position is determined in the $array$ attribute. When this occurs the non-terminal <V> is invoked, finally allowing the specification of the value to be placed at this previously determined position. The mapping process finishes with the output of the final phenotypic binary string.

$<S>_{n.array}$ ::= ϵ
Cond : $<S>_n=0$;
| $<P>_{n.array}$
Cond : $<S>_n>0$;

$<V>_n$ ::= 0
| 1
| :
| n-1

$\langle P\rangle_{n.array}$ ::= $\langle F\rangle_{c.array.val.n.v}$
$\langle F\rangle_{v}$:=0;
$\langle F\rangle_{c}$:=0;
$\langle F\rangle_{array}$:= $\langle P\rangle_{array}$;
$\langle F\rangle_{n}$:= $\langle P\rangle_{n}$;
$\langle F\rangle_{val}$:=0;
| $\langle F\rangle_{c.array.val.n.v}$
$\langle F\rangle_{v}$:=0;
$\langle F\rangle_{c}$:=0;
$\langle F\rangle_{array}$:= $\langle P\rangle_{array}$;
$\langle F\rangle_{n}$:= $\langle P\rangle_{n}$;
$\langle F\rangle_{val}$:=1;
:
| $\langle F\rangle_{c.array.val.n.v}$
$\langle F\rangle_{v}$:=0;
$\langle F\rangle_{c}$:=0;
$\langle F\rangle_{array}$:= $\langle P\rangle_{array}$;
$\langle F\rangle_{n}$:= $\langle P\rangle_{n}$;
$\langle F\rangle_{val}$:=n-1;

$\langle F\rangle_{c.array.val.n.v}$::= $\langle F\rangle_{c}$ $\langle V\rangle\langle S\rangle_{n.array}$
$\langle S\rangle_{n}$ = $\langle F\rangle_{n}$-1;
$\langle S\rangle_{array}$ = $\langle F\rangle_{array}$
$\langle S\rangle_{array}$ = $\langle S\rangle_{array[\langle F\rangle_c]}$
Cond : $(\langle F\rangle_{c} >= \langle F\rangle_{v})\&\&(\langle F\rangle_{array[\langle F\rangle_c]} = -1)$
| $\langle F\rangle_{c.array.val.n.v}$
$\langle F\rangle_{c}$:=$\langle F\rangle_{c}$+1;$\langle F\rangle_{v}$:=$\langle F\rangle_{c}$+1;
Cond : $\langle F\rangle_{array[\langle F\rangle_c]} = -1$
| $\langle F\rangle_{c.array.val.n.v}$
$\langle F\rangle_{c}$:=$\langle F\rangle_{c}$+1;
Cond : $\langle F\rangle_{array[\langle F\rangle_c]} != -1$

Example Individual

Consider the following individual, using four bit numbers to specify genes :

```
0110 0000 1000 0001 1011 0000 0001 0000
```

For clarity, we rewrite these as decimal pairs:

```
<6,0> <8,1> <11,0> <1,0>
```

Although we have used four bit genes to code for the value in this example, most Gauge experiments simply use a single bit when producing binary strings as output.

The first position gene has the value 6, which is **mod**ed by the number of vacant positions, 4, to give the value 2. This means bit position 2 will get the value 0, giving us a phenotpye of :

```
* * 0 *
```

A similar operation is performed on the next position gene, which is this time **mod**ed by 3, as there are now three vacant positions. This also gives 2, so we count through the vacant positions until we arrive at that position, which means that the phenotype becomes

```
* * 0 1
```

This continues until there are no vacant positions left. There is always one pair per bit position in the final string, so there is no chance of underspecification or overspecification.

Genetic Operators

The underlying representation is a simple binary string, so virtually all GA operators can be applied without fear of violating the structure. Because Gauge is built on top of GE, it inherits a crucial effect of GE's crossover operator, the *ripple* (Keijzer *et al.*, 2001) effect, as described in Section 2.0.

The ripple effect occurs in GE due to the left to right dependency that exists in a genome. All codons are evaluated in context, thus, a change of context can change the meaning of a codon, and any such change can ripple through the rest of the genome.

Similarly, there is also a left to right dependency in Gauge. For these examples, we adopt the following notation for clarity :

$$N_x$$

where N is decimal value of the position gene, and x an associated label to identify which location it actually maps to. Note that at this stage, we are not concerned with what the actual values code for. Thus, 37_a will attempt to put a into position 37, **mod**ed to a legal position, of course.

Thus, the string $3_a, 2_b, 1_c, 0_d$ will map to d, c, b, a, while $1_a, 3_b, 0_c, 3_d$ will map to b, a, c, d. These individuals are fundamentally different at the representation level, so a crossover between them will cause a dramatic ripple effect after the crossover point. However, individuals that code for the same positions at the start of the chromosome can be crossed over with less disruption.

Consider crossing $3_a, 2_b, 0_c, 1_d$ from above with $3_a, 1_b, 2_c, 1_d$, which decodes to c, d, b, a. Depending on where one chooses the crossover point, the level of disruption varies. The closer to the start, that is, to the *region of similarity*, the less disruption one will encounter. Thus, if the crossover point is chosen immediately after the common area, the two children will be identical to the parents.

This is one of the primary motivations for using a GP like representation because, although one can expect a diverse array of representations at the start of evolution, one can also expect the population to evolve its representation at the same time it evolves solutions to the problem.

Due to the position independent nature of the encoding, the regions of similarity do not have to be identical. Consider $3_a, 2_b, 1_c, 0_d$ and $7_a, 5_b, 1_c, 0_d$. Both these individuals produce the same phenotype, despite having a different pair of genes at the start. Further, the region of similarity includes both of these genes, so areas do not have to have exactly the same representation.

Clearly, the greater the diversity of *representation* in the population, the more likely genes are to change in their meaning. We postulate that as the representation starts to converge, the ripple effect will become less pronounced. Theoretically, it is possible that all individuals will end up with the same representation, but the speed of optimisation (of the problem) is often faster than that of parameter adaptation (where the parameter undergoing adaptation is the representation). Initial results (Ryan *et al.*, 2002) show that the speed of optimisation is accelerated by this elastic representation.

4. Experiments

Gauge has been applied successfully to a number of problems, such as ordering and symbolic regression. Recent work, however, has been more concerned with the dynamics of the system. Specifically, does Gauge prioritise the information as envisaged? The problem examined in this work is InvBinInt, a variant of BinInt.

BinInt

Perhaps the most well known saliency problem is the BinInt problem, as defined by the formula:

$$f(x) = \sum_{i=0}^{l-1} x_i 2^{l-i-1} \qquad x_i \in \{1, 0\}$$

where l is the string length, and x_i the allele at position i (with positions in a phenotype string ranging from 0 to $l-1$). This problem has the interesting characteristic that the salience of each allele is higher than the combined marginal fitness contributions of all the following alleles. This means that within an individual (phenotype), the fitness contributions decrease from left to right. Table 12-1 shows some examples of fitness evaluations.

This problem was first introduced by Rudnick (Rudnick, 1992), where it was used to investigate the phenomenon of *domino convergence* (the convergence speed of building blocks). Its convergence time complexity was later measured by Thierens et al. (Thierens *et al.*, 1998), to illustrate the idea of the temporal-salience structure of problems.

Table 12-1. Example fitness evaluations for a series of binary strings of length 8 using the BinInt problem, illustrating the salience of leftmost allele.

Binary string	Fitness value
01011111	95
01111101	125
01111111	127
10000000	128
11111111	255

InvBinInt

For this analysis, we use the InvBinInt problem, a variation of the BinInt problem, in which an individual is evaluated as an inverted binary number, i.e. encoded from right to left. It is defined by the formula:

$$f(x) = \sum_{i=0}^{l-1} x_i 2^i \qquad x_i \in \{1, 0\}$$

InvBinInt is of more interest to us because of the manner in which individuals are built up. That is, although the most salient allele is that in position $l - 1$, the closer to position 0 in the genotype that individuals have it, the more likely it is to be passed on to offspring.

Experimental Setup

In each of these experiments, the same setup has been used as was employed previously (Ryan *et al.*, 2002; Nicolau and Ryan, 2002). That is, a steady-state replacement strategy with roulette-wheel selection and probabilities of one-point crossover and (point) mutation of 0.9 and 0.01 respectively. The position fields were encoded using eight bits while the value fields were encoded using just one bit. Several population sizes were used, of 50, 100 and 800 individuals, and all experiments were run 100 times, with different random seeds. The phenotype length was 64 bits. Each population size yielded similar results, although the problem was more difficult for the smallar population. All results reported here are for a population size of 50.

Observed Characteristics

Figure 12-2 shows that Gauge achieves a 100% success rate on this problem by generation 200. There is clearly some settling down going on in the population prior to generation 100, as, at that stage there is a success rate of just 5%, followed by a jump to 100%.

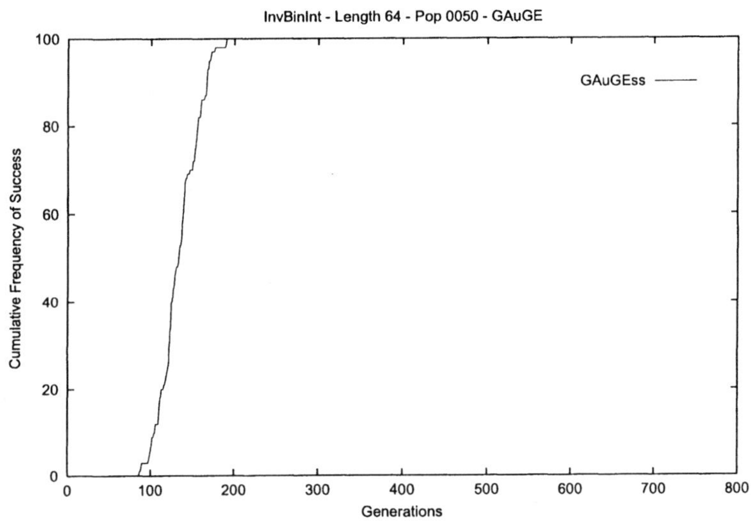

Figure 12-2. Cumulative Probability of success.

Our main concern here, however, is the manner in which Gauge evolves the representation. Recall from section 12.2 that Gauge uses ripple crossover, so the further to the left a gene is in the genotype, the more stable it will be, so we would expect the more salient genes to tend to be on the left.

The key word here is *tend.* The position independent nature of the representation is such that there is a large number of different ways of representing the same phenotype. While there is no distinction fitness-wise between individuals that have evolved the ideal representation and those that have a less than perfect one, the probability of producing fit offspring is directly related to this.

Individuals with a good representation, that is, those that have the more salient genes towards the left of the genome, are less likely to disrupt their sequences of good genes during crossover. Genes towards the end of the chromosome, on the other hand, are more likely to be effected by the ripple effect.

To measure the quality of representation of an individual, we plot each of its locus specifications and measure the slope of the least-squares regression line calculated from the average position specifications. Because the more salient genes for the InvBinInt problem are those towards the end of the phenotype, the optimal representation for Gauge would be to reverse the chromosome. That is, the last gene should be stored first, so that it will be retained after crossover. The ideal representation is shown in figure 12-3, and has a fitness of -1.

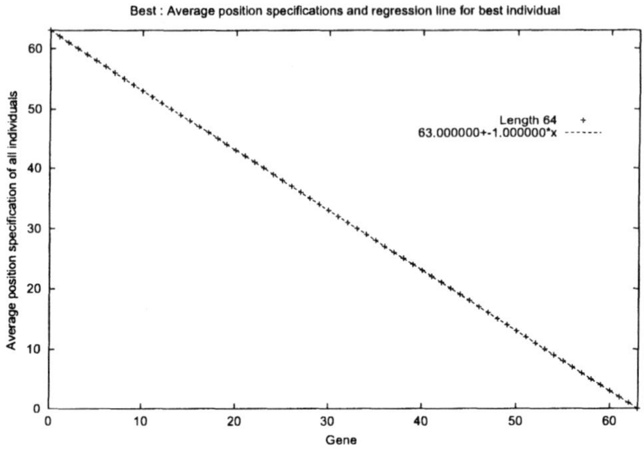

Figure 12-3. The ideal individual, with the phenotype reveresed in the genotype.

In the following section, we track the average quality of representation of the *best* individual of each generation and the average quality of all individuals in a population at each generation. Although all experiments were run for 800 generations, populations invariably contained all perfectly fit individuals by generation 200. The use of steady state selection precluded the replacement of any more individuals after this, so only results up to generation 150 are shown.

Results

The initial population contains entirely random individuals, so, as one would expect, the slopes of the plots are close to zero.

By generation 20, there is already a noticeable downward slope, even in the average case, which is down to -0.07. The slope continues to get steeper until generation 100 when, somewhat surprisingly, it starts to level out again. This continues until generation 200, where the average quality stays at -0.16. Table 12-2 summarises the changes in the slopes.

If we recall the cumulative probability of success measure as reported in figure 12-2 we see that there is an enormous increase in performance at approximately the same time. Our hypothesis is that the speed of optimisation of the problem is increased when a suitable representation has been evolved. The representation isn't optimal simply because it doesn't need to be. Once it has reached a certain level the problem becomes trivial for Gauge.

The fact that the discovery of the optimal representation is not a prerequisite for solving the problem is key to the success of Gauge, and this is the case because *all* individuals produce legal phenotypes.

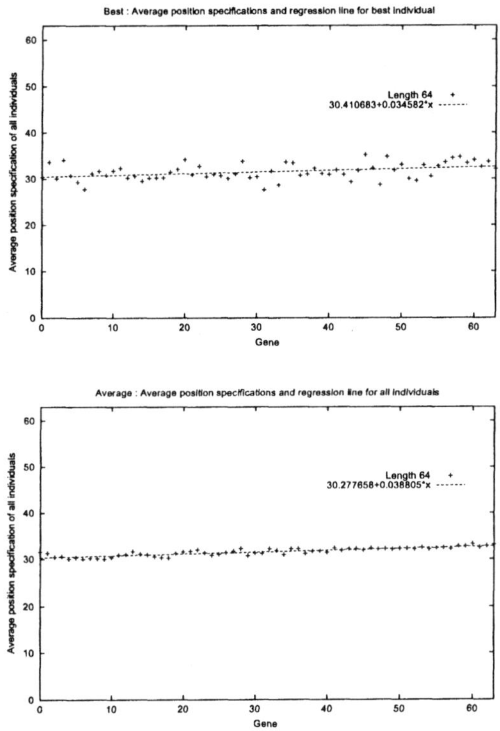

Figure 12-4. Representation quality in the initial generation.

5. Conclusions

We have described the Gauge system, a hierarchical Genetic Algorithm which has been strongly influenced by the manner in which GP both evolves individual trees and loses structural diversity in the population.

We take the view that the loss of structural diversity is not necessarily a bad thing, and attempt to harness this dynamic. We discover that, although the initial population is comprised of individuals with radically different structures, the representation evolves during the run. Once the quality of representation has reached a certain level, solving the problem is a much easier task.

Clearly, the next step is to identify some suitable salience problems for Gauge, and to devise measures to determine whether Gauge is capable of moving entire building blocks around the genome. If this proves to be the case, particularly with building blocks that would otherwise be geographically disparate, then Gauge could prove to be a very useful tool.

Table 12-2. Evolution of the quality of representation.

Generation	Best	Average
0	0.03	0.03
20	-0.11	-0.07
40	-0.18	-0.14
60	-0.24	-0.20
80	-0.27	-0.24
100	-0.24	-0.24
120	-0.21	-0.22
140	-0.18	-0.19
160	-0.18	-0.17
180	-0.17	-0.16
200	-0.17	-0.16

6. Speculation

The floating nature of the representation, at the very least, doesn't harm the GA. However, the next issue is to determine how much it is actually helping. Initial attempts at this have measured how far apart linked genes end up on the chromosome. For example, if there is a linkage between four genes, *A, B, C* and *D*, one would expect that Gauge would attempt to minimise the distance between them. Early results suggest that this is not necessarily the case, and that the there can be quite a considerable distance between them. This seems to support the finding in section 12.4, that it isn't necessary for this sytem to produce the optimal representation in order to solve a problem. This may have implications for scaling of the system.

The high level of dependency present in the representation raises the question of how building blocks are combined in Gauge, particularly when the start of each parent chromosome code for different locations. The more homogenous the representation is, the more constructive one would expect crossover to be, so there must be some element of convergence. In particular, there may be an element of *competitive* building blocks in addition to the co-operative building blocks often assumed to exist in GAs. Future work will examine the notion of competitive building blocks, and investigate whether this phenomenon can be harnessed.

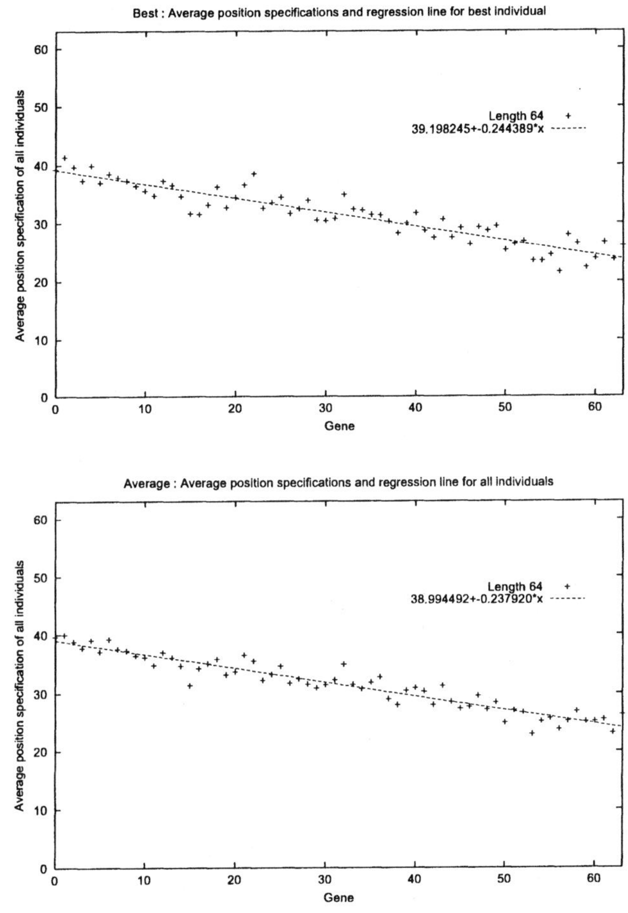

Figure 12-5. Representation quality in generation 100.

References

Keijzer M., Ryan C., O'Neill M., Cattolico M. and Babovic V. (2001). Ripple Crossover in Genetic Programming. In *Proceedings of the Fourth European Conference on Genetic Programming*, J. Miller *et al.*., (Eds.), pp. 74-86. Springer Lecture Notes in Computer Science, Vol. 2038. Springer-Verlag.

McPhee N. F. and Hopper N. J. (1999). Analysis of Genetic Diversity through Population History. In *Proceedings of the Genetic and Evolutionary Computation Conference (GECCO 1999)*, Orlando, Florida, pp. 1112-1120. Morgan Kaufmann.

Nicolau, M. and Ryan, C.(2002). LINKGAUGE: Tackling hard deceptive problems with a new linkage learning genetic algorithm. In *Proceedings of the Genetic and Evolutionary Compu-*

tation Conference (GECCO 2002), W. Langdon *et al.* (Eds.), pp 488-494. Morgan Kaufmann Publishers, San Francisco.

O'Neill, M. (2001). Automatic Programming in an Arbitrary Language: Evolving Programs with Grammatical Evolution. Doctoral Dissertation, University of Limerick.

Rudnick M. Genetic Algorithms and Fitness Variance with an Application to the Automated Design of Articial Neural Networks. Unpublished Doctoral Dissertation, Oregon Graduate Institute of Science and Technology (1992).

Ryan, C., Collins, J.J. and O'Neill, M. (1998). Grammatical Evolution: Evolving Programs for an Arbitrary Language. In *Proceedings of the First European Workshop on Genetic Programming, EuroGP'98*, Banzhaf *et al.*. (Eds.). Lecture Notes in Computer Science (LCNS) Vol. 1391, pp. 83–95. Springer-Verlag.

Ryan, C., Nicolau, M. and O'Neill, M. (2002). Genetic Algorithms using Grammatical Evolution. In *Proceedings of EuroGP-2002*, J. Foster *et al.* (Eds.). Lecture Notes in Computer Science, Vol. 2278, pp 278-287. Springer-Verlag.

Thierens, D., Goldberg, D. E. and Pereira, A.G. (1998). Domino convergence, drift, and the temporal-salience structure of problems. In *Proceedings of the 1998 IEEE World Congress on Computational Intelligence*, pp 535-540.

Chapter 13

PROBABILISTIC MODEL BUILDING AND COMPETENT GENETIC PROGRAMMING

Kumara Sastry, David E. Goldberg
Illinois Genetic Algorithms Laboratory (IlliGAL)
Department of General Engineering
University of Illinois, Urbana-Champaign, IL
104 S. Mathews Ave, Urbana, IL 61801

Abstract This paper describes probabilistic model building genetic programming (PM-BGP) developed based on the extended compact genetic algorithm (eCGA). Unlike traditional genetic programming, which use fixed recombination operators, the proposed PMBGA adapts linkages. The proposed algorithms, called the extended compact genetic programming (eCGP) adaptively identifies and exchanges non-overlapping building blocks by constructing and sampling probabilistic models of promising solutions. The results show that eCGP scales-up polynomially with the problem size (the number of functionals and terminals) on both GP-easy problem and boundedly difficult GP-hard problem.

Keywords: Competent genetic program, Linkage adaptation, Linkage learning, Probabilistic model building, Extended compact genetic algorithm

1. Introduction

In the last few decades, there has been a significant progress in the theory, design and development of genetic algorithms (GAs). A design methodology for understand the mechanics of genetic algorithms and developing better genetic and evolutionary algorithms in a principled manner has been proposed (Goldberg *et al.*, 1992). Based on the design decomposition, different facets of selectorecombinative GAs have been modeled using simple dimensional and facetwise models. These models have not only enabled a better understanding of the complex mechanisms of GAs, but also have provided GA practitioners with guidelines for tuning GA parameters. Finally, design decomposition has led to the creation of *competent genetic algorithms*—GAs that solve hard problems, *quickly*, *reliably*, and *accurately* (Goldberg, 1999; Goldberg, 2002). The competent GAs have been very effective in solving a wide variety of problems—including hierarchical problems—of *bounded* difficulty.

On the other hand, even though the growth in application of genetic programming (GP) to problems of practical and scientific importance has been remarkable (Koza, 1992; Spector *et al.*, 1999), there has been limited attention given to the development of competent operators that adapt linkage. Most of the studies on GP employ fixed crossover operators such as sub-tree crossover. Analysis of fixed recombination operators in selectorecombinative GAs suggests that fixed operators are highly inadequate and suggests operators that adapt linkage are essential for solving GA-hard problems in tractable time (Thierens and Goldberg, 1993; Thierens, 1999; Sastry and Goldberg, 2003).

Therefore, the purpose of this paper is to develop a competent GP—called the extended compact genetic programming (eCGP)—that adaptively identifies and propagates important subsolutions of a search problem. The proposed algorithm is similar to probabilistic incremental program evolution (PIPE) (Salustowicz and Schmidhuber, 1997). Unlike PIPE, our algorithm is not only capable of handling univariate variable interactions, but also can capture fairly complex multivariate interactions among the variables of the search problem. The multivariate interactions are captured in a similar fashion as in the extended compact genetic algorithm (eCGA) (Harik, 1999a). Initial scalability results show that eCGP solves problems of bounded difficulty in polynomial time, as opposed to exponential time required by simple GP.

This paper is structured as follows. The next section presents a brief review of related work followed by a synopsis of the extended compact genetic algorithm in section 3. The key features of the proposed algorithm (eCGP) are described in section 4 followed by an outline of test problems used to compare eCGP and simple GP in section 5. Section 6 describes the results followed by a note on the future work and key conclusions.

2. Literature Review

The last few decades have witnessed great strides in the development of so-called *competent* genetic algorithms—GAs that solve hard problems, quickly, reliably, and accurately (Goldberg, 1999). Since the development of the *messy genetic algorithm* (Goldberg *et al.*, 1989), a number of competent GAs have been constructed using different mechanisms and can be broadly classified into three classes:

Perturbation techniques include the fast messy GA (fmGA) (Goldberg *et al.*, 1993), gene expression messy GA (GEMGA) (Kargupta, 1996), and linkage identification by nonlinearity check GA (LINC GA), and linkage identification by monotonicity detection GA (LIMD GA) (Munetomo and Goldberg, 1999).

Linkage adaptation techniques such as linkage learning GA (LLGA) (Harik and Goldberg, 1997; Harik, 1999b; Chen and Goldberg, 2002).

Probabilistic model building techniques such as population-based incremental learning (PBIL) (Baluja, 1994), the bivariate marginal distribution algorithm (BMDA) (Pelikan and Mühlenbein, 1999), the extended compact GA (eCGA) (Harik, 1999a), iterated distribution estimation algorithm (IDEA) (Bosman and Thierens, 1999), Bayesian optimization algorithm (BOA) (Pelikan *et al.*, 2000), and hierarchical Bayesian optimization algorithm (hBOA) (Pelikan and Goldberg, 2001). For details on probabilistic model building GAs, see (Pelikan, 2002b; Pelikan, 2002a; Larrañaga and Lozano, 2002).

While most of the above competent GAs operate on binary and real valued variables, there exist very few competent genetic programming designs that operate on program trees. An exception is probabilistic incremental program evolution (PIPE) (Salustowicz and Schmidhuber, 1997). PIPE is a univariate probabilistic model building technique and is based on Baluja's PBIL (Baluja, 1994). In PIPE, computer programs or mathematical functions are evolved as in genetic programming. Programs are represented by trees where each internal node represents a function/instruction and leaves represent terminals. In PIPE, probabilistic representation of the program trees is used and the probabilities of each instruction in each node in a maximal possible tree are used to model promising trees and generate new programs.

Similar to other univariate PMBGAs, PIPE can only handle problems where there are no interactions among variables. However, there is a large class of problems where variables do interact and it is essential to capture those interactions and the design operators that respect these higher-order interactions. This study uses ECGA to adaptively model higher-order interactions among variables (nodes of the program tree).

3. Extended Compact Genetic Algorithm (eCGA)

The extended compact GA Extended compact genetic algorithm (eCGA) proposed by (Harik, 1999a) is based on a key idea that the choice of a good probability distribution is equivalent to linkage learning. The measure of a good distribution is quantified based on minimum description length (MDL) models. The key concept behind MDL models is that all things being equal, simpler distributions are better than more complex ones. The MDL restriction penalizes both inaccurate and complex models, thereby leading to an optimal probability distribution. Thus, MDL restriction reformulates the problem of finding a good distribution as an optimization problem that minimizes both the probability model as well as population representation. The probability distribution used in eCGA is a class of probability models known as marginal product models

(MPMs). MPMs are formed as a product of marginal distributions on a partition of the genes and are similar to those of the compact GA (CGA) (Harik *et al.*, 1998) and PBIL (Baluja, 1994). Unlike the models used in CGA and PBIL, MPMs can represent probability distributions for more than one gene at a time. MPMs also facilitate a direct linkage map with each partition separating tightly linked genes. For example, the following MPM, `[1,3] [2] [4]`, for a four-bit problem represents that the 1st and 3rd genes are linked and the 2nd and 4th genes are independent. Furthermore, the MPM consists of the following marginal probabilities $\{p(x_1 = 0, x_3 = 0),\ p(x_1 = 0, x_3 = 1),\ p(x_1 = 1, x_3 = 0),\ p(x_1 = 1, x_3 = 1),\ p(x_2 = 0),\ p(x_2 = 1),\ p(x_4 = 0),\ p(x_4 = 1)\}$, where x_i is the value of the ith gene.

The eCGA can be algorithmically outlined as follows:

1 Initialization: The population is usually initialized with random individuals. However, other initialization procedures can also be used.

2 Evaluate the fitness of each individual

3 Selection: The eCGA uses s-wise tournament selection (Goldberg *et al.*, 1989). However, other selection procedures can be used instead of tournament selection.

4 Build the probabilistic model: In eCGA, both the structure and the parameters of the model are searched. A greedy search heuristic is used to find an optimal model of the selected individuals in the population.

5 Create new individuals: In eCGA, new individuals are created by sampling the probabilistic model.

6 Replace the parental population with the offspring population.

7 Repeat steps 2–6 until some convergence criteria are met.

Two things need further explanation, one is the identification of the MPM using the MDL model and the other is the creation of a new population based on the MPM. The identification of the MPM in every generation is formulated as a constrained optimization problem,

$$\text{Minimize} \quad C_m + C_p \tag{13.1}$$

$$\text{Subject to}$$

$$\chi^{k_i} \leq n \ \ \forall i \in [1, m]. \tag{13.2}$$

where C_m is the model complexity which represents the cost of a complex model. In essence, the model complexity, C_m, quantifies the model representation size in terms of number of bits required to store all the marginal probabilities. Let, a given problem of size ℓ with alphabet cardinality χ, have

m partitions with k_i genes in the ith partition, such that $\sum_{i=1}^{m} k_i = \ell$. Then each partition i requires $\chi^k - 1$ independent frequencies to completely define its marginal distribution. Furthermore, each frequency is of size $\log_2(n)$, where n is the population size. Therefore, the model complexity (or the model representation size), C_m, is given by

$$C_m = \log_2(n) \sum_{i=1}^{m} \left(\chi^{k_i} - 1\right). \tag{13.3}$$

The compressed population complexity, C_p, represents the cost of using a simple model as opposed to a complex one. In essence, the compressed population complexity, C_p, quantifies the data compression in terms of the entropy of the marginal distribution over all partitions. Therefore, C_p is evaluated as

$$C_p = n \sum_{i=1}^{m} \sum_{j=1}^{\chi^{k_i}} -p_{ij} \log_2\left(p_{ij}\right) \tag{13.4}$$

where p_{ij} is the frequency of the jth gene sequence of the genes belonging to the ith partition. In other words, $p_{ij} = N_{ij}/n$, where N_{ij} is the number of chromosomes in the population (after selection) possessing bit-sequence $j \in [1, \chi^{k_i}]$ [1] for the ith partition. The constraint (Equation 13.2) arises due to the finite population size.

The following greedy search heuristic is used to find an optimal or near-optimal probabilistic model:

1 Assume each variable is independent of each other. The model is a vector of probabilities.

2 Compute the model and population complexity for the current model.

3 Consider all possible $\frac{1}{2}\ell(\ell - 1)$ merges of two variables.

4 Evaluate the model and compressed population complexity values for each model structure.

5 Select the merged model with lowest combined complexity.

6 If the combined complexity of the best merged model is better than the combined complexity of the model evaluated in step 2, replace it with the best merged model and go to step 2.

[1] Note that a BB of length k has χ^k possible sequences where the first sequence is denoted $00\cdots0$ and the last sequence is $(\chi - 1)(\chi - 1)\cdots(\chi - 1)$

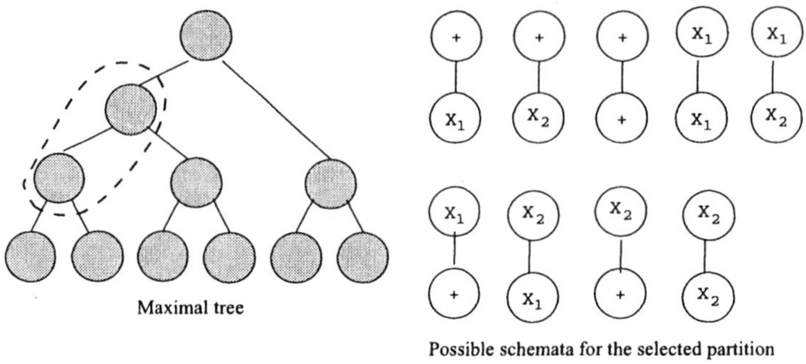

Figure 13-1. Illustration of all possible subtrees in a given partition.

7 If the combined complexity of the best merged model is less than or equal to the combined complexity of the model evaluated in step 2, the model cannot be improved and the model of step 2 is the probabilistic model of the current generation.

A new population is generated based on the optimal MPM as follows: A population of size $n(1 - P_c)$ where P_c is the crossover probability, is filled by the best individuals in the current population. The remaining nP_c individuals are generated by randomly choosing subsets from the current individuals according to the probabilities of the subsets as calculated in the probabilistic model.

4. Extended Compact Genetic Programming (eCGP)

The previous section outlined the key features of extended compact genetic algorithm. This section combines the features of eCGA (Harik, 1999a) and PIPE (Salustowicz and Schmidhuber, 1997) to create multivariate probabilistic model building genetic programming. The proposed algorithm, called extended compact genetic programming (eCGP), adaptively identifies both the building blocks and their structure. eCGP also exchanges the building blocks from different partitions effectively, and therefore drastically improves BB mixing when compared to fixed recombination operators. The eCGP can be viewed as an extension of both ECGA and PIPE. While eCGA operates on only fixed-length bitstrings, eCGP operates on variable-size program trees. While PIPE uses only fixed-structure univariate probabilistic models to create new offspring, eCGP utilizes adaptive-structure multivariate probabilistic models.

The eCGP algorithm is similar to that of eCGA described in the previous section (Section 3). Similar to PIPE, the probabilistic model is built on a maximal tree. It should be noted that we use a maximal tree only to build the probabilistic model. The trees that are generated from the probabilistic model need not, and indeed are not, maximal trees. Unlike PIPE which considers each node in the tree to be independent of each other, eCGP accommodates multivariate interactions between nodes. In other words, eCGP decomposes or partitions the maximal tree into subtrees and simultaneously builds the probabilistic models for each subtree. Therefore, eCGP not only searches for the structure of the probabilistic model, but also learns the parameters of the model. The structure of the model, similar to eCGA, is an MPM, and the parameters are the respective frequencies of all possible subtrees in a given partition. For example, figure 13-1 shows all possible subtrees, for a 3-variable problem (1 functional, 2 terminals) (Sastry *et al.*, 2003).

A greedy search heuristic is used to find an optimal probabilistic model and the MDL metric is used as a measure of the model quality. Finally, the "best" model is used to sample new individuals (program trees). Similar to PIPE, a maximal tree is created using the probabilistic model for each offspring, and unused portions of the tree are pruned before the evaluation of the candidate solution. We emphasize once more that even though we sample new offspring from the maximal-tree probabilistic model, the final pruned tree need not be a maximal tree. This is because, during the sampling process, both functionals and terminals are candidates and if a terminal is chosen at a particular internal node all the other nodes that are below it and connected to it are discarded during the pruning process.

5. Test Problems

Our approach in designing pilot experiments for investigating the scalability of eCGP is to *design* bounding *adversarial problems* that exploit one or more dimensions of problem difficulty. Particularly, our pilot test problems should possess the following properties:

- Building-block identification should be critical for successful innovation. That is, if the BBs of the problem are not identified and exchanged, it should be impossible to attain the global solution.
- Building-block structure and interactions of the problem should be known to the researchers, but not to the problem solver (search method). Ensuring that the BB identification methods work on such problems, provides assurance that they would also identify BBs of real-world problems, where the BBs are not known a priori.
- The properties such as building-block size, and problem difficulty should be tunable without significantly changing the functional.

This adversarial and systematic design method contrasts sharply with the common practices of using historical, randomly generated, or ad hoc test functions (Goldberg, 2002).

In this study we employ two classes of problems: (1) OneMax-like GP-easy problem, and (2) Deceptive trap problem. They are respectively described in the following sections.

GP-Easy Problem: ORDER

ORDER is a GP version of the OneMax problem in GAs (Goldberg and O'Reilly, 1998; O'Reilly and Goldberg, 1998), and therefore is a GP-easy problem. The primitive set of ORDER consists of the primitive JOIN of arity two and complimentary primitive pairs $(X_i, \bar{X}_i)$, $i = 0, 1, \cdots, \ell$ of arity one. A candidate solution of the ORDER problem is a binary tree with JOIN primitive at the internal nodes and either X_i's or $\bar{X}_i$'s at its leaves. The candidate solution's output is determined by parsing the program tree inorder (from left to right). The program expresses the value X_i if, during the inorder parse, an X_i leaf is encountered before its complement $\bar{X}_i$ and neither X_i nor its complement are encountered earlier. That is, the program only expresses unique primitives during the inorder parse.

For each unique X_i (or $\bar{X}_i$) that a program expresses, an equal unit of fitness value is accredited. That is,

$$f_1(x_i) = \begin{cases} 1 & if x_i \in \{X_1, X_2, \cdots, X_\ell\} \\ 0 & otherwise \end{cases} . \tag{13.5}$$

The fitness function for ORDER is then defined as

$$F(\mathbf{x}) = \sum_{i=1}^{\ell} f_1(x_i) \tag{13.6}$$

where $\mathbf{x}$ is the set of primitives output by the program. The expression for optimal solution of a ℓ-primitive ORDER problem is $\{X_1, X_2, \cdots, X_\ell\}$, and its fitness value is ℓ.

For example, consider a candidate solution for a 4-primitive ORDER problem as shown in figure 13-2. The sequence of leaves for the tree is $\{X_1, \bar{X}_1, \bar{X}_1, X_4, X_1, \bar{X}_2\}$, the expression during inorder parse is $\{X_1, \bar{X}_2, X_4\}$, and its fitness is 2.

For more details, motivations, and analysis of the ORDER problem, the interested reader should refer elsewhere (Goldberg and O'Reilly, 1998; O'Reilly and Goldberg, 1998).

GP-Hard Problem: Deceptive Trap

Another test problem used in this study is the deceptive *trap* function (Deb and Goldberg, 1993; Deb and Goldberg, 1994) which consists of additively separable *deceptive* functions (Goldberg, 1987). Deceptive functions are designed to thwart the very mechanism of selectorecombinative search by punishing

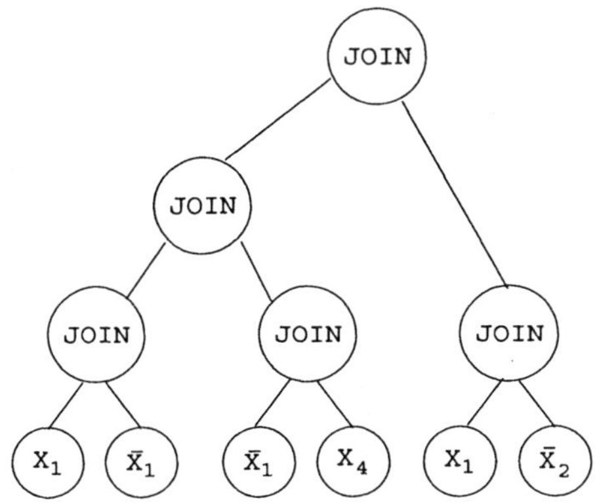

Figure 13-2. A candidate solution for a 4-primitive ORDER problem. The output of the program is $\{X_1, \bar{X}_2, X_4\}$ and its fitness is 2.

any localized hillclimbing and requiring mixing of whole building blocks at or above the order of deception. Using such *adversarially* designed functions is a stiff test—in some sense the stiffest test—of algorithm performance. The idea is that if an algorithm can beat an adversarially designed test function, it can solve other problems that are equally hard or easier than the adversary. Furthermore, if the building blocks of such deceptive functions are not identified and respected by a selectorecombinative GAs, then they almost always converge to the local optimum.

The expression mechanism of the program for deceptive trap function is identical to that of ORDER, and the difference is in the fitness evaluation procedure. Unlike ORDER, a deceptive trap function divides the expressed primitives into subgroups of k-primitives, and the fitness value for a k-primitive subgroups is defined as follows:

$$f_k\left(u(x_1, x_2, \cdots, x_k)\right) = \begin{cases} 1.0 & u = k \\ (1.0 - \delta)\left(1 - \frac{u}{k-1}\right) & u < k \end{cases}, \tag{13.7}$$

where u is the unitation (number of primitives), X_i, in a portion of the tree:

$$u\left(x_1, x_2, \cdots, x_k\right) = \sum_{i=1}^{k} f_1(x_i), \tag{13.8}$$

where, x_i is the ith unique primitive, δ is the difference in the functional value between the good BB (all ones) and its deceptive attractor (all zeros). The difficulty of a trap function can be adjusted by modifying the values k, and δ. The problem becomes more difficult as the value of k is increased and that of δ is decreased. A 4-primitive deceptive trap function is illustrated in figure 13-3.

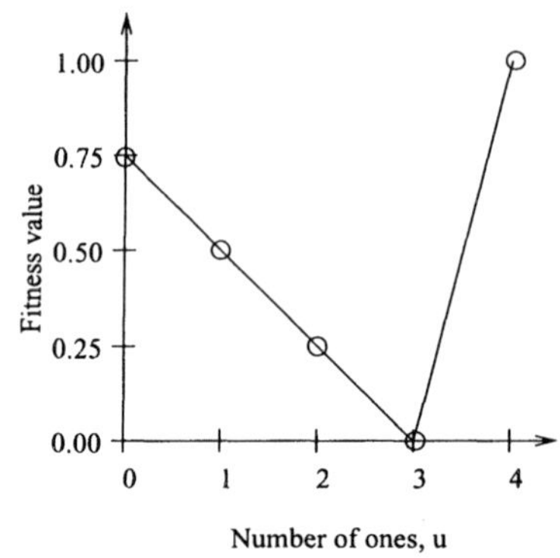

Figure 13-3. A fully deceptive trap function with $k = 4$, and $\delta = 0.25$.

The fitness function of the trap function is then defined as

$$F(\mathbf{x}) = f_k(u(x_1, x_2, \cdots, x_k)) + f_k(u(x_{k+1}, \cdots, x_{2k})) + \cdots + f_k\left(u(x_{(m-1)k+1}, \cdots, x_{mk})\right), \tag{13.9}$$

where, F is the fitness function, $\mathbf{x}$ is the expressed primitives (expressed leaf nodes of the tree), m is the number of BBs, and $\ell = mk$.

The important feature of additively separable trap functions is that if the good BB (all ones) in any particular partition is not identified, then the GA tends to converge to the deceptive attractor (all zeros) in that partition. Therefore, BB identification and mixing is critical to innovation success. Furthermore, notice that the problems are of bounded difficulty $k < \ell$. If the trap was of length ℓ, nothing could work better than enumeration or random search without replacement. However, given that the difficulty is bounded, a GA that identifies BBs and mixes them well has the opportunity to solve the problem in polynomial time (Goldberg, 2002).

6. Results and Discussion

This section compares the performance of eCGP and a simple GP for both ORDER and 3-primitive deceptive trap function (k = 3). Specifically, we investigate how eCGP and simple GP scale-up with the problem size (number of terminals and functionals) for both ORDER, which is a GP-easy problem, and the deceptive trap function which is a boundedly GP-hard problem.

A simple GP consists of a tree-swap crossover and s-wise tournament selection. Crossover probability of 1.0 and tournament size of 2 was used for both eCGP and simple GA. Mutation is not considered in this study. The initial population for both eCGP and simple GP were generated using the ramped

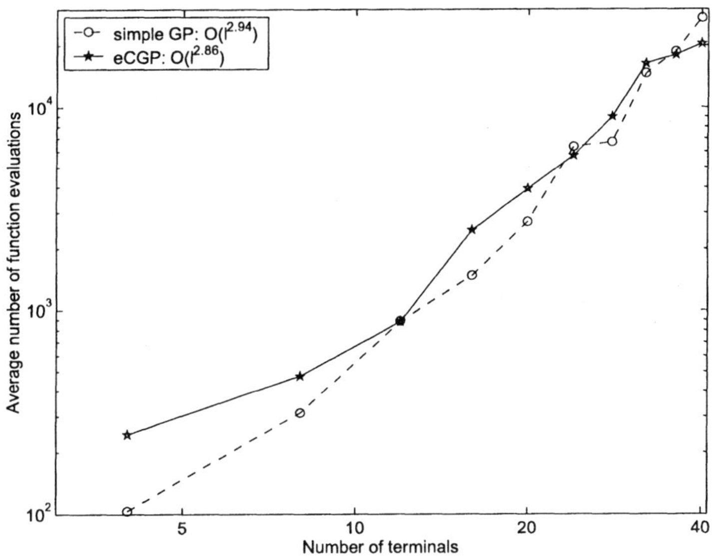

Figure 13-4. Comparison of eCGP and simple GP for ORDER. The figure plots the average number of function evaluations as a function of problem size. The results are averaged over 50 independent runs.

half-and-half method and maximum tree depth of $2\log_2 \ell$ was used. A GP run was stopped when the fitness variance of the population was less than ℓ^{-2} for ORDER and less than m^{-2} for the deceptive trap function. The number of function evaluations required in order to obtain a population with at least $m-1$ correct building blocks are empirically computed. All the results presented in this section are averaged over 50 independent runs.

Figure 13-4 compares the scalability of eCGP and simple GP for ORDER, plotting the number of function evaluations against problem size. The results indicate that even though eCGP requires a slightly larger number of function evaluations, the slope of the curve is slightly less than that for simple GP. This result is consistent with the performance of competent GAs on GA-easy problems (Pelikan *et al.*, 2000; Pelikan, 2002a) and is caused by the spurious dependencies that are introduced by the selection mechanism and fixed population size (Pelikan, *et al.*, 2001; Pelikan *et al.*, 2003). However, as the problem size increases, simple GP will suffer from mixing effects (Sastry and Goldberg, 2003) and the slope of the number of function evaluations will increase.

Figure 13-5 compares the scalability of the eCGP and simple GP for the deceptive trap problem. The building block size, k is set to 3 and the signal difference, δ is set to 0.25. The figure plots the number of function evaluations

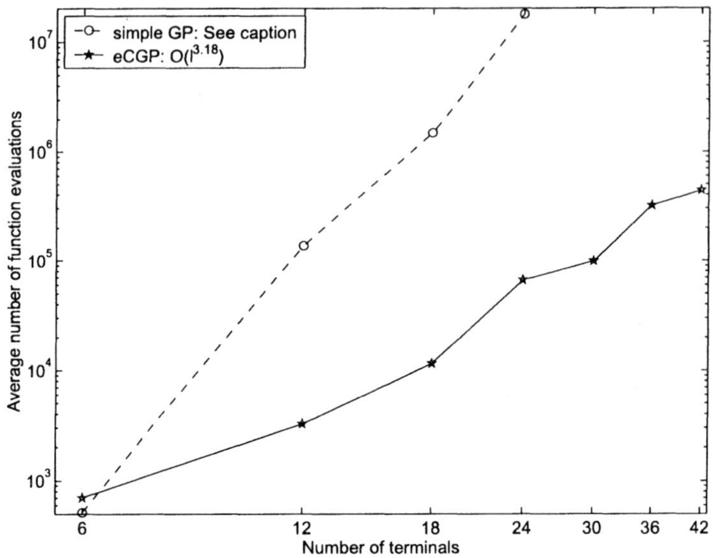

Figure 13-5. Comparison of eCGP and simple GP for the deceptive trap function (k = 3). The figure plots the average number of function evaluations as a function of problem size. The results are averaged over 50 independent runs. Simple GP was not able to solve problems beyond 24 primitives. For example, for a deceptive trap function with 30 primitive, simple GP did not converge to the optimal solution even after 5×10^8 function evaluations. For the problem sizes that we were able to obtain the optimal solution via a simple GP, the scale-up is $O(\ell^{8.91})$. But its failure to converge to the optimal solution for larger problems suggests that the scale-up is indeed much larger than $O(\ell^{8.91})$.

as a function of problem size. The results clearly indicate the effectiveness of linkage-adaptive recombination operator of eCGP over the fixed recombination operator of simple GP. The results indicate that eCGP scales-up polynomially (cubic) with problem size. Furthermore, the scalability of eCGP is about the same for both ORDER and deceptive trap function. Finally, the number of function evaluations required for simple GP appears to be very high for problems with more than 24 terminals, and therefore the problems become intractable quickly. For example, for deceptive trap problem with 30 terminals, simple GP was not able to converge to the optimal solution even after 5×10^8 evaluations. However, for the problem sizes where simple GP was able to converge to the optimal solutions, the scale-up with problem size is $O(\ell^9)$.

We recognize that both ORDER and the deceptive trap problems have non-overlapping building blocks. That is, an expressed primitive is a part of only one building block. As mentioned earlier, eCGP partitions the search problem into linkage groups and therefore is very effective in identifying non-overlapping building blocks. Even though the success of eCGP in identifying, propagating, and exchanging of non-overlapping building blocks, enables it to solve a broad class of additively decomposable problems, we acknowledge that an even broader class of problems can be tackled if complex interactions such as overlapping BBs and hierarchically interacting BBs can also be identified. The successful adaptation of eCGA into the genetic programming domain, makes the adaptation of other powerful competent GAs such as the Bayesian optimization algorithm (BOA) (Pelikan *et al.*, 2000) and the hierarchical Bayesian optimization algorithm (hBOA)(Pelikan and Goldberg, 2001) into the GP domain straightforward and promising.

7. Summary, Future Work, and Conclusions

This paper developed a competent GP, called extended compact GP (eCGP) using the ideas of the extended compact genetic algorithm (ECGA) and probabilistic incremental program evolution (PIPE). The proposed algorithm adaptively identifies, propagates, and exchanges important subsolutions of a search problem. The subsolutions are identified by building a multivariate probabilistic model of promising solutions and the subsolutions are exchanged by creating new offspring by sampling the probabilistic model. The results of eCGP was compared to those of a simple GP for two test problems: (1) ORDER: A GP-easy problem, and (2) Deceptive trap: A GP-hard problem. The results show that eCGP scales up polynomially with the problem size (number of terminals) for both GP-easy and GP-hard problems. On the other hand, as expected, a simple GP scales-up polynomially for a GP-easy problem, and exponentially for GP-hard problems.

This paper presents initial results of eCGP, and there are many avenues of future work, some of which are listed as follows:

- The eCGP can only identify non-overlapping BBs, and we need to develop competent operators that not only identify complex BB structures, but also tackle BBs with hierarchical interactions.

- Two adversarially designed test problems were considered in this study, so more tests should be conducted on different classes of problems and the scalability of eCGP should be analyzed.

- We need to investigate the convergence time and population-sizing requirement of eCGP, both analytically through facetwise and dimensionless models, and also empirically for different classes of problems.

- This study considered problems with only one functional and many more terminals. However, many GP problems consider many functionals and only a few terminals (assuming the ephemeral random constant is one type of terminal). The performance of eCGP on such a class of problems needs to be investigated.

Acknowledgments

This work was sponsored by the Air Force Office of Scientific Research, Air Force Materiel Command, USAF, under grant F49620-00-0163 and F49620-03-1-0129, and the National Science Foundation under grant DMI-9908252, and CSE fellowship, UIUC. The U.S. Government is authorized to reproduce and distribute reprints for government purposes notwithstanding any copyright notation thereon.

The views and conclusions contained herein are those of the authors and should not be interpreted as necessarily representing the official policies or endorsements, either expressed or implied, of the Air Force Office of Scientific Research, the National Science Foundation, or the U.S. Government.

References

Baluja, S. (1994). Population-based incremental learning: A method of integrating genetic search based function optimization and competitive learning. Technical Report CMU-CS-94-163, Carnegie Mellon University.

Bosman, P. and Thierens, D. (1999). Linkage information processing in distribution estimation algorithms. In *Proceedings of the Genetic and Evolutionary Computation Conference*, Banzhaf, W. (Ed.), pp. 60–67. Orlando, FL, July 1999. Morgan Kauffmann.

Chen, Y.-P. and Goldberg, D. E. (2002). Introducing start expression genes to the linkage learning genetic algorithm. In *Parallel Problem Solving from Nature - PPSN VII, Lecture Notes in Computer Science Vol. 2439*, pp. 351–360. Springer, Berlin. (Also IlliGAL Report No. 2002007.)

Deb, K. and Goldberg, D. E. (1993). Analyzing deception in trap functions. In *Foundations of Genetic Algorithms 2*, D. Whitley (Ed.), pp. 93–108. Morgan Kaufmann.

Deb, K. and Goldberg, D. E. (1994). Sufficient conditions for deceptive and easy binary functions. *Annals of Mathematics and Artificial Intelligence 10*: 385–408. (Also IlliGAL Report No. 92001.)

Goldberg, D. E. (1987). Simple genetic algorithms and the minimal, deceptive problem. In *Genetic algorithms and simulated annealing*, Davis, L. (Ed.), pp. 74–88. Los Altos, CA: Morgan Kaufmann.

Goldberg, D. E. (1999). The race, the hurdle, and the sweet spot: Lessons from genetic algorithms for the automation of design innovation and creativity. In *Evolutionary Design by Computers*, Bentley, P. (Ed.), pp.105–118. San Mateo, CA: Morgan Kaufmann.

Goldberg, D. E. (2002). *The design of innovation: Lessons from and for competent genetic algorithms*. Boston MA: Kluwer Academic Publishers.

Goldberg, D. E., Deb, K. and Clark, J. H. (1992). Genetic algorithms, noise, and the sizing of populations. *Complex Systems 6(4):* 333–362. (Also IlliGAL Report No. 91010.)

Goldberg, D. E., Deb, K., Kargupta, H. and Harik, G. (1993). Rapid, accurate optimization of difficult problems using fast messy genetic algorithms. In *Proceedings of the Fifth International Conference on Genetic Algorithms*, S. Forrest (Ed.), pp 56–64. Morgan Kaufmann. (Also IlliGAL Report No. 93004.)

Goldberg, D. E., Korb, B. and Deb, K. (1989). Messy genetic algorithms: Motivation, analysis, and first results. *Complex Systems 3(5):* 493–530. (Also IlliGAL Report No. 89003.)

Goldberg, D. E. and O'Reilly, U.-M. (1998). Where does the good stuff go, and why? How contextual semantics influences program structure in simple genetic programming. In *Proceedings of the First European Workshop on Genetic Programming (EuroGP '98)*, W. Banzhaf *et al.* (Eds.), pp. 16–36. Springer, Berlin.

Harik, G. (1999a). *Linkage learning via probabilistic modeling in the ECGA.* IlliGAL Report No. 99010. Urbana, IL: University of Illinois at Urbana-Champaign.

Harik, G. R. (1999b). *Learning gene linkage to efficiently solve problems of bounded difficulty using genetic algorithms.* Doctoral dissertation, University of Michigan, Ann Arbor, MI. (Also IlliGAL Report No. 97005.)

Harik, G. and Goldberg, D. E. (1997). Learning linkage. In *Foundations of Genetic Algorithms 4*, R.Belew and M.Vose (Eds.), pp. 247–262. Morgan Kaufmann. (Also IlliGAL Report No. 96006.)

Harik, G., Lobo, F. and Goldberg, D. E. (1998). The compact genetic algorithm. In *IEEE Conference on Evolutionary Computation*, pp. 523–528. (Also IlliGAL Report No. 97006.)

Kargupta, H. (1996). The gene expression messy genetic algorithm. In *Proceedings of the International Conference on Evolutionary Computation*, pp 814–819. Springer Verglag.

Koza, John (1992). *Genetic Programming: On the Programming of Computers by Natural Selection.* MIT Press, Cambridge, MA, USA.

Larrañaga, P. and Lozano, J. A. (Eds.) (2002). *Estimation of distribution algorithms.* Boston, MA: Kluwer Academic Publishers.

Munetomo, M. and Goldberg, D. E. (1999). Linkage identification by non-monotonicity detection for overlapping functions. *Evolutionary Computation 7(4)*: 377–398.

O'Reilly, U.-M. and Goldberg, D. E. (1998). How fitness structure affects subsolution acquisition in genetic programming. In *Genetic Programming 1998: Proceedings of the Third Annual Conference*, W. Banzhaf *et al.* (Eds.), pp. 269–277.

Pelikan, M. (2002a). *Bayesian optimization algorithm: From single level to hierarchy.* Doctoral dissertation, University of Illinois at Urbana-Champaign, Urbana, IL.

Pelikan, M. (2002b). A survey of optimization by building and using probabilistic models. *Computational Optimization and Applications 21:* 5–20. (Also IlliGAL Report No. 99018).

Pelikan, M. and Goldberg, D. E. (2001). Escaping hierarchical traps with competent genetic algorithms. In *Proceedings of the Genetic and Evolutionary Computation Conference (GECCO 2001)*, pp. 511–518.

Pelikan, M., Goldberg, D. E. and Cantú-Paz, E. (2000). Linkage learning, estimation distribution, and Bayesian networks. *Evolutionary Computation 8(3):* 314–341.

Pelikan, M., Goldberg, D. E. and Sastry, K. (2001). Bayesian optimization algorithm, decision graphs, and Occam's razor. In *Proceedings of the Genetic and Evolutionary Computation Conference (GECCO 2001)*, pp. 519–526. (Also IlliGAL Report No. 2000020.)

Pelikan, M. and Mühlenbein, H. (1999). The bivariate marginal distribution algorithm. In *Advances in Soft Computing - Engineering Design and Manufacturing*, Roy, R., Furuhashi, T. and Chawdhry, P. K. (Eds.), pp. 521–535. London: Springer-Verlag.

Pelikan, M., Sastry, K. and Goldberg, D. E. (2003). Scalability of the Bayesian optimization algorithm. *International Journal of Approximate Reasoning 31(3):* 221–258. (Also IlliGAL Report No. 2002024.)

Salustowicz, R. P. and Schmidhuber, J. (1997). Probabilistic incremental program evolution. *Evolutionary Computation 5(2)*: 123–141.

Sastry, K. and Goldberg, D. E. (2003). Scalability of simple selectorecombinative genetic algorithms on problems with tight linkage. In *Proceedings of the Genetic and Evolutionary Computation Conference (GECCO 2003).* (Also IlliGAL Report No. 2002013.)

Sastry, K., O'Reilly, U.-M., Goldberg, D. E. and Hill, D. (2003). Building-block supply in genetic programming. In *Genetic Programming, Theory and Practice*, B. Worzel and R. Riolo (Eds.). Kluwer. (Also IlliGAL Report No. 2003011.)

Spector, L., Langdon, W. B., O'Reilly, U.-M. and Angeline, P. J. (Eds.) (1999). *Advances in genetic programming 3*. Cambridge, MA: MIT Press.

Thierens, D. (1999). Scalability problems of simple genetic algorithms. *Evolutionary Computation 7(4):* 331–352.

Thierens, D. and Goldberg, D. E. (1993). Mixing in genetic algorithms. In *Proceedings of the Fifth International Conference On Genetic Algorithms*, S. Forrest (Ed.), pp. 38–45. Morgan Kaufmann.

Chapter 14

AUTOMATED SYNTHESIS BY MEANS OF GENETIC PROGRAMMING OF COMPLEX STRUCTURES INCORPORATING REUSE, PARAMETERIZED REUSE, HIERARCHIES, AND DEVELOPMENT

John R. Koza,[1] Matthew J. Streeter,[2] and Martin A. Keane[3]
[1]*Stanford University, Stanford, California*
[2]*Genetic Programming Inc., Mountain View, California*
[3]*Econometrics Inc., Chicago, Illinois*

Abstract: Genetic programming can be used as an automated invention machine to synthesize designs for complex structures. In particular, genetic programming has automatically synthesized complex structures that infringe, improve upon, or duplicate the functionality of 21 previously patented inventions (including analog electrical circuits, controllers, and mathematical algorithms). Genetic programming has also generated two patentable new inventions (involving controllers). Genetic programming has also generated numerous additional human-competitive results involving the design of quantum computing circuits as well as other substantial results involving antennae, networks of chemical reactions (metabolic pathways), and genetic networks. We believe that these results are the direct consequence of a group of techniques—many unique to genetic programming—that facilitate the automatic synthesis of complex structures. These techniques include automatic reuse, parameterized reuse, parameterized topologies, and developmental genetic programming. The paper describes these techniques and how they contribute to automated design.

Key words: Hierarchy, reuse, development, parameterized topologies, architecture-altering operations, automatically defined functions, automatically defined iterations, automatically defined loops, automatically defined recursions, automatically defined stores, circuits, controllers

1. Introduction

Genetic programming is an automatic method for solving problems. It is an extension of the genetic algorithm (Holland, 1975). Genetic programming starts from a high-level statement of the requirements of a problem and attempts to produce a computer program that solves the problem. Specifically, genetic programming starts with a primordial ooze of thousands of randomly created computer programs and uses the Darwinian principle of natural selection; analogs of recombination (crossover), mutation, gene duplication, gene deletion; and certain mechanisms of developmental biology to progressively breed an improved population over a series of generations (Koza, 1992; Koza, 1994; Koza, Bennett, Andre, and Keane, 1999; Koza, Keane, Streeter, Mydlowec, Yu, and Lanza, 2003; Banzhaf, Nordin, Keller, and Francone, 1998; Langdon, 1998; Langdon and Poli, 2002).

Genetic programming can be used as an automated invention machine to synthesize designs for complex structures. In particular, genetic programming has automatically synthesized complex structures that infringe, improve upon, or duplicate the functionality in a novel way of 21 previously patented inventions including analog electrical circuits, controllers, and mathematical algorithms (Table 14.1). In addition, genetic programming generated two patentable new inventions (controllers) for which patent applications are pending (Keane, Koza, and Streeter, 2002). Genetic programming has also generated numerous additional human-competitive results involving the design of antennae, networks of chemical reactions (metabolic pathways), genetic networks (Koza, Keane, Streeter, Mydlowec, Yu, and Lanza, 2003) and the design of quantum computing circuits that are superior to those designed by human designers (Spector, Barnum, and Bernstein, 1998, 1999; Spector, Barnum, Bernstein, and Swamy, 1999; Barnum, Bernstein, and Spector, 2000; Spector and Bernstein, 2003).

We believe that these results are the consequence of a group of techniques that facilitate the automatic synthesis of complex structures by genetic programming. These techniques include developmental genetic programming, reuse, parameterized reuse, and parameterized topologies.

Section 2 describes developmental genetic programming. When genetic programming is used to synthesize the design of complex structures, it is often advantageous to use a developmental process that preserves syntactic validity and locality.

We believe that reuse is an essential precondition for scalability in automated design. Section 3 describes several aspects of reuse, including:

- reuse of useful substructures by means of the ordinary crossover (recombination) operation (section 3.1),

- reuse of useful substructures by means of subroutines (automatically defined functions) (section 3.2),
- parameterized reuse (section 3.3),
- creation of reuse by the architecture-altering operations (section 3.4),
- reuse of code by means of automatically defined iterations, automatically defined loops, and automatically defined recursions (section 3.5), and
- reuse of the results of executing code by means of automatically defined stores (memory) (section 3.6).

Section 4 describes the automatic creation of parameterized topologies. Section 5 describes the automatic creation of parameterized topologies containing conditional operators. Section 6 contains a discussion of future possibilities for genetic programming theory and practice.

2. Developmental Genetic Programming

When genetic programming is used to automatically create computer programs, the programs are ordinarily represented as program trees (i.e., rooted, point-labeled trees with ordered branches).

When genetic programming is used to synthesize the design of a cyclic graphical structure (such as an electrical circuit), a developmental process may be advantageously used to establish a mapping between the program trees (acyclic graphs) and the cyclic graphs. The developmental process may begin with a simple embryo consisting of a single modifiable wire that is not initially connected to the inputs or outputs of the to-be-synthesized circuit. The final circuit is developed by progressively applying the functions in a circuit-constructing program tree (evolved by genetic programming) to the embryo's initial modifiable wire (and to succeeding modifiable wires and modifiable components).

Table 14.1 Twenty-one previously patented inventions reinvented by genetic programming

Invention	Date	Inventor	Place	Patent
Darlington emitter-follower section	1953	Sidney Darlington	Bell Telephone Laboratories	2,663,806
Ladder filter	1917	George Campbell	American Telephone and Telegraph	1,227,113
Crossover filter	1925	Otto Julius Zobel	American Telephone and Telegraph	1,538,964
"*M*-derived half section" filter	1925	Otto Julius Zobel	American Telephone and Telegraph	1,538,964

Cauer (elliptic) topology for filters	1934–1936	Wilhelm Cauer	University of Gottingen	1,958,742, 1,989,545
Sorting network	1962	Daniel G. O'Connor and Raymond J. Nelson	General Precision, Inc.	3,029,413
Computational circuits	NA	Numerous	Numerous	Numerous
Electronic thermometer	NA	Numerous	Numerous	Numerous
Voltage reference circuit	NA	Numerous	Numerous	Numerous
60 dB and 96 dB amplifiers	NA	Numerous	Numerous	Numerous
Second-derivative controller	1942	Harry Jones	Brown Instrument Company	2,282,726
Philbrick circuit	1956	George Philbrick	George A. Philbrick Researches	2,730,679
NAND circuit	1971	David H. Chung and Bill H. Terrell	Texas Instruments Incorporated	3,560,760
PID (proportional, integrative, and derivative) controller	1939	Albert Callender and Allan Stevenson	Imperial Chemical Limited	2,175,985
Negative feedback	1937	Harold S. Black	American Telephone and Telegraph	2,102,670, 2,102,671
Low-voltage balun circuit	2001	Sang Gug Lee	Information and Communications University	6,265,908
Mixed analog-digital variable capacitor circuit	2000	Turgut Sefket Aytur	Lucent Technologies Inc.	6,013,958
High-current load circuit	2001	Timothy Daun-Lindberg and Michael Miller	International Business Machines Corporation	6,211,726
Voltage-current conversion circuit	2000	Akira Ikeuchi and Naoshi Tokuda	Mitsumi Electric Co., Ltd.	6,166,529
Cubic function generator	2000	Stefano Cipriani and Anthony A. Takeshian	Conexant Systems, Inc.	6,160,427
Tunable integrated active filter	2001	Robert Irvine and Bernd Kolb	Infineon Technologies AG	6,225,859

The functions in the circuit-constructing program trees may include

(1) topology-modifying functions that alter the topology of a developing circuit (e.g., series division, parallel division, via between nodes, via to ground, via to a power supply, via to the circuit's input signal, via to the circuit's output points),

(2) component-creating functions that insert components (i.e., resistors, capacitors, and transistors) into a developing circuit, and

(3) development-controlling functions that control the developmental process (e.g., cut, end).

To illustrate the developmental process, Figure 14.1 shows a circuit-constructing program tree that produces the circuit shown in Figure 14.2.

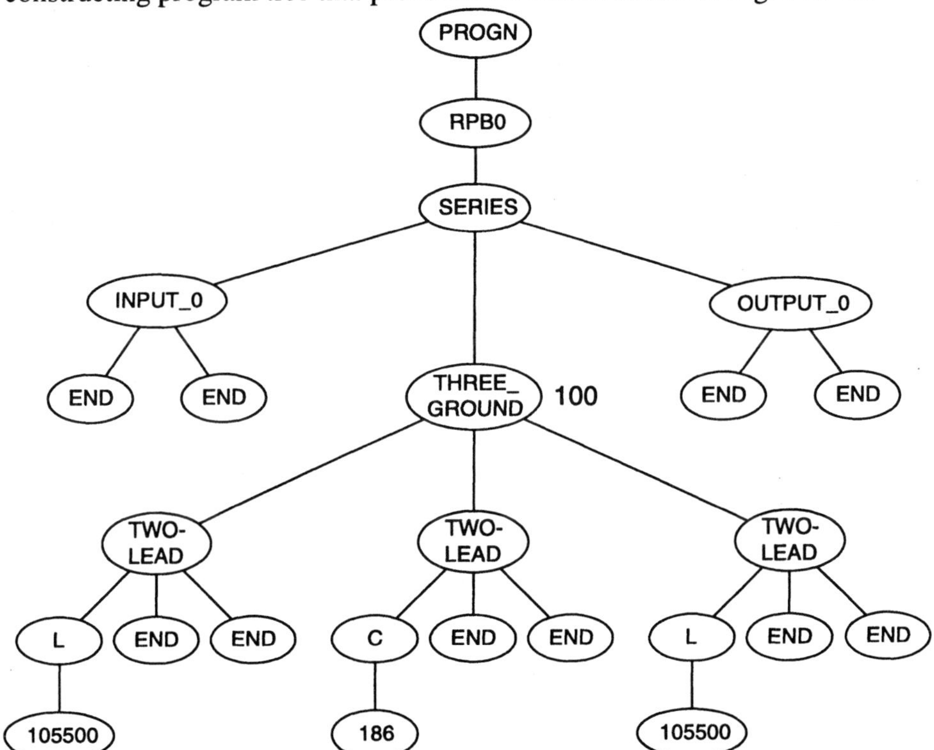

Figure 14.1 Circuit-constructing program tree that develops into one T-section

The fully developed circuit of Figure 14.2 consists of one T-section containing two 105,500-micro-Henry inductors (one on each arm of the "T") and a 186-nanofarad capacitor on the vertical segment of the "T." In Figure 14.2, the incoming signal VSOURCE and the source resistor RSOURCE at the far left as well as the load resistor RLOAD at the far right are part of a fixed test fixture that is not subject to evolutionary change.

In the circuit-constructing program tree of Figure 14.1, the second argument of the THREE_GROUND topology-modifying function (described in detail in Koza, Bennett, Andre, and Keane 1999) is a TWO_LEAD component-creating function that creates a capacitor C1 connected between node 0 in Figure 14.2 and node 6 in Figure 14.2 (an intermediate node created by the operation of the THREE_GROUND function).

The first argument of the three-argument SERIES topology-modifying function in Figure 14.1 is a two-argument INPUT_0 topology-modifying function that makes a connection between the incoming signal and L1. The third argument of the SERIES function in Figure 14.1 is a two-argument OUPTUT_0 topology-modifying function that makes a connection between the circuit's output point and L2. The development-controlling END functions terminate the developmental process.

The first argument of the three-argument THREE_GROUND topology-modifying function in Figure 14.1 is a TWO_LEAD function that creates the 105,500-micro-Henry inductor L1 connected between nodes 2 (one end of the embryo's now-replaced original modifiable wire) and 6 of Figure 14.2. The values of components, such as capacitors and inductors, are established by value-setting subtrees that typically contain a multiplicity of arithmetic-performing functions and numerical values. However, for simplicity, Figure 14.1 shows only the component's final numerical value. The second argument of the THREE_GROUND function in Figure 14.1 is a TWO_LEAD function that creates the 186-nanofarad capacitor C1 connected between nodes 6 and 0 of Figure 14.2. The third argument of the THREE_GROUND function in Figure 14.1 is a TWO_LEAD function that creates the 105,500-micro-Henry inductor L2 connected between nodes 3 (the other end of the embryo's now-replaced modifiable wire) and 6 of Figure 14.2.

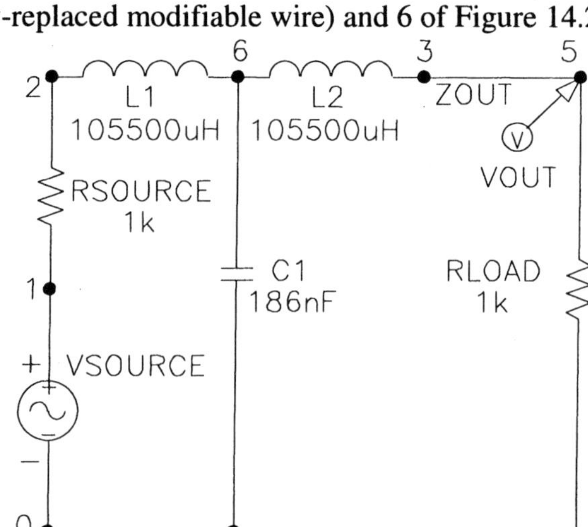

Figure 14.2 Circuit consisting of one T-section

3. Reuse

Anyone who has ever looked at a blueprint for a building, an electrical circuit, a protein molecule, a corporate organizational chart, a musical score, a city map, or a computer program will be struck by the ubiquitous reuse of certain basic substructures within the overall structure. Indeed, complex structures are almost always replete with modularities, symmetries, and regularities. Reuse can accelerate automated learning by avoiding "reinventing the wheel" on each occasion requiring a particular sequence of already-learned steps. We believe that reuse is the cornerstone of meaningful machine intelligence.

3.1 Reuse by Means of Crossover

The crossover (recombination) operation is one mechanism by which genetic programming can advantageously reuse useful substructures. For example, one way to construct a lowpass filter with a passband boundary of 1,000 Hz is by means of a cascade of identical T-sections. See Koza, Keane, Streeter, Mydlowec, Yu, and Lanza, 2003 for additional details of this illustrative problem of filter design.

A circuit (such as Figure 14.2) consisting of one T-section is a lowpass filter, albeit an extremely poor one in that it does not sharply suppress frequencies higher than 1,000 Hz. The crossover operation may create a circuit-constructing program tree (such as the one shown in Figure 14.4) by reusing genetic material from Figure 14.1. The subtree rooted at the point labeled 200 in Figure 14.4 corresponds to the subtree rooted at the THREE_GROUND function labeled 100 in Figure 14.1. That is, the subtree that was responsible for the filtering ability of the single T-section of Figure 14.2 is now embedded inside another individual that itself produces a T-section. The result of the execution of the circuit-constructing program tree in Figure 14.4 is two identical T-sections (Figure 14.3). A cascade consisting of these two identical T-sections (Figure 14.3) is a slightly better lowpass filter than a single T-section. Note that the tree depicted in Figure 14.4 could be the result of either a crossover of two individuals that each develop into a T-section or the result of a crossover of an individual with itself.

3.2 Reuse by Means of Automatically Defined Functions

Subroutines (automatically defined functions) provide another mechanism by which genetic programming can automatically reuse code—either exactly or with different instantiations of dummy variables or formal parameters. Genetic programming is capable of automatically creating reusable subroutines along with main programs (result-producing

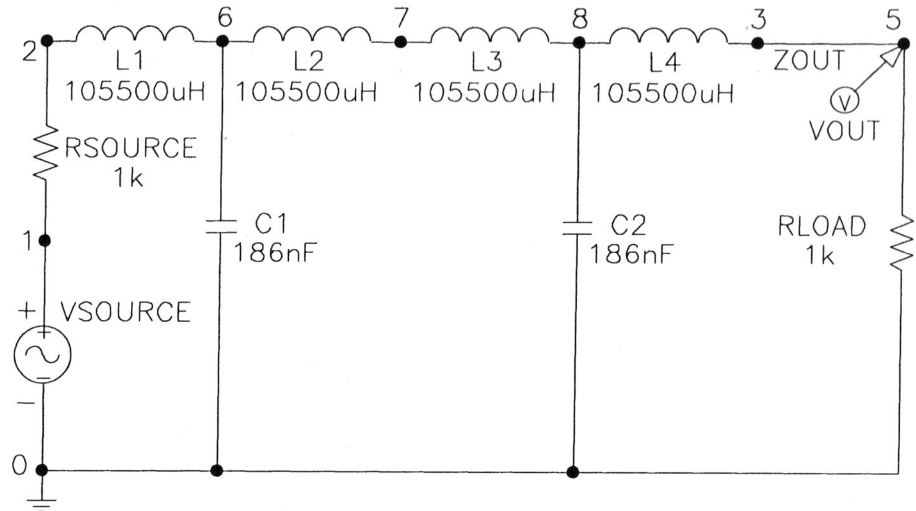

Figure 14.3 Circuit consisting of a cascade of two identical T-sections

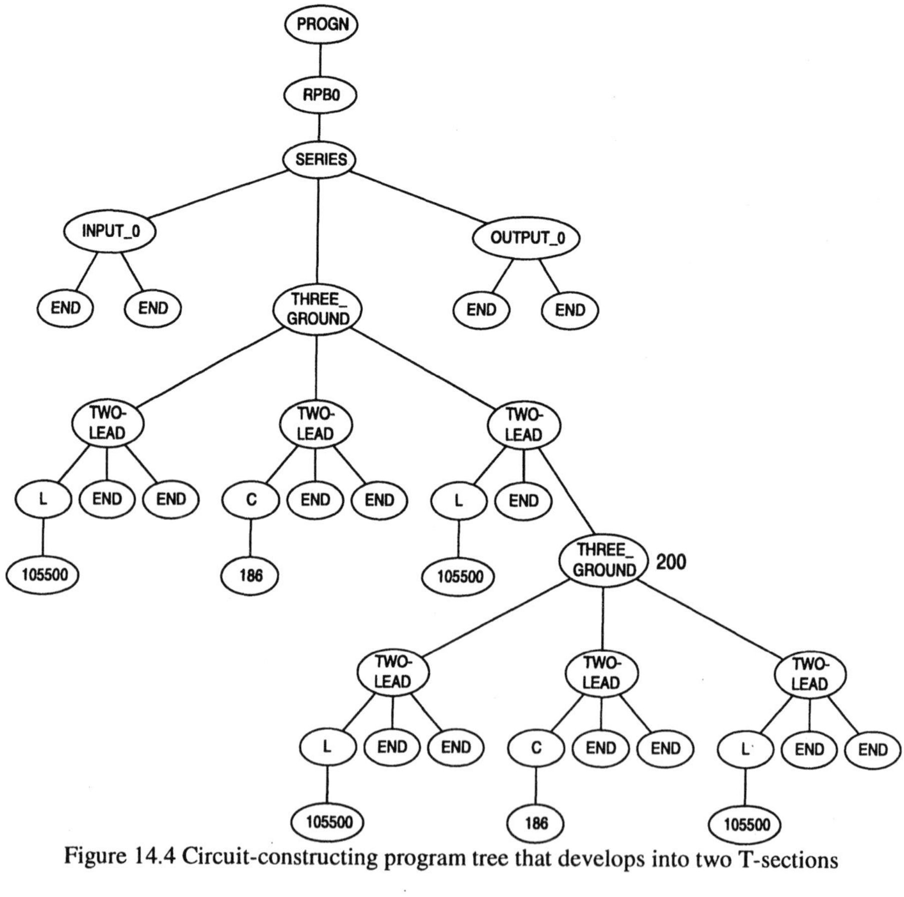

Figure 14.4 Circuit-constructing program tree that develops into two T-sections

branches) dynamically during a run. Figure 14.5 shows a circuit-constructing program tree that contains an automatically defined function (ADF0) that develops into the two T-sections shown in Figure 14.3. ADF0 develops into one T-section. The result-producing branch (RPB0) invokes ADF0 twice, thereby creating a circuit consisting of two identical T-sections.

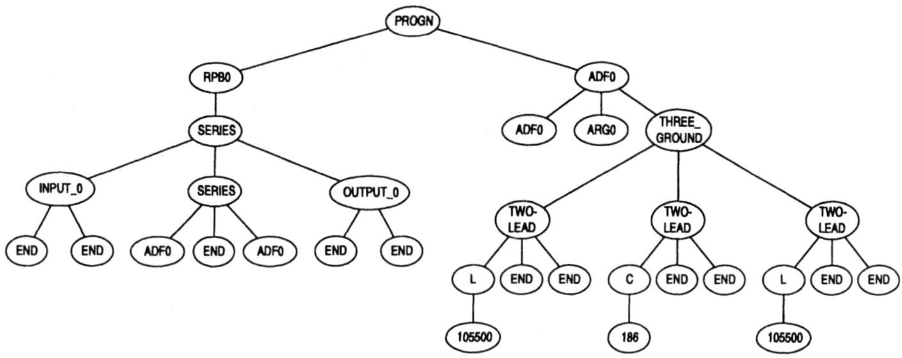

Figure 14.5 Circuit-constructing program tree containing automatically defined function ADF0 that develops into two T-sections

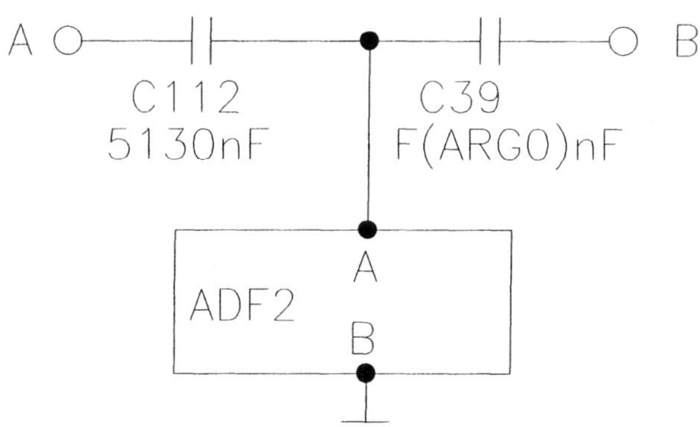

Figure 14.6 Subcircuit produced by three-ported automatically defined function ADF3

3.3 Parameterized Reuse

In genetic programming, different instantiations of a dummy variable (formal parameter) of an automatically defined function may be passed to an automatically defined function. This aspect of genetic programming can be illustrated by a portion of a genetically evolved circuit for a two-band crossover (woofer-tweeter) filter (described in detail in Koza, Bennett, Andre, and Keane, 1999). Figure 14.6 shows the subcircuit that is created by the execution of a three-ported automatically defined function. The execution of automatically defined function (ADF3) has the following three consequences.

- It inserts a (not noteworthy) fixed 5,130 nanofarad capacitor in the upper left of the Figure.
- It inserts a (noteworthy) parameterized capacitor C39 whose component value is dependent on the dummy variable `ARG0`.
- It invokes an automatically defined function `ADF2`. The dummy variable `ARG0` is passed to `ADF2`. In turn, `ADF2` creates a (noteworthy) parameterized inductor whose component value is dependent on the dummy variable `ARG0` that is passed to `ADF2`.

Parameters may be similarly passed into automatically defined iterations, automatically defined loops, and automatically defined recursions that possess dummy variables (as described in later sections of this paper).

3.4 Creation of Reuse by Means of the Architecture-Altering Operations

The genome of *Mycoplasma genitilium* contains only 470 different genes (Fraser *et al.*, 1995), whereas the human genome contains approximately two orders of magnitude more. The question arises as to how new genes—that is, new biological functions—originate in nature. Occasionally, a gene may be duplicated, thereby creating two places on the chromosome that manufacture the same protein. After such a gene duplication, one of the two initially identical genes may remain intact and continue to manufacture the original protein (thus conferring the gene's presumably survival-related function on the organism). Meanwhile, over many generations, the second gene may harmlessly accumulate changes and diverge. Eventually the second gene may come to manufacture a new protein with an entirely new function. Thus, new biological functions emerge in nature as part of the evolutionary process (Ohno, 1970).

Genetic programming uses architecture-altering operations (described in detail in Koza, Bennett, Andre, and Keane, 1999) to automatically determine program architecture in a manner that parallels gene duplication, and the related operation of gene deletion, in nature. The architecture-altering operations are performed (usually with a low probability in the neighborhood of 1%) during each generation of a genetic programming run (along with the genetic operations of crossover, mutation, and reproduction). When the architecture-altering operations are used, the overall architecture and hierarchy of the evolved computer program are part of the output produced by genetic programming—not part of the input supplied by the human user.

The subroutine duplication operation duplicates a preexisting subroutine in an individual program, gives a unique new name to the copy, and randomly divides the preexisting calls to the old subroutine between the two.

This operation changes the program architecture by broadening the hierarchy of subroutines in the overall program. As with gene duplication in nature, this operation preserves semantics when it first occurs. The two subroutines typically diverge later—sometimes yielding specialization.

The argument duplication operation duplicates one argument of a subroutine, randomly divides internal references to it, and preserves overall program semantics by adjusting all calls to the subroutine. This operation enlarges the dimensionality of the subspace on which the subroutine operates.

The subroutine creation operation can create a new subroutine from part of a main result-producing branch (main program), thereby deepening the hierarchy of references in the overall program by creating a hierarchical reference between the main program and the new subroutine. `ADF0` in Figure 14.5, for example, could possibly have been created in this way. The subroutine creation operation can also create a new subroutine from part of an existing subroutine by creating a hierarchical reference between a preexisting subroutine and a new subroutine, thus creating a deeper and more complex overall hierarchy.

The subroutine deletion operation deletes a subroutine from a program thereby making the hierarchy of subroutines narrower or shallower.

The argument deletion operation deletes an argument from a subroutine, thereby (usually) reducing the amount of information available to the subroutine—a process that can be viewed as generalization.

The architecture-altering operations rapidly create an architecturally diverse population containing programs with different numbers of subroutines, different numbers of arguments, and different hierarchical arrangements of the subroutines. Programs with architectures that are well-suited to the problem at hand will tend to grow and prosper in the competitive evolutionary process, while programs with inappropriate architectures will tend to wither away under the relentless selective pressure of the problem's fitness measure.

3.5 Reuse of Code with Iterations, Loops, and Recursions

Other architecture-altering operations add and delete iterations, loops, and recursions. Automatically defined iterations, automatically defined loops, and automatically defined recursions provide additional mechanisms that enable genetic programming to automatically reuse code (Koza, Bennett, Andre, and Keane, 1999).

The circuit-constructing functions responsible for a useful subcircuit may reside in an automatically defined copy (ADC). An automatically defined copy is an iteration specialized to problems of circuit synthesis. Multiple occurrences of the subcircuit result when the copy body branch (CBB) is repeatedly invoked by the copy control branch (CCB) of the automatically defined copy (Koza, Bennett, Andre, and Keane, 1999).

3.6 Reuse of the Results of Executing Code with Automatically Defined Stores

Automatically defined functions, automatically defined iterations, automatically defined loops, and automatically defined recursions provide a mechanism for reusing code at different levels of granularity in a complex system.

In contrast, automatically defined stores (memory) provide a way for automatically reusing the remembered results produced by the execution of code (Koza, Bennett, Andre, and Keane, 1999).

4. Automatic Synthesis of Parameterized Topologies

Genetic programming can automatically create, in a single run, a general (parameterized) solution to a problem in the form of a graphical structure whose nodes or edges represent components and where the values of the components are specified by mathematical expressions containing free variables. The free variables represent parameters of the design problem (e.g., the passband boundary frequency for a lowpass filter). By producing a graphical structure whose component values are specified by mathematical equations containing the free variables, genetic programming produces a general solution to the problem (e.g., a design for a collection of lowpass filters, each of which has a different passband boundary) as opposed to a design for a specific single problem (i.e., the design of a lowpass filter having a specific single passband boundary).

The genetically evolved individual represents a complex structure (e.g., electrical circuit, controller, network of chemical reactions, antenna, genetic network). In the automated design process, genetic programming determines the gross size of the graphical structure (i.e., the number of nodes in the graph) as well as the graph's connectivity (i.e., a specification of the nodes that are connected). Genetic programming also assigns component types to various edges (or nodes) of the graphical structure. In a circuit, the components may be transistors, resistors, and capacitors. In a controller, they may be integrators, differentiators, gain blocks, leads, lags, delays, adders, and subtractors. Genetic programming also creates mathematical expressions that establish the parameter values for the components (e.g., the capacitance of a capacitor in a circuit, the amplification factor of a gain block in a controller). The free variables in the mathematical expressions confer generality on the genetically evolved solution. The free variables enable a single genetically evolved graphical structure to represent a parameterized solution of an entire category of problems. In creating parameterized topologies, genetic programming can do all of the above in an automated way in a single run.

The capability of genetic programming to create parameterized topologies for design problems is illustrated by the automatic creation of a general-purpose controller (Figure 14.7). The purpose of a controller is to force a system—conventionally called the *plant*—to produce an output signal that matches a specified *reference signal.* This genetically evolved controller outperforms PID controllers tuned using the widely used 1942 Ziegler-Nichols tuning rules (Ziegler and Nichols, 1942) and the recently developed 1995 Astrom and Hagglund tuning rules on an industrially representative set of plants in terms of the criteria described in Astrom and Hagglund 1995. Some of the blocks of this genetically evolved controller are parameterized by constants whereas others are parameterized by mathematical expressions containing the problem's four free variables. These free variables are readily measurable behavioral characteristics of plants. In particular, they are the plant's time constant, T_r, ultimate period, T_u, ultimate gain, K_u, and dead time, L. This controller's overall topology consists of three adders, three subtractors, four gain blocks parameterized by a constant, two gain blocks parameterized by non-constant mathematical expressions containing free variables, and two lead blocks parameterized by non-constant mathematical expressions containing free variables. For purposes of illustration, we mention that gain block 730 of the Figure has a gain of

$$\left| \log \left| T_r - T_u + \log \left| \frac{\log\left(|L|^L\right)}{T_u + 1} \right| \right| \right|$$

and gain block 760 of the Figure has a gain of

$$\left| \log \left| T_r + 1 \right| \right| .$$

The genetically evolved mathematical expressions for the other blocks in this controller and other details of this problem can be found in Koza, Keane, Streeter, Mydlowec, Yu, and Lanza 2003.

The techniques above have been applied to a variety of additional problems (Koza, Keane, Streeter, Mydlowec, Yu, and Lanza, 2003), including synthesis of

- a parameterized circuit-constructing program tree containing free variables that yields a Zobel network,

- a parameterized circuit-constructing program tree that yields a passive third-order elliptic lowpass filter whose modular angle is specified by a free variable,
- a parameterized circuit-constructing program tree that yields a passive lowpass filter whose passband boundary is specified by a free variable,
- a parameterized circuit-constructing program tree that yields an active lowpass filter whose passband boundary is specified by a free variable, and
- a parameterized controller for controlling a three-lag plant whose time constant is specified by a free variable.

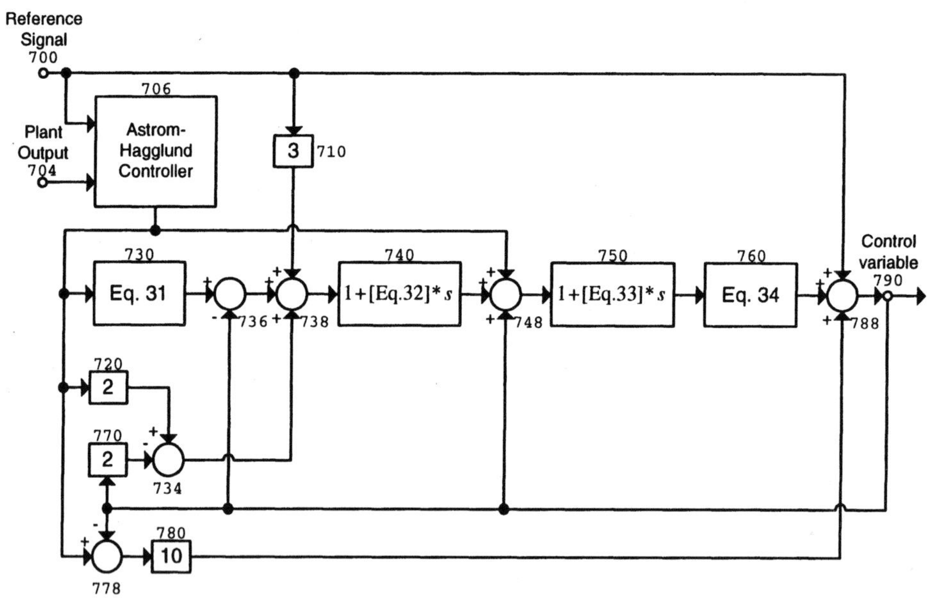

Figure 14.7 Genetically evolved general-purpose controller.

5. Automatic Synthesis of Parameterized Topologies Containing Conditional Operators

If the genetically evolved program for a parameterized topology contains conditional developmental operators as well as free variables, a different graphical structure may be produced for different instantiations of the free variables. That is, the genetically evolved program for the parameterized topology may operate as a genetic switch. Each genetically evolved program has inputs, namely the problem's free variables. Depending on the values of the free variables, different graphical structures may result from the execution of the program.

For example, a single genetically evolved circuit-constructing program tree with free variables (representing two frequencies, `F1` and `F2`) and conditional operators may yield either

- a lowpass filter with a variable passband boundary determined by the free variables (Figure 14.8), or
- a highpass filter with a variable passband boundary determined by the free variables (Figure 14.9).

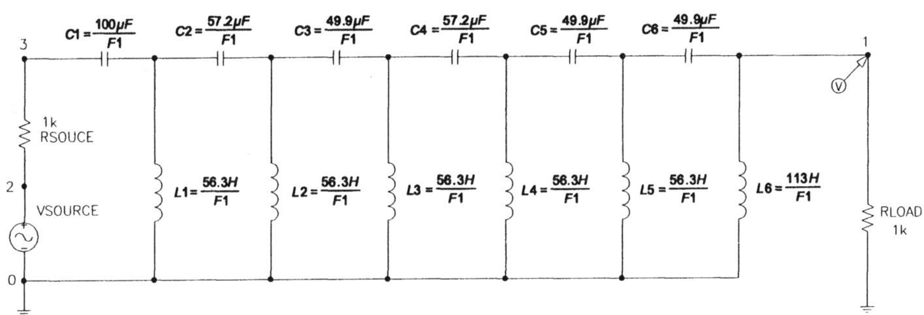

Figure 14.8 Genetically evolved generalized circuit when free variables call for a highpass filter (that is, `F1` > `F2`)

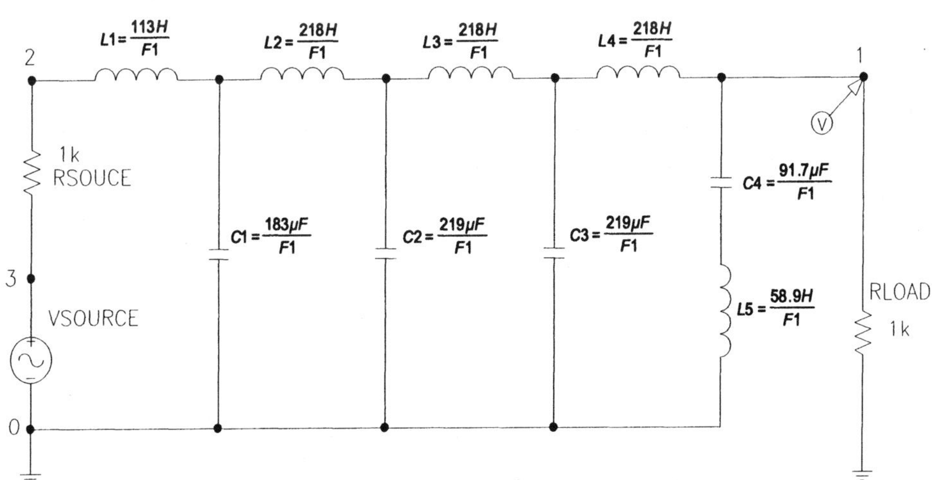

Figure 14.9 Genetically evolved generalized circuit when free variables call for a lowpass filter.

Automatically created mathematical expressions parameterize the capacitors and inductors in these automatically created circuits.

The technique above has also been applied to additional problems (Koza, Keane, Streeter, Mydlowec, Yu, and Lanza 2003), including the synthesis of

- a circuit-constructing program tree containing conditional developmental operators and free variables that yields either a quadratic or cubic function, and
- a circuit-constructing program tree containing conditional developmental operators and free variables that yields either a 40 dB or 60 dB amplifier.

6. Discussion

In this paper we have discussed how genetic programming can be used to synthesize a variety of complex structures. Our approach to the synthesis of complex structures has itself evolved over the years, with changes to methodology, representations, and fitness functions made primarily based on practical experience, intuition, and judgments derived from the theoretical foundations of genetic algorithms. Examples include our decision to use representations that we would expect to preserve building blocks in the face of crossover. In some cases—notably our design of the architecture-altering operations and our use of a developmental process for growing electrical circuits—the technique enhancements found their inspiration in biology. However, there is a clear need for more rigorous guidance and theory. Given that a representation is "crossover-friendly" (i.e., there exist within the representation local structures that contribute to fitness), the question naturally arises as to how can we further characterize the effectiveness of a representation. To what extent is the scalability of optimization algorithms affected by their ability to exploit opportunities for reuse? When is the use of higher-level constructs, such as iterations, loops, and recursions, essential for efficient solution of a problem? We believe the answers to these and similar questions are essential to the development of truly competent genetic programming (in the sense advocated by David Goldberg in his 2002 book *The Design of Innovation: Lessons from and for Competent Genetic Algorithms*).

References

Astrom, Karl J. and Hagglund, Tore. (1995). *PID Controllers: Theory, Design, and Tuning.* Second Edition. Research Triangle Park, NC: Instrument Society of America.

Banzhaf, Wolfgang, Nordin, Peter, Keller, Robert E., and Francone, Frank D. (1998). *Genetic Programming – An Introduction.* San Francisco, CA: Morgan Kaufmann and Heidelberg: dpunkt.

Barnum, H., Bernstein, H.J. and Spector, Lee (2000). Quantum circuits for OR and AND of ORs. *Journal of Physics A: Mathematical and General 33(45)*: 8047–8057.

Fraser, C. M., Gocayne, J. D., White, O., Adams, M. D., Clayton, R. A., Fleischmann, R. D., Bult, C. J., Kerlavage, A. R. , Sutton, G., Kelley, J. M., *et al.* (1995). The Minimal Gene Complement of Mycoplasma genitalium. *Science 270(5235)*: 397-403.

Goldberg, David E. (2002). *The Design of Innovation: Lessons from and for Competent Genetic Algorithms.* Boston: Kluwer Academic Publishers.

Holland, John H. (1992). *Adaptation in Natural and Artificial Systems: An Introductory Analysis with Applications to Biology, Control, and Artificial Intelligence.* Ann Arbor, MI: University of Michigan Press, 1975. Second edition. Cambridge, MA: The MIT Press.

Keane, Martin A., Koza, John R., and Streeter, Matthew J. (2002). *Improved General-Purpose Controllers.* U.S. patent application filed July 12, 2002.

Koza, John R. (1992). *Genetic Programming: On the Programming of Computers by Means of Natural Selection.* Cambridge, MA: MIT Press.

Koza, John R. (1994). *Genetic Programming II: Automatic Discovery of Reusable Programs.* Cambridge, MA: MIT Press.

Koza, John R., Bennett III, Forrest, H., Andre, David, and Keane, Martin A. (1999). *Genetic Programming III: Darwinian Invention and Problem Solving.* San Francisco, CA: Morgan Kaufmann.

Koza, John R., Keane, Martin A., Streeter, Matthew J., Mydlowec, William, Yu, Jessen, and Lanza, Guido. (2003). *Genetic Programming IV. Routine Human-Competitive Machine Intelligence.* Kluwer Academic Publishers.

Langdon, William B. (1998). *Genetic Programming and Data Structures: Genetic Programming + Data Structures = Automatic Programming!* Amsterdam: Kluwer.

Langdon, William B. and Poli, Riccardo. (2002). *Foundations of Genetic Programming.* Springer-Verlag.

Ohno, Susumu. (1970). *Evolution by Gene Duplication.* New York: Springer-Verlag.

Spector, Lee, Barnum, Howard, and Bernstein, Herbert, J. (1998) Genetic programming for quantum computers. In Koza, John R., Banzhaf, Wolfgang, Chellapilla, Kumar, Deb, Kalyanmoy, Dorigo, Marco, Fogel, David B., Garzon, Max H., Goldberg, David E., Iba, Hitoshi, and Riolo, Rick. (editors). Genetic Programming 1998: In *Proceedings of the Third Annual Conference*, 365-373. San Francisco, CA: Morgan Kaufmann.

Spector, Lee, Barnum, Howard, and Bernstein, Herbert J. (1999). Quantum computing applications of genetic programming. In Spector, Lee, Langdon, William B., O'Reilly, Una-May, and Angeline, Peter (editors). *Advances in Genetic Programming 3.* Cambridge, MA: The MIT Press, 1999. 135–160.

Spector, Lee, Barnum, Howard, Bernstein, Herbert J., and Swamy, N. (1999). Finding a Better-than-classical Quantum AND/OR Algorithm Using Genetic Programming. In IEEE. In *Proceedings of 1999 Congress on Evolutionary Computation*, pp. 2239-2246. Piscataway, NJ, IEEE Press.

Spector, Lee, and Bernstein, Herbert J. (2003). Communication Capacities of Some Quantum Gates, Discovered in part through Genetic Programming. In Shapiro, Jeffery H. and Hirota, Osamu (editors). In *Proceedings of the Sixth International Conference on Quantum Communication, Measurement, and Computing*, 500-503. Princeton, NJ: Rinton Press.

Ziegler, J. G. and Nichols, N. B. (1942). Optimum Settings for Automatic Controllers. *Transactions of ASME 64*: 759–768

Chapter 15

INDUSTRIAL STRENGTH GENETIC PROGRAMMING

Empirical Modeling and Symbolic Regression via GP: Integrated Methodologies, Best Practices, Lessons Learned

Mark Kotanchek[1], Guido Smits[2], and Arthur Kordon[3]

[1]*Dow Chemical, Midland, MI USA*

[1]*Dow Benelux, Terneuzen, NV*

[1]*Dow Chemical, Freeport, TX USA*

Abstract Since the mid-1990's, symbolic regression via genetic programming (GP) has become a core component of a multi-disciplinary approach to empirical modeling at Dow Chemical. Herein we review the role of symbolic regression within an integrated empirical modeling methodology, discuss symbolic regression system design issues, best practices and lessons learned from industrial application, and present future directions for research and application

Keywords: Genetic Programming, Empirical Modeling, Symbolic Regression, Support Vector Machines, Neural Networks

In theory, there is no difference between theory and practice. In practice, there is.
– Jan L.A. van de Snepscheut

1. Introduction

Our objective is to summarize our experience in industrial application of genetic programming to empirical modeling and to transfer key lessons with respect to real-world application. Hence, there are three themes which are explored in the following pages: (1) the role of GP in empirical modeling, (2) symbolic regression system design and (3) open issues and research suggestions.

Our interest in symbolic regression is motivated by the industrial need to quickly and efficiently convert multivariate data sets and data streams into actionable insight and knowledge. Although we use a variety of technologies in our analysis methodology, GP offers some unique capabilities which are complementary to those of neural networks, support vector machines, linear statistics, etc. Hence, we explore the framework of industrial empirical modeling, practical issues, and how modeling success may be best achieved. In general, success is achieved by using a combination of tools and techniques and analysis discipline.

We have developed two symbolic regression environments in support of our modeling efforts. We will explore the system design objectives, issues, components, experiences, and lessons learned in the development and exploitation of those environments. Some tricks-of-the-trade for achieving efficiency, accuracy and simplicity – from the human as well as genetic programming perspective – are presented.

Although symbolic regression via GP has proven to be a valuable tool, there remain open issues which are worthy of further research to improve the speed, robustness and effectiveness of GP and empirical modeling. For example, truth is an evasive term in the realm of empirical modeling. Hence, we often assemble ensembles of models and use their concurrence, or lack thereof, as an indicator of model confidence. The validity of this strategy depends upon the ensemble containing diverse models with comparable complexities. Hence, we have a fundamental need to characterize both *diversity* and *complexity* in a population of models. Unfortunately, quantifying either of these is a difficult task. Another research area of potentially large impact is the conversion of hard problems into easy ones – either via integration with other techniques, incorporation of heuristics and *a priori* knowledge or identification of metavariables to simplify the solution discovery.

2. Industrial Application of GP to Empirical Modeling

In this section we review the importance of empirical modeling to industry, the unique capabilities of symbolic regression via GP for empirical modeling, and how an integrated methodology exploits the strengths of GP and complementary analysis techniques.

Empirical Modeling

In an industrial setting, the objective of empirical modeling is to convert data to knowledge which can then be transformed into money. As such, it tends to be goal-driven rather than exploratory as is typically the case with data mining. In other words, we typically have a response (or responses) which we seek to model as a prelude to understanding underlying mechanisms or system optimization. Reality, however, often requires that empirical modeling efforts include an exploratory aspect to identify key parameters and variable

Table 15-1. Industrial applications of empirical models

Application	*Comment*
Research Acceleration	Understanding variable relationships to each other as well as to the response of interest accelerates research due to providing cues to underlying physical mechanisms or areas of parameter space to explore. These models facilitate exploration of highly multivariate problems. (Kordan *et al.*, 2002)
Inferential Sensors	Also known as soft sensors, these empirical models combine one or more parameters to infer a system state (response) which may be difficult to measure directly. Predictive models can give advance warning of impending product quality issues or other operational issues. (Kordon and Smits, 2001)
Emulators	System optimization can be difficult for industrial systems due to the cost and difficulty of acquiring data points. An empirical model may be used as a system emulator to provide insight as a prelude to more focused experimental validation. Additionally, an emulator may be used as a lower-fidelity approximation of a first-principles model for coarse optimization or online deployment. (Mercure *et al.*, 2001)
Variable Transforms	Identifying combinations of variables which have a physical meaning can lead to simpler and more robust system models. These metavariables can also provide insight into underlying mechanisms or convert a nonlinear problem into a linear one.(Kordan *et al.*, 2002; Castillo *et al.*, 2002)
Nonlinear Design-of-Experiments	In environments where data points are expensive to collect, the ability to understand the underlying system behavior while simultaneously facilitating focusing data collection in critical regions. An example of this might include design of model discrimination experiments to distinguish between competing models of chemical kinetics. (Castillo *et al.*, 2002)

relationships. Empirical modeling may have a first-principles aspect if *a priori* information provides models which should be fitted to the available data.

Industrial Applications of Empirical Models. Some of the industrial applications of empirical models are captured in Table 15-1. In general, we seek either system insight or the ability to more effectively and efficiently monitor, control or optimize a system

Requirements for a Successful Model. Achieving an empirical model may be accomplished using a number of technologies in addition to symbolic regression: linear statistics, neural networks, support vector machines, particle swarm optimizers, genetic algorithms, GMDH, fuzzy rules, etc. A quality model must have aspects of:

credibility	the model matches the observed behavior
robustness	the model accuracy is able to withstand minor changes in the targeted system
extrapolation	closely related to robustness, this is the ability to predict outside the training range (within reason)
self-assessment	the model is able to assess the quality of its output after deployment
interpretability	humans are able to explore and qualitatively validate the "reasonableness" of the model"
cost effectiveness	in addition to development cost, the total cost of ownership includes ease-of-use and effective maintenance and support

Model quality is not a static notion; rather, in addition to raw accuracy, it must consider the cost and timeliness of the development process as well as performance and maintainability during its lifetime.

The Role of Symbolic Regression. As we shall see when we review the integrated methodology, the primary analysis techniques used are linear statistics, neural networks, support vector machines and symbolic regression via genetic programming. Each of these foundation technologies have their strengths and weaknesses which are complementary (Kotanchek *et al.*, 2002). In general, symbolic regression is more computationally intensive and produces less accurate models than the other nonlinear techniques. In compensation, it has unique benefits which include those summarized in Table 15-2.

Integrated Methodology

Combining engineering discipline and rigor with nonlinear and classical data analysis techniques leads to efficient and reliable empirical model development. The components of the methodology are summarized in Table 15-3. As indicated, the primary nonlinear modeling tools used are: neural networks, support vector machines and symbolic regression via genetic programming. Neural networks are universal approximators and are, therefore, capable of modeling any system. Of course, they can also model noise very well so caution must be used in their development since it is often difficult to interpret the developed models due to their black box nature.

Support vector machines (SVMs) (Jordaan, 2002; Hastie *et al.*, 2001) are a recent introduction to the empirical modeling toolbox. The basic idea is that a subset of the data samples — the support vectors — will capture the response function of the data if we locate appropriate kernels (e.g, radial basis functions or polynomials) on the support vectors. SVMs have a reasonably rigorous theoretical foundation founded in statistical learning theory and the theory specifically addresses small data sets. The benefit is SVMs can recognize sets which are

Table 15-2. The unique benefits of symbolic regression to empirical modeling include:

Benefit	*Discussion*
Physical Insight & Understanding	Examining the models returned by a symbolic regression may be interpreted by an expert with first-principles insight to identify underlying mechanisms or to suggest new outside-the-box approaches.
Visualization	Visualizing highly multivariate spaces is inherently difficult. A symbolic representation is, effectively, a visual representation which summarizes the interaction between variables. Of course, the model may also be converted into graphical form if desired.
Simple Summary Models	Evaluating developed symbolic models may require fewer computing resources and complexity than those derived by their nonlinear brethren such as neural networks or support vector machines. This can facilitate online model implementation in manufacturing plants.
Identify Transforms & Metasensors	Natural combinations of variables may appear as part of the symbolic regression model building. These (typically nonlinear) synthetic metavariables can be used to capture variable dependencies as well as facilitate data transforms so that classical linear statistics may be applied and metrics such as lack-of-fit used for model validation. Additionally, incorporating these metavariables may increase the efficiency and accuracy of additional symbolic regressions by transforming the problem into a different space which is easier for model development.
Variable Selection	Often more variables are available to the empirical modeling than are warranted by the underlying physics. GP can identify these significant variables since they will, typically, appear in the more fit models. Focusing the subsequent modeling on these variables can increase the efficiency of the modeling effort.
Variable Sensitivity	Once a model (or models) have been developed, they can be numerically or symbolically differentiated to understand the sensitivity to parameter changes. This can be useful in applications such as robust design.

oversampled with respect to their information content (or specified complexity). This also leads to the truly spectacular capability of SVMs — their ability to perform nonlinear outlier detection. They may also be used for downselection in the sample space to ensure that the input to other modeling tools are balanced in terms of their information content.

GP (Jacob, 2001; Banzhaf *et al.*, 1998) offers an escape from the black box modeling techniques and provides a link between the empirical and the fundamental. Although there are continuing theoretical debates regarding the mechanisms of effective function evolution, this is not a major issue from a practical perspective since effective application has been demonstrated. GP is attractive for a number of reasons including the natural selection of the most important inputs, relatively parsimonious analytical expressions which are evolved and the associated ease of implementation and the lack of *a priori* model assumptions.

Table 15-3. The components of the integrated empirical modeling methodology

Component	*Discussion*
Problem & Success Definition	Although often neglected, *defining* success is *the* key ingredient to achieving success. This involves understanding the system to avoid context-free analysis (which tends to lead to confidently wrong answers) as well as specifying the requisite levels of model accuracy and robustness.
Data Review & Preparation	There is no substitute for understanding the input data and its quality. Bad data points should be removed and appropriate data transforms and combinations performed to synthesize metavariables. Data should be partitioned into test and training sets. If system changes have occurred, the data should be partitioned to avoid mode mixing. Often, data samples should also be downsampled to reflect the actual information content (this avoids redundant data points skewing the evaluation). Regression-oriented support vector machines are especially appropriate for downsampling data as well as identifying outliers.
Select Variables & Metavariables	Often many more variables are available than required. Downselecting variables to the critical parameters helps to focus the models onto the underlying physics rather than system noise. Additionally, many modeling techniques make an implicit assumption that variables are uncorrelated; violating that assumption can invalidate the resulting models. Genetic programming is useful for identifying possible metavariables. Although GP can be used for variable downselection, neural networks are generally more computationally efficient.
Build Models	Many techniques are available for model building These include model fitting simple mathematical models (e.g., polynomials) or first-principles models as well as nonlinear data-driven techniques. Support vector machines generally produce the most robust models, neural networks the most accurate, and symbolic regression the most insightful and simplest to deploy. The metavariables identified via GP can enable model linearization which is generally preferable from a robustness perspective.
Analyze & Validate Models	Any empirical model should be viewed with suspicion until proven valid. The possibility of overfitting must be explored as well as the physical reasonableness of results. If necessary, additional data partitioning and data cleansing should be done and the modeling exercise repeated.
Select & Combine Models	The best model may not be the most accurate depending upon the definitions of success. For example, simplicity of implementation or model parsimony with "good enough" performance may be preferred. Stacking or boosting models (Hastie *et al.*, 2001) can result in improved performance as well as an indication of operation in unknown regions of parameter space. However, using the consensus of models as an indicator of model validity makes an implicit assumption of independent constituent models which may be difficult to verify.
Exploit Models	In an industrial setting, the modeling effort is not a success unless the models are being exploited either by providing system insight, enabling optimization or deployed in an operational system.

As a final note, irrespective of modeling technique, correlation (quality of fit of a model) does not imply causality. Causality always needs to be verified independently.

3. System Design

In this section we first discuss some of the key design issues for a symbolic regression environment. Following that, the components which have been implemented into our *Mathematica* (Kotanchek, 2003) and MATLAB implementations are reviewed and key features discussed. Finally, we close with a review of tricks-of-the-trade in the practical use of the system capabilities.

The current implementations are currently restricted to evolving sample-oriented models (i.e., no recursive functions or operations across the entire data space). The typical symbolic regression will operate on a data set of 3–10 variables and 20–3000 data points. (Recall that we often use neural net and SVM-based preprocessing to reduce the size of the input data set.) However, we have used GP to perform sensitivity analysis of systems with up to 100 variables.

Implementation & Design Issues

Better. Faster. Cheaper. Those are essentially the objectives of any industrial design effort. In the context of symbolic regression system design, these translate into:

- speed of model development and selection,
- robust parsimonious models, and
- human efficiency of model development and selection.

Achieving these objectives requires a system level approach to provide an environment efficient for both the computer and the user. Since garbage in results in garbage out, input data quality assessment tools must be provided to assist in the data set design process. The initiation and execution of the evolutionary process also must not be labor intensive. From the computer standpoint, the evolutionary process must be computationally viable either through efficient code or via parallel execution. For maximum efficiency the symbolic regression should be autonomous to minimize demands on the modeler. The evolved solutions also must balance complexity and performance. Finally, given the plethora of solutions which will typically be evolved, tools must be provided for the user to sort through the entities (candidate expressions), evaluate them, extract insights, and facilitate model deployment.

System Components

On the order of 200 functions have been defined as part of the *Mathematica* symbolic regression and data analysis environments; hence, of necessity, we will only cover the highlights of the design functionality.

Define Success: Fitness & Parsimony Penalty Functions. The first step in the symbolic regression is to define a fitness function to characterize the quality of the fit of the evolved expression to the available data. Generally, we opt for an absolute correlation metric since a translation and scale coefficient can be calculated in post processing — thereby simplifying the search problem by eliminating the need to evolve those terms. There are multiple aspects of fitness to consider when evaluating an entity. The raw (or absolute) fitness characterizes the ability of the expression to predict the targeted response behavior.

We generally prefer parsimonious (simple) functions. A parsimony adjustment which penalizes complexity also prevents explosive growth of function complexity due to the inclusion of introns. Although we continue to explore alternate strategies, the default, which is generally appropriate, is to use a multiplicative penalty function which penalizes the raw fitness according to the number of nodes in the genome. A sigmoid function,

$$ParsimonyFunction \rightarrow 1 - \frac{1}{1 + e^{-\frac{\mathrm{nodes}[genome]-a}{b}}}$$

is the default adjustment to the raw fitness. Adjusting the sigmoid coefficients, a and b, allows tweaking of permissible complexity and complexity sensitivity.

Initial Population Creation. Functions need to be provided to define entities as a starting point for the evolutionary process. These can either be created randomly from the genetic code components or drawn from a pool of previously evolved entities. Additionally, functions must be provided to seed the populations with *a priori* expressions; hence, the ability to inversely map from defined mathematical expressions (the phenotypes) into their genetic codes (genotypes) is critical — at least to give the user the feeling of control!

Genetics. Although not strictly required, our implementation of genetic programming distinguishes between the genetic code abstraction (genotype) and the mathematical expression (phenotype) upon which the raw fitness is calculated. This abstraction leads to considerable flexibility in the application of the symbolic regression.

The Genetic Code: Building Blocks for Evolution. The genetic code for symbolic regression requires three components:

- *Function Patterns*: codings for mathematical and logical operators

- *Variables*: the independent parameters in the data matrix
- *Constants*: constant integers and reals (our implementation synthesizes random constants upon demand)

The problem is to assemble these into a structure which describes the observed behavior. We can simplify the evolutionary process by including function patterns appropriate for the problem at hand as well as downselecting the variables to an appropriate subset. In addition to controlling the maximum arity (number of slots or parameters) of functions like Plus[ ] or Times[ ] which can handle an arbitrary number of arguments, we also control the complexity and structure by weighting the likelihood of any given building block (or class of building block) being selected. To illustrate, the default function patterns are:

$$\text{FunctionPatterns} \rightarrow \{\Sigma[\_\_], \mathbb{S}[\_,\_], \Pi[\_\_], \mathbb{D}[\_,\_], \mathbb{P}[\_,\_]\}$$

which, when converted into the phenotype, leads to summation, subtraction, multiplication, division, and power, respectively. Summation and multiplication are arbitrary arity functions (denoted by two underscores) whereas the others are explicitly limited to two (possibly compound) slots. The default is to automatically synthesize variable names ($x_1, x_2, \ldots$) from a user-defined symbol or explicit symbols, as desired.

The evolution can also be accelerated by inclusion of proper metavariables (combinations of variables or transforms) which simplify the task of discovering a quality genome. Currently, metavariables are defined by users examining evolved structures or using physical insight; however, a natural extension is to adopt a coevolution strategy for automated definition.

In the ensuing discussion, it is important to note that any genome can be represented as a tree-based structure with terminals (constants or variables or metavariables) at the end of the branches and functions at the other nodes. The term *node* refers to any specific location in the genetic tree — function or terminal.

Complexity Control. In addition to the indirect means of complexity control by applying parsimony pressure and limiting the arity of arbitrary arity functions, we have explicit limits on the maximum allowable tree depth and leafcounts for the genome structures. These are safety regulators since, if unfettered by parsimony controls, expressions can rapidly evolve with millions of elements — which likely do not meet the engineering definition of success!

Phenotypes. The genetic code must be converted into its phenotype (physical representation) for fitness evaluation.. This is the representation which can be evaluated by *Mathematica* or MATLAB at each of the evaluation points and the predicted result compared to the actual observed response. As with natural selection, it is the phenotype (entity) which is judged and the underlying genetics which are propagated.

The phenotype can also learn from experience to improve its performance. We shall address this more as part of the genetic engineering discussion.

Representation. In addition to the genetic code, any given entity also has three fitness metrics: absolute, adjusted and relative. The absolute fitness is the quality independent of parsimony pressure whereas the adjusted fitness penalizes this value based upon its complexity. The relative fitness is the quality relative to its peers in the breeding pool. Additionally, we track miscellaneous attributes as part of a personality aspect for analytic flexibility.

Evaluating Entities. Evaluating entities is a relatively straightforward process in that the genetic code is converted into an executable form and evaluated. The mapping can be straightforward, e.g., $\mathbb{SR}[x] \rightarrow \sqrt{x}$, or they can be *protected*, e.g., $\mathbb{SR}[x] \rightarrow \sqrt{|x|}$, which can handle the situation when x is negative. The advantage of protected functions is that they avoid pathologies in the phenotypes and, therefore, may result in a faster solution. The disadvantage is an increase in complexity of the evolved expressions, although not necessarily in the underlying genetic code.

Each entity is evaluated at each of the sample points and the result judged relative to the observed response behavior. This is where design of the data set comes into play relative to correspondence of the calculated vs. true quality of the correspondence. If the data set is *unbalanced* in the sense that much of the data is localized in one region of parameter space, then the evolved entities will be judged primarily on how well they fit *that* region. SVMs are especially useful for addressing this problem and produce *balanced* data sets which avoid this problem. Alternately, human insight, experimental design, genetic algorithms or other techniques may be used to preserve *diversity* in the data space.

Propagation. The purpose of a phenotype is to earn the right to propagate its genetic code into the next generation. As with nature, there are a plethora of competition models which can be used to assign breeding rights; these include:

- **Proportional** – Each entity is awarded breeding rights proportional to its contribution to the cumulative fitness of the population. This is also known as the roulette wheel model since the size of each entity's bin corresponds to it's component of the total population fitness (*Mathematica* default).

Table 15-4. Diversity introduction mechanisms include:

Mechanism	*Description*
MutateSubtree	Replace an arbitrary node and all of its branches with a new subtree
MutateNode	Replace an arbitary node with a different function or terminal which is a valid replacement.
Crossover	Replace subtrees from an arbitrary node with those from another entity. Although analogous to sexual reproduction, the genetic exchange is not restricted to being pairwise.
Clone	Sometimes reproduction is asexual without an introduction of diversity
Hoist	Create a new entity from a subtree of the donor entity.

- **Rank-Based** – The breeding rights are proportional to the rank of an entity (the lowest quality entity gets a rank of one and the rank increments by one from there). (MATLAB default).
- **Elitist** – The top fraction of entities are treated as equals from a breeding rights perspective and the lower ranked entities have no chance to propagate.
- **Tournament** – Subsets of the population are placed into pools for competition with the winners achieving propagation rights. The intent here is, again, to preserve the diversity in the population.

Our preference is to run many independent evolutions; hence, premature convergence and loss of genetic diversity isn't a major problem. We have also used mixtures of selection strategies.

Successful evolution requires only two things: a fitness-based pressure and a means to introduce diversity into the population. If we recall that the genome is represented as a tree, possible operators to introduce this diversity include those summarized in Table 15-4.

Although analogous to cloning, we also have found the presence of a Methuselah individual to be effective wherein the best entity is given a new lease on life and propagates unchanged into the next generation. This leads to a more continuous fitness behavior during evolution which simplifies termination criteria and related issues. The others in a generation are killed off and replaced by their progeny.

Genetic Engineering: Optimizing for Performance & Simplicity. When Darwin proposed the theory of evolution based in survival of the fittest, there was not yet an awareness of genetics and their role in evolution. One alternate theory was proposed by Lamarck wherein the experiences collected by an entity would

Table 15-5. Diversity introduction mechanisms include:

Operation	*Description*
Remove Introns	Remove nonfunctional genetic code to produce a simplified genome
Simplify	Evaluate a phenotype expression and simplify it using mathematical rules. This is relatively easy in an environment such as *Mathematica* which provides support for such symbolic manipulations. The simplifed form is then converted into it genome equivalent and becomes the new genetic code of the entity.
Align	Scale and translate the entity to maximize the alignment to the observed response behavior. This linear operation is generally fairly efficient computationally.
Optimize	Optimize the constants in the phenotype to maximize alignment to the response behavior. Note that this is typically a nonlinear optimization problem and relatively time consuming.
Optimize Structure	Prune the structure to eliminate branches and leaves which do not provide a benefit in terms of increased phenotypic fitness. Alternately, prune out selected variables.
Local Optimize	Randomly select a constant in the genome and optimize its value.

be passed on to the next generation. Thus, a giraffe stretching its neck reaching for leaves would result in progeny with longer necks. Although Lamarckian genetics are not realistic in natural systems, the implicit genetic engineering is easily achieved in symbolic regression via genetic programming. To achieve this, we need inverse mappings from phenotypes to genotypes as well as tools to manipulate and optimize both phenotypes and genotypes. The basic classes of genetic engineering which we have found to be effective include those listed in Table 15-5.

With the exception of the local optimization which is relatively efficient and occasionally included as a propagation operator, the other genetic engineering operations are performed offline on the best entities resulting from a round of evolution.

Working with Populations: Isolated and Cascaded Evolutions. Although, in principle, it should be possible to evolve high-quality expressions in a single monolithic evolution, maintaining sufficient genetic diversity and avoiding premature convergence on suboptimal solutions has proven to be very difficult in our experience. A more robust strategy has been to spawn many isolated and relatively small populations which evolve in parallel. The best entities out of each of these *worlds* are placed in a pool and randomly selected as the seed population for a subsequent round of evolution of many parallel worlds. This cascaded evolution model has proven to be very effective and robust.

Table 15-6. Tools to review evolution results include:

Tool	*Description*
Genome Tree Plot	Treeform graphic of the entity genetic structure
Entity Selection Table	Tablular view of top entity expressions, their fitnesses (raw, absolute, and relative), and included variables
Response Surface Plot	Response surface or curve over a specified or automatic region of parameter space. Unvaried variables may either be specified or automatically calculated from the data matrix.
Entity Evaluation Plot	The predicted vs. observed response. Ideally, all points should fall on the diagonal
Entity Residual Plot	Residual (error between predicted and observed) vs. sample number
Variable Prediction Plot	Actual vs. predicted values as a function of specified variable
Entity Regression Report	Summary statistics relative to the quality of fit of the entity to the observed behavior.

The term *efficient* in genetic programming is a relative term since GP is, typically, very time and compute resource intensive. Towards that end, eliminating unecessary variables from the modeling exercise simplifies the task of high quality expression discovery. Although we generally have pruned the number of variables using a neural network based sensitivity analysis as part of the data processing, a good practice is to examine the presence and frequency of variables in the best evolved structures to ascertain whether additional variable elimination is possible. Additionally, identification of metavariables from the evolution results can help to convert a hard analysis problem into a much easier one.

Given the investment in time and computation in the evolutionary process, a viable system design needs to include tools for archiving evolutionary results, retrieving those results, and merging populations. Our utility functions in this space also provide for automatic file naming which facilitates project organization and execution for analysis, and avoids inadvertent file overwrites.

Reviewing Entities. Machine intelligence should not be confused with intelligence. At this stage of the technology curve, humans must be involved in reviewing evolutionary results, validating performance and selecting models for deployment and exploitation. From a system design standpoint, the human interaction must be efficient and insightful. Towards that end, we provide a number of tools in our systems to facilitate that interaction; these include those summarized in Table 15-6.

Tricks of the Trade

Although our symbolic regression environments are actively being used in empirical analysis projects, they are also the subjects of continued research and development. Despite the success of the application efforts, it would also be hubris to believe that the choices of evolutionary operators, population sizes, population operators, selection strategies, complexity controls, etc., are optimal. However, they are good enough — which is a sufficient level of performance from an evolutionary (and engineering) viewpoint. That said, key features of the symbolic regression system designs are:

- **Multiple Worlds** — many smaller evolutions minimize issues around maintaining genetic diversity
- **Cascaded Evolutions** — cascading evolutions using the best entities from prior rounds initializes the subsequent rounds with genetically diverse but high quality entities
- **Intron Removal** — removing the genetic noise helps to ensure that the key genetic building blocks are propagated
- **Entity Optimization** — optimizing the entity structure or coefficients in post-processing improves the quality of the models
- **Absolute Correlation** — using absolute correlation as the fitness metric eliminates the need to evolve scaling and translation coefficients
- **Protected Functions** — this speeds up the discovery of fit models, albeit at the cost of increased phenotype complexity and difficulty in post-processing simplification and optimization
- **Variable Downselection** — trimming the number of variables to those which most drive the response behavior helps to reduce the search space and accelerates the model discovery process
- **Metavariables** — the right synthetic variables also reduce search space and accelerate the discovery process
- **Evolution, Evaluation & Review Tools** — ease of use and analysis is critical for success since humans provide the contextual insight and real-world validation

4. Open Issues & Research Topics

Although the industrial value of symbolic regression and genetic programming have been established, the optimal strategies and parameter settings remain a topic of ongoing research. Additionally, there are deeper conceptual and theoretical explorations which remain as significant issues to fully capture the value of GP in industry. Some of these are discussed in the following sections.

How to control complexity and characterize smoothness?

The quality of an entity is determined by how well it describes the observed response behavior *at the sample points*. An open question is: What is the behavior between the sample points and in the surrounding region? To some extent, we attempt to control this via parsimony pressure which penalizes complexity. However, parsimony pressure does not necessarily eliminate pathologies and singularities which are not physically reasonable. Occam's razor implies that we want "smooth" behavior unless we have information to the contrary. The question is how do we *efficiently* recognize and characterize *global* and *local* smoothness (complexity) over parameter space? The ability to answer this question should lead to more robust expressions.

How to blend heuristics and prior knowledge?

Currently we have three basic strategies for including *a priori* insight and knowledge into a symbolic regression: (1) choice of function set and variables, (2) inclusion of metavariables, and (3) seeding the evolution with likely expressions. Improving the inclusion of prior knowledge should lead to greater physical interpretability of evolved expressions and, presumably, more robustness and more efficient and faster evolution.

How do we identify metavariables?

In the ideal case we could identify transforms which would convert nonlinear analysis problems to ones which would be tractable via linear techniques. Currently, identification of metavariables and transforms is done by "eyeballing" the evolved fit solutions. There has got to be a better way! Coevolution of metavariables is one possible approach, or we could also use a pattern analysis post-processing of the best solutions. Making the metavariable selection more automated and robust should improve the speed of analysis and interpretation.

How do we detect diversity in a population?

A strategy which we have found to be effective is to *stack* (very similar to boosting) models and to use their average or median as a robust predictor and their consensus as an indicator of prediction confidence. The validity of this approach depends upon the *diversity* of the population. Obviously, stacking minor variations on the same theme will not add a great deal of robustness to the composite estimate.

To assemble a collection of diverse models we need a metric to characterize diversity. To some extent, we can evolve a diverse population by using many isolated evolutions or by evolving expressions using different sets of building block functions and variables. But, how can we be sure that we have truly achieved the desired diversity objective?

Related to the diversity issue is the ability to measure expression complexity since we would like to balance the complexity of the expressions used in the composite model.

Are we using the wrong paradigm?

A cultural metaphor (e.g., particle swarm optimization, PSO) appears to hold a significant advantage in effectiveness over their genetic brethren (e.g., genetic algorithms) for optimization problems and has replaced GAs in our color additive formulation and appearance engineering systems. As with human culture, the interchange of *memes* (ideas) operates in a far faster and more effective fashion than the genetic transfer operating in parallel. This leads us to wonder whether the *genetic programming* paradigm is passé and should be replaced or augmented with *memetic programming*? The question, of course, is how to to rephrase the symbolic regression search problem from a genetic analogy into a cultural framework?

5. Summary

GP-based symbolic regression has become a key component in Dow Chemical's empirical modeling capability due to its unique ability to provide physical insight, generate computationally efficient models, and identify key variable and associated transforms and combinations. Used with other analysis techniques as part of an integrated methodology, the result is improved modeling accuracy, robustness, and development efficiency. Continuing improvement of the symbolic regression foundations used by Dow and their integration with other linear and nonlinear analytic techniques continues to be a significant research effort at Dow Chemical with the expectation of a corresponding return on investment.

References

Banzhaf, W., Nordin, P., Keller, R. and Francone, F. (1998). *Genetic Programming - An Introduction.* Morgan Kaufmann, San Francisco, CA.

Castillo, F. A., Marshall, K., Green, J. and Kordon, A. K. (2002). Symbolic Regression in Design of Experiments: A Case Study with Linearizing Transformations. In *Proceedings of the Genetic and Evolutionary Computation Conference (GECCO 2002)*, W. B. Langdon, *et al.* (Eds.), pp. 1043–1048. New York: Morgan Kaufmann.

Hastie, T., Tibshirani, R. and Friedman, J. (2001). *The Elements of Statistical Learning: Data Mining, Inference, and Prediction.* New York: Springer-Verlag.

Jacob, C. (2001). *Illustrating Evolutionary Computation with Mathematica*. San Francisco: Morgan Kaufmann.

Jordaan, E. Maria. (2002). *Development of Robust Inferential Sensors: Industrial Application of Support Vector Machines for Regression*. Eindhoven: Universiteitsdrukkerij TU Eindhoven.

Kordon, A. K., Pham, H. T., Bosnyak, C. P. and Kotanchek, M. E. (2002). Accelerating Industrial Fundamental Model Building with Symbolic Regression: A Case Study with Structure-Property Relationships. In *GECCO 2002: Presentations in the Evolutionary Compution in Industry Track*, D. Davis and R. Roy (Eds.), pp. 111–116. New York: GECCO 2002 Conference.

Kordon, A. K. and Smits, G. F. (2001). Soft Sensor Development using Genetic Programming. In *Proceedings of the Genetic and Evolutionary Computation Conference (GECCO 2001)*, L. Spector, *et al.*. (Eds.), pp. 1346–1351. New York, Morgan Kaufmann.

Kotanchek, M. E., et al. (2002). Evolutionary Computing in Dow Chemical. In *GECCO 2002 Presentations in the Evolutionary Computation in Industry Track*, D. Davis and R. Roy (Eds.), pp. 101–110. New York: GECCO 2002 Conference.

Kotanchek, M. E., (2003). Industrial Strength Symbolic Regression: Evolving Empirical Models from Industrial Data. *2003 Mathematica Developers Conference*, Champaign, IL : Presentation.

Mercure, P. Kip., Smits, G. F. and Kordon, A. K. (2001). Empirical Emulators for First Principle Models. *Fall 2001 AIChE Meeting*. Reno: Presentation.

Chapter 16

OPERATOR CHOICE AND THE EVOLUTION OF ROBUST SOLUTIONS

Terence Soule
Department of Computer Science
University of Idaho, Moscow, ID, 83844-1010
tsoule@cs.uidaho.edu

Abstract

This research demonstrates that evolutionary pressure favoring *robust* solutions has a significant impact on the evolutionary process. More robust solutions are solutions that are less likely to be degraded by the genetic operators. This pressure for robust solutions can be used to explain a number of evolutionary behaviors. The experiments examine the effect of different types and rates of genetic operators on the evolution of robust solutions. Previously robustness was observed to occur through an increase in inoperative genes (introns). This work shows that alternative strategies to increase robustness can evolve. The results also show that different genetic operators lead to different strategies for improving robustness. These results can be useful in designing genetic operators to encourage particular evolutionary behaviors.

Keywords: Code growth, code bloat, operators, introns, design, robust solutions

1. Introduction and Background

The goal of this paper is to demonstrate that pressure for robust solutions is a general phenomenon effecting more than just tree-based GP and that growth is not the only evolutionary response to this pressure. The results show that evolution may favor any of several different strategies to improve robustness and that the strategy adopted depends significantly on the operators involved.

It is worth emphasizing that in this discussion robustness refers only to resistance to change from variation operators such as crossover and mutation. In a different context robustness might refer to other factors such as resistance to noise or to changing environments. Further, pressure for robustness is not the *only* factor influencing evolution. Clearly selection produces considerable pressure for more fit solutions and there may be other, as yet unidentified, pressures as well.

In this paper the robustness (or stability) of a solution is a measure of the solution's average change in fitness under genetic operations such as crossover and mutation. It has been theorized that there is significant evolutionary pressure in favor of solutions that are more robust (Soule, 2002; Soule, 2002; Streeter, 2002). The most outstanding evidence of pressure towards stability is the phenomenon of code growth (or bloat) in genetic programming (GP) (Koza, 1992; Blickle and Thiele, 1994; Nordin and Banzhaf, 1995; McPhee and Miller, 1995; Soule, 1996; Soule, 1998; Luke, 2000c; Nordin and Banzhaf, 1995; Nordin *et al.*, 1997; Nordin, 1997). Code bloat is a rapid increase in code size that does not result in fitness improvements. The extra code usually consists of inoperative code or introns (code that does not contribute to the program's fitness). It is generally accepted that in large part programs generated with GP grow as a means of protecting the useful code within good solutions against the potentially negative effects of crossover. By adding introns the useful code (commonly known as operative code or exons) is less likely to be affected by crossover.

Luke has argued that introns themselves are not the cause of code growth (Luke, 2000c). Smith and Harries have shown that growth can occur in code that does influence fitness if the code has only a negligible effect on performance (Smith and Harries, 1998). The author has recently shown that code growth can also occur in exons that have a significant effect on fitness (Soule, 2002; Soule, 2002). It has also been shown that some forms of mutation can encourage code growth (Langdon *et al.*, 1999; Soule and Heckendorn, 2002). Finally, theoretical and experimental research strongly suggests that code growth is a factor in any evolutionary system using a variable sized representation (McPhee and Miller, 1995; Soule, 1998; Langdon, 1997).

Whereas early code growth research focused on tree based GP, introns and crossover, these recent results point to a more general phenomenon relating to any evolutionary system with variable length representations, any genetic operators and both introns and exons. This supports the hypothesis that code growth in GP is only one symptom of a general underlying evolutionary pressure favoring robust individuals. To avoid the focus on tree based representations a linear representation is used in these experiments.

The results presented here show that different operators create pressure for different forms of solutions. In particular, two versions of crossover are compared: with one version the pressure for robust solutions leads to growth and with the other version it does not. It is also shown that with some genetic operators a reduction in solution size is favored to make solutions more robust.

Finally, the results suggest rules for determining how a particular operator will influence the evolutionary process by favoring robust solutions. The rules will be particularly useful to a practitioner designing a new representation and operators for a novel problem. These rules will make it possible for the practitioner to begin to predict how the new operators will influence the evolutionary process and thus will be useful in designing evolutionary algorithms and operators that guide the evolutionary process in specific directions.

Other Causes of Growth

In addition to protective growth, two other causes of growth in GP have also been proposed: *removal bias* and *drift*. Because growth is an important strategy for evolving robustness these additional causes need to be considered in examining the data presented here.

Removal bias hypothesizes that growth occurs in part because the size of the region removed during crossover is much more significant in determining the offspring fitness than is the size of the added region (Soule and Foster, 1998). In particular, offspring created by removing a small region and adding a larger region (such offspring will be larger than their parents) are much more likely to maintain their fitness than offspring created by removing a large region and adding a small region (such offspring will be smaller than their parents). Thus, in the selection step the larger than average individuals are more likely to survive. Experimental data has confirmed that removal bias occurs and is a significant factor in GP growth (Soule and Heckendorn, 2002; Luke, 2000c). Theoretical analysis suggests that removal bias will be most significant when inviable and 'near' inviable regions (regions that have no effect or a negligible effect on fitness when modified) are common.

The drift theory of code growth is based on the structure of the search spaces. It has been experimentally observed that for many problems the number of solutions of a given fitness that are larger than a given size is much greater than the number of solutions with the same fitness that are smaller than the given size (Langdon *et al.*, 1999; Langdon, 1999). E.g. given a solution of fitness X and size Y there are many more solutions with fitness X that are larger than Y than that are smaller than Y.

Because larger programs are more common it has been proposed that an unbiased search is more likely to find larger solutions of a given fitness than it is to find smaller solutions of that fitness, simply because there are more larger solutions within the search space to be found. The general idea is that "... any stochastic search technique, such as GP, will tend to find the most common programs in the search space of the current best fitness" (Langdon, 1999) and for any given fitness, larger programs are more common.

It's worth noting that these three views of growth (protection, removal bias or drift) are not mutually exclusive. Nor would proving that any one of them is correct invalidate the general notion of selective pressure for robust solutions. In fact, as noted above, the protective view of growth can be taken as a special case of the general pressure for robustness.

2. Experiment

A very simple problem is used to examine the importance of code and operator types on a variable length, linear evolutionary algorithm. The goal is to find a set of integers that sum to a given target value T. The allowed integers are 0, 1 and 4. Individuals are variable length strings consisting of those three integers.

The fitness of an individual is the absolute value of the difference between the sum of the integers and the target value, i.e. fitness = |sum - T|. For the individual:

$$00141001104 \tag{16.1}$$

the sum is

$$0+0+1+4+1+0+0+1+1+0+4=12 \tag{16.2}$$

and the fitness is

$$|12-T|. \tag{16.3}$$

Clearly, a lower fitness is better.

Note that for these experiments there are effectively three types of "genes." Genes with a large effect on fitness: 4's, genes with a smaller effect on fitness: 1's and genes with no effect on fitness (inoperative genes): 0's. This problem is similar to those used in previous experiments, although a linear rather than a tree-based representation is used here (Soule, 2002; Soule and Heckendorn, 2002).

The evolutionary algorithm is generational, with a population size of 500, elitism of 2 and is run for 1000 generations. The genetic operators, crossover and mutation, are described below. Other details of the evolutionary algorithm are shown in Table 16-1.

Crossover

Two-point crossover is used. Because the size of the individuals are variable the two crossover points are chosen independently in each parent. Thus, in general the sizes of the exchanged regions will be different. E.g.

Parents		Offspring
00|14|1001104	$\rightarrow$	00|0041|1001104
0140|0041|04	$\rightarrow$	0140|14|04

Two methods of picking the crossover points are used. In the first method both crossover points in an individual are chosen randomly. This is referred to as *proportional crossover* because the length of the exchanged region is proportional to the length of the parent. In general, for two randomly chosen crossover points 33% of each parent will be selected for crossover. This method is analogous to typical GA two-point crossover.

In second method the length l of the crossed region is chosen according to the following algorithm:

```
l = 2
While(l < L/2 AND random real < 0.5)
        l = l * 2
```

where L is the length of the parent individual. Thus, the size of the region selected for crossover is 2 "genes" long 50% of the time, 4 genes long 25% of the time, 8 genes long 12.5% of the time, etc. Thus, crossing short regions is very common and crossing longer regions happens infrequently. Once the length of the crossed region is generated, the left-hand crossover point is randomly selected and the right-hand point is l beyond the left-hand point. This form of crossover is referred to as *constant crossover* because the distribution of lengths of the crossed region is constant, regardless of the parents' sizes.

Constant crossover is analogous to crossover in tree-based GP. In standard tree-based GP crossover a random point is chosen for crossover. For full binary trees this results in an average crossover branch consisting of two nodes regardless of the tree size (Rosca and Ballard, 1996a; Soule and Foster, 1997). Larger branches are exponentially less likely to be chosen for crossover. In practice, GP usually leads to randomly shaped trees rather than full trees (Poli and Langdon, 1997; Langdon, 1999). However, the distribution of crossover points still heavily favors small branches (Soule and Heckendorn, 2002) and using the 90/10 rule (choosing leaf nodes for crossover only 10% of the time) only slightly shifts the distribution towards larger branches (Soule and Heckendorn, 2002). Thus, the distribution of crossover sizes with constant crossover is comparable to those seen in tree-based GP; both emphasize exchanging small branches.

Mutation

For a typical, fixed-length GA the mutation rate is proportional to the inverse of the length of the individuals. However, with a variable-length representation the inverse of the length is not the same for all members of the population.

Table 16-1. Summary of the evolutionary algorithm parameters.

Objective	Find integers whose sum is T
Integer values	0, 1, 4
Population Size	500
Crossover Probability	0.9
Mutation Probability	0, 0.0001, 0.001, 0.005 or 0.01
Selection	3 member tournament
Run Time	1000 Generations
Maximum Size	None
Elitism	2 copies of the best individual are preserved
Initial Population	Random individuals of length 5 to 59
Number of trials	200
Crossover	Proportional and Constant (see text)

Thus, several fixed mutation rates are chosen (0, 0.0001, 0.001, 0.005, 0.01) and compared.

3. Results

Three separate sets of results are presented in this paper. First, the effect of the crossover operator on size is examined. It is shown that the different types of crossover have distinctly different effects on size. Next, the evolved number of gene's (0's, 1's and 4's) is examined. These results show that the pressure for robustness favors 1's over 4's. Finally, the effect of incorporating mutation is examined.

Growth

We begin by examining growth, as growth is the best understood strategy for increasing robustness. Figure 16-1 shows the average size (length) of individuals evolved using proportional and constant crossover with no mutation. Growth is quite pronounced with constant crossover and non-existent with proportional crossover.

The explanation for these very different results seems fairly straightforward. With proportional crossover the two crossover points are chosen randomly. Thus, as noted previously, the exchanged regions of the chromosome are proportional to the overall lengths of the chromosomes. The protective effect of growth observed with tree-based GP doesn't apply because adding inoperative code (0's) simply causes a larger region of the chromosome to be exchanged

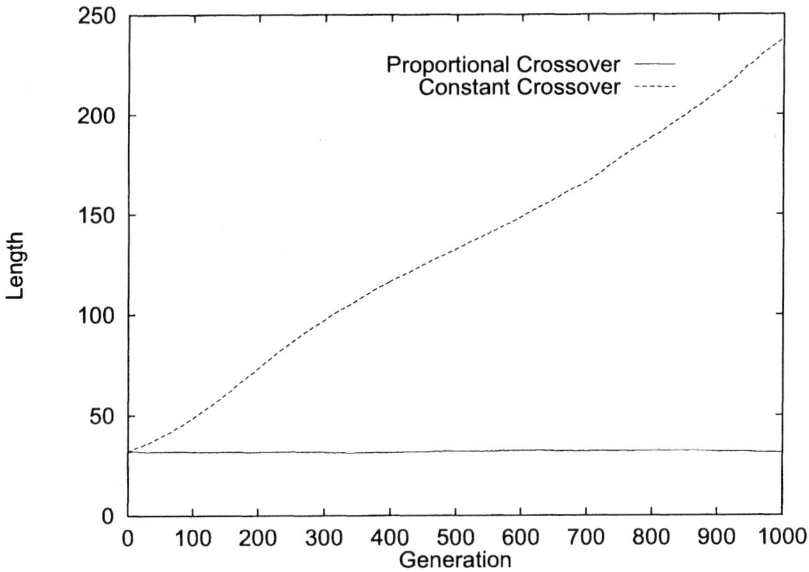

Figure 16-1. Average size (length) for individuals evolved using proportional and constant crossover.

during crossover. Thus, the probability of effecting operative code (1's and 4's) doesn't change when inoperative code (0's) is added.

In contrast, with constant crossover the expected size of the exchanged regions is constant regardless of the chromosome's overall length. Thus, in this case additional inoperative code does decrease the probability of effecting operative code. This is exactly what is observed in GP, where the average size of the exchanged branches during crossover does not grow proportionally with the overall tree size and rapid growth is observed.

These results explain why growth has not been a significant problem in most versions of evolutionary computation using variable length, linear representations. Typical variable length, linear representations use the proportional version of crossover, where the crossover points are picked randomly with a uniform distribution. As these results demonstrate, this method of picking crossover points does not encourage growth.

It is worth considering the other two proposed causes of growth: removal bias and drift. Removal bias is not expected to cause growth in either of the trials described above. The primary assumption of the removal bias hypothesis is that growth occurs because of a bias regarding the size of the removed region that does not apply to the added region during crossover. Normally this bias occurs because there is a probability that the added region will fall within an inviable

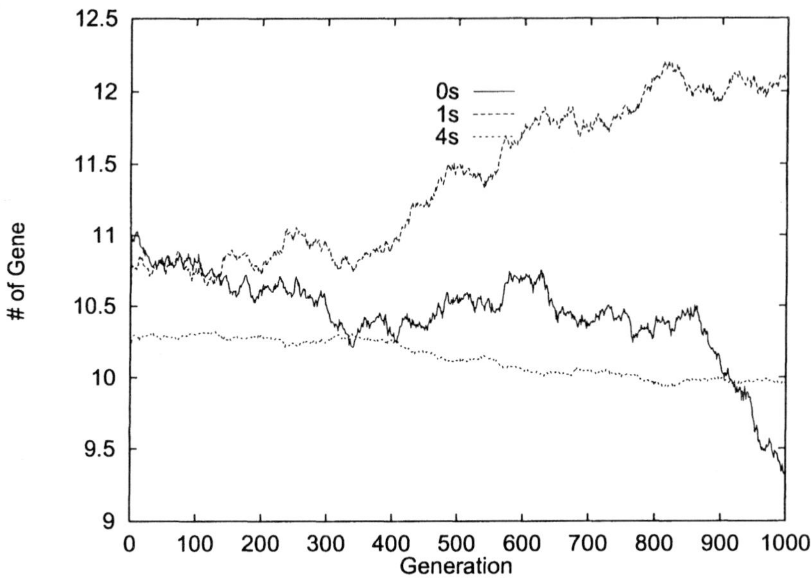

Figure 16-2. Number of 'genes' (0's, 1's or 4's) of each type when using proportional crossover. The number of 1's increases slowly, with a corresponding decrease in 4's.

region of the parent it is being added to. (For example, the added section may be within an "if(false)" branch or may be multiplied by zero.) However, with the current encoding there are no inviable regions; every added 1 or 4 has an effect. Thus, adding a large region is more likely to add 1's and 4's and thus effect fitness just as removing a large region is more likely to effect fitness by removing 1's or 4's. With no bias between removed and added regions removal bias does not apply to these results.

In contrast, drift should occur with either type of crossover. The primary assumption of the drift hypothesis is that there are more larger solutions of a given fitness than there are smaller solutions of the same fitness. Thus, the search naturally stumbles upon the larger solutions and growth occurs. For this experiment this assumption certainly holds. (Consider a solution such as 010401, an infinite number of larger solutions of the same fitness can be made by adding 0's and, to a lesser extent, by replacing 4's with 1's. However, there are only a few smaller solutions with the same fitness.) The fact that growth does not occur with proportional crossover casts considerable doubt upon the drift hypothesis.

Gene Choice

Although proportional crossover does not lead to growth, the pressure for robust solutions still influences evolution. Given that the size of the crossed regions is proportional to the length of the chromosome a more robust solution can still be achieved by limiting the importance of the "genes" exchanged during crossover. Clearly 1's represent a less important gene than 4's as an extra or missing 1 has a much less significant effect on fitness than an extra or missing 4. Thus, we expect to see an increase in the number of 1's and a decrease in the number of 4's.

Figure 16-2 shows the numbers of 0's, 1's and 4's when proportional crossover is used. As predicted there is a (slight) increase in the number of 1's and a decrease in the number of 4's. Although this difference is small, on average there are two more 1's than 4's in the final generation, the difference in the final generation is significant at the 1% level (Student's 2-tailed, t-test, $t = 2.6$). This shows that there is pressure for more robust solutions with proportional crossover. The system is simply using a different (non-growth based) strategy to increase robustness.

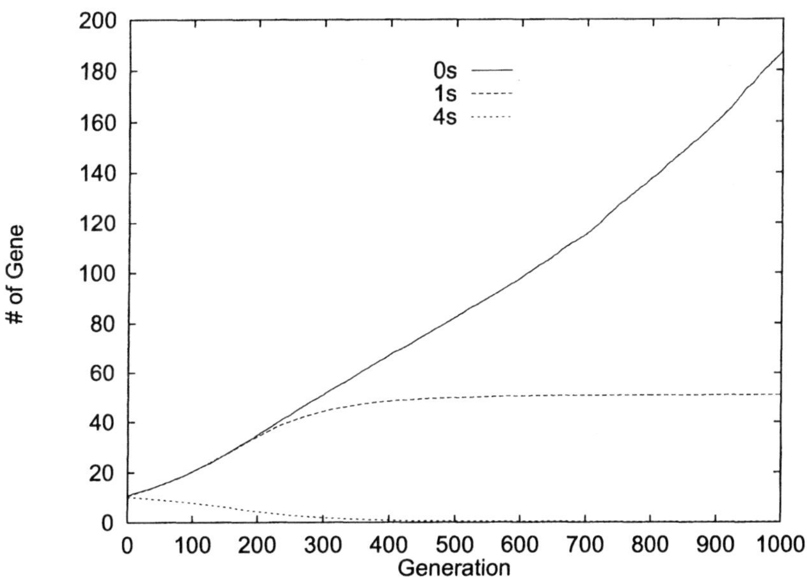

Figure 16-3. Number of "genes" (0's, 1's or 4's) of each type when using constant crossover. The number of 1's increases, with a corresponding decrease in 4's.

In fact the same robustness strategy is also adopted with constant crossover. Figure 16-3 shows the number of each type of genes with constant crossover; the results are much more pronounced than with proportional crossover. There is a pronounced shift away from 4's and toward 1's. In fact, by the final generation several of the trials showed no 4's at all. The reason for the shift from 4's to 1's is the same as with proportional crossover-individuals with relatively more 1's are more robust with respect to crossover. Possibly there is stronger pressure for the shift with constant crossover. Alternatively, constant crossover or the growth it creates may make it "easier" to exchange 1's for 4's; the pressure is the same in both cases, but the shift occurs more readily with the larger individuals.

As expected Figure 16-3 shows that the growth observed with constant crossover is primarily created by an increase in 0's. Thus, with constant crossover two evolutionary strategies are adopted to increase robustness, grow by adding 0's and increase the stability of the operative code by shifting from 4's to 1's.

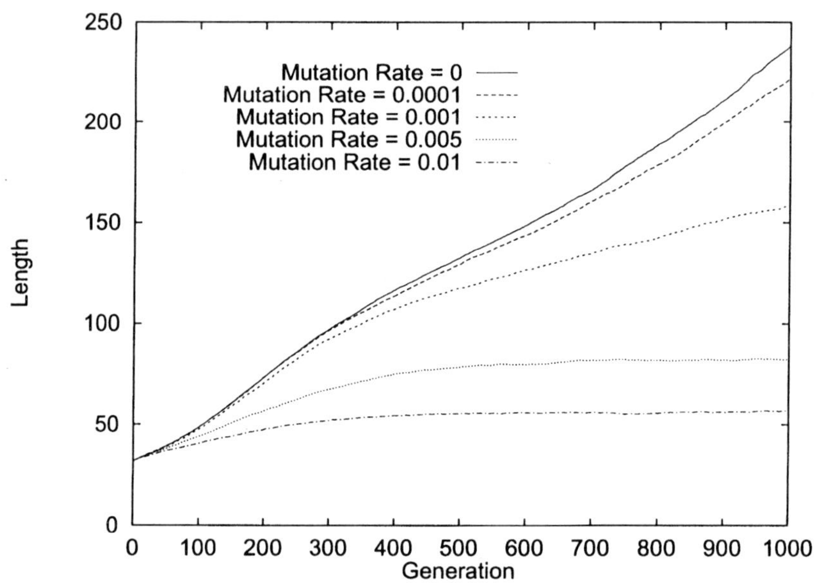

Figure 16-4. Average size (length) for individuals evolved using constant crossover with a variety of different mutation rates.

Mutation and Reduced Size

The previous experiments looked only at the effect of crossover. Next we examine the influence of mutation when constant crossover is used. Because the length of the individuals varies using a fixed mutation rate of 1/length is not possible. Instead several mutation rates are tried: 0.0001, 0.001, 0.005, 0.01.

Table 16-2. Average sizes and standard deviations in the final generation (1000) when using constant crossover with a variety of different mutation rates.

Mutation Rate	Average Size	Std. Dev.
0.0	238.0	126.5
0.0001	221.9	90.36
0.001	158.2	38.20
0.005	82.50	7.967
0.01	56.93	5.253

Each of these are the chance per "gene" of a mutation occurring. Figure 16-4 shows the results of these experiments on the growth rate of the evolving individuals. Table 16-2 shows the average sizes and standard deviations for the final generation. These differences are all significant at the 1% level.

Clearly the higher mutation rates limit growth. Again this can be explained in terms of robustness. Because the mutation rate is per gene (or per site), additional genes make the occurrence of a mutation event more likely. Because there are no inviable genes every mutation has an effect on fitness. Thus a short individual (one with fewer genes) is less likely to have a mutation event and is more robust with regards to mutation.

This increased robustness with regards to mutation for shorter individuals has to be balanced against the greater robustness for longer individuals with regards to crossover. Thus, as the mutation rate increases the ideal length for maximum robustness decreases. These results are particularly significant in that they appear to be the first time that pressure for robustness has been shown to lead to a smaller overall solution length.

4. Discussion and Conclusions

These results further support the idea the pressure for robustness is an important factor in the evolutionary process. The evolutionary process must balance this pressure for more robust solutions with the pressure for more fit solutions introduced by selection. Previous research has shown that robustness is often increased via the growth of inoperative and/or inviable code. The present results show that other evolutionary strategies are also adopted to increase robustness. In particular, we observed that increased robustness is achieved through gene choice, in this case replacing 4's with 1's. Robustness also increased by reducing the overall chromosome length (attenuation) in some cases.

Based on the results presented here it is possible to make predictions regarding the effects of different types and frequencies of operations with regards to "code" growth and gene selection. In general, changes that increase robustness (decrease the probability of fitness changes) will be favored.

These results show that increased growth will lead to increased robustness (and thus be evolutionarily favored) with respect to a given operator *if* that operator is applied with a per individual, and not a per site, probability. (Where "site" means nodes in a tree representation or loci in a linear representation.) I.e., when the operator is applied to individuals uniformly regardless of size. In these cases adding additional sites decreases the probability of any specific site being effected.

Examples of per individual, rather than per site operations (which encourage growth), include: GP crossover (one crossover per individual and the size of the crossed branches is not directly proportional to the number of nodes in the tree), mutation that is applied to a fixed number of sites per individual regardless of the individuals length, and the constant crossover operation described in this paper.

Examples of per site operations (which do not encourage growth) include: mutations with a per site probability (e.g. standard GA mutation) and the proportional two-point crossover described above (as individuals get longer-have more sites-the number of sites changed by crossover increases proportionally).

Even in cases where the operators are designed or chosen so that they will not encourage growth to increase robustness operators may have more subtle effects. E.g. these experiments demonstrated that pressure for robustness can effect "gene" choice.

The rules and examples presented in this paper can be used to predict the effect of novel operators on the evolutionary process and thus can improve our ability to design useful genetic operators. This is particularly beneficial to a practitioner who must design a novel representation and associated operators to solve a new problem. The rules described above can be used to design operators that guide evolution in a desired direction.

References

Blickle, T. and Thiele, L. (1994). Genetic programming and redundancy. In *Genetic Algorithms within the Framework of Evolutionary Computation (Workshop at KI-94, Saarbrücken)*, Hopf, J. (Ed.), pp. 33–38, Im Stadtwald, Building 44, D-66123 Saarbrücken, Germany. Max-Planck-Institut für Informatik (MPI-I-94-241).

Igel C. and Chellapilla K. (1999). Investigating the Influence of Depth and Degree of Genotypic Change on Fitness. In *Proceedings of the Genetic and Evolutionary Computation Conference (GECCO 1999)*, pp 1061-1068, Orlando, Florida. Morgan Kaufmann.

Koza, John (1992). *Genetic Programming: On the Programming of Computers by Natural Selection*. MIT Press, Cambridge, MA, USA.

Langdon, W. B. (1997). Fitness causes bloat: Simulated annealing, hill climbing and populations. Technical Report CSRP-97-22, The University of Birmingham, Birmingham, UK.

Langdon, W. B. (1999). Size fair and homologous tree genetic programming crossovers. In *Proceedings of the Genetic and Evolutionary Computation Conference 1999*, Banzhaf, W. *et al.* (Eds.), pp. 1092–1097. Morgan Kaufmann.

Langdon, W. B., Soule, T., Poli, R. and Foster, J. A. (1999). The evolution of size and shape. In *Advances in Genetic Programming III*, Spector, L. *et al.* (Eds.), pp. 163–190. Cambridge, MA: The MIT Press.

Luke, S. (2000c). Code growth is not caused by introns. In *Late Breaking Papers, Proceedings of the Genetic and Evolutionary Computation Conference (GECCO-2000)*, pp. 228–235.

McPhee, N. F. and Miller, J. D. (1995). Accurate replication in genetic programming. In *Proceedings of the Sixth International Conference on Genetic Algorithms*, Eshelman, L. J. (Ed.), pp. 303–309. Morgan Kaufmann.

Nordin, P. (1997). *Evolutionary Program Induction of Binary Machine Code and its Application*. Muenster: Krehl Verlag.

Nordin, P. and Banzhaf, W. (1995). Complexity compression and evolution. In *Proceedings of the Sixth International Conference on Genetic Algorithms*, Eshelman, L. J. (Ed.), pp. 310–317. San Francisco, CA: Morgan Kaufmann.

Nordin, P., Banzhaf, W. and Francone, F. D. (1997). Introns in nature and in simulated structure evolution. In *Proceedings Bio-Computing and Emergent Computation*, Lundh, D., Olsson, B., and Narayanan, A. (Eds.), pp. 22–35. Springer.

Poli, R. and Langdon, W. B. (1997). A new schema theory for genetic programming with one-point crossover and mutation. In *Genetic Programming 1997: Proceedings of the Second Annual Conference*, J Koza *et al.* (Eds.), pp 278–285. San Francisco, CA: Morgan Kaufmann.

Rosca, J. P. and Ballard, D. H. (1996a). Complexity drift in evolutionary computation with tree representations. Technical Report NRL5, University of Rochester, Computer Science Department, Rochester, NY, USA.

Smith, P. and Harries, K. (1998). Code growth, explicitly defined introns, and alternative selection schemes. *Evolutionary Computation 6(4):* 339–360.

Soule, T. (1998). *Code Growth in Genetic Programming*. PhD thesis, University of Idaho, University of Idaho.

Soule, T. (2002). Exons and code growth in genetic programming. In *Genetic Programming, 5th European Conference, EuroGP 2002*, J. Foster *et al.* (Eds.), pp. 142–151.

Soule, T. and Foster, J. A. (1997). Code size and depth flows in genetic programming. In *Genetic Programming 1997: Proceedings of the Second Annual Conference*, J. Koza *et al.* (Eds.), pp. 313–320. San Francisco, CA: Morgan Kaufmann.

Soule T. and Foster J. A. (1998). Removal Bias: a New Cause of Code Growth in Tree Based Evolutionary Programming. In *IEEE International Conference on Evolutionary Computation*, pp. 178-186, Anchorage, Alaska, IEEE Press.

Soule, T. and Heckendorn, R. B. (2002). An analysis of the causes of code growth in genetic programming. *Genetic Programming and Evolvable Machines 3(3):* 283–309.

Soule, T., Foster, J. A. and Dickinson, J. (1996). Code growth in genetic programming. In *Genetic Programming 1996: Proceedings of the First Annual Conference*, J. Koza *et al.* (Eds.), pp. 215–223. Cambridge, MA: MIT Press.

Soule, T., Heckendorn, R. and Shen, J. (2002). Solution stability in evolutonary computation. In *Proceedings of the 17th International Symposium on Computer and Information Systems*, I. Cicekli *et al.* (Eds.), pp. 237–241.

Streeter, M. J. (2002). The root causes of code growth in genetic programming. In *Genetic Programming, 6th European Conference, EuroGP 2003*, C. Ryan *et al.* (Eds.), pp. 443–454.

Chapter 17

A HYBRID GP-FUZZY APPROACH FOR RESEVOIR CHARACTERIZATION

With a Gentle Introduction to Oil Exploration and Production

Tina Yu, Dave Wilkinson and Deyi Xie

ChevronTexaco Information Technology Company & ChevronTexaco Exploration and Production Technology Company

6001 Bollinger Canyon Road, San Ramon CA 94583 U.S.A.

Abstract: A hybrid GP-fuzzy approach to model reservoir permeability is presented. This approach uses a two-step divide-and-conquer process for modelling. First, GP is applied to construct classifiers that identify permeability ranges. Within each range, ANFIS is employed to build a Takagi-Sugeno-Kang fuzzy inference system that gives permeability estimation. We applied this method to five well log data sets. The results show that this hybrid system gives more accurate permeability estimation than other previous works.

Key words: Oil Exploration and Production, Reservoir Characterization, Soft Computing, Permeability Estimation, Fuzzy Logic, Genetic Programming, Fuzzy Modelling.

1. Introduction

Where has all the oil gone? The image of oil shooting from the ground in some of the old western films gives us an illusion that if you drill it, it will come. However, this is quite far from reality. Not only is the oil reservoir difficult to find, sometimes the discovered reserve can "disappear" to the neighbouring properties. Oil hunting is therefore an adventure that continues to attract many people from different disciplines.

Oil exploration and production is a business of oil hunting. Currently, less than one third of proven reserve is recovered by oil and gas companies. How to reduce the amount of oil left behind (bypassed oil) is of great importance to the world's energy supply. One key step to increasing oil recovery is using data integration (such as seismic and well logs) intelligently to provide accurate reservoir characterization and modelling. With such information, better reservoir management decisions can be made to optimize oil production.

We first give a gentle introduction to the process of oil exploration and production (Section 2). The data integration step for reservoir characterization is then explained (Section 3). After that, we present a hybrid GP-fuzzy approach that uses well log and core data to model reservoir permeability (Section 4 & 5). The experimental results are then presented and analyzed (Section 6 & 7). Finally, we conclude the paper and give avenues of future work (Section 8).

2. Oil Exploration and Production

Oil exploration and production is the process of discovering and developing new petroleum or natural gas reservoirs. Typically, the process starts with a thorough analysis of the geology in a potential region, particularly the probability of finding hydrocarbons (oil or gas), and the economic factors such as risk and investment needed. If it is decided to explore for resources in the region, a seismic exploration survey is conducted. In a survey, a controlled sound source is used to set off sound waves. These waves penetrate the earth, propagate, reflect, refract and then reach back to the surface of the earth where they are recorded by geophones or hydrophones. Figure 17.1 gives the schematic of the seismic method. Typically, the sound wave is generated by a vibrator truck. A computer database system is connected to the geophones lying on the ground to record the data.

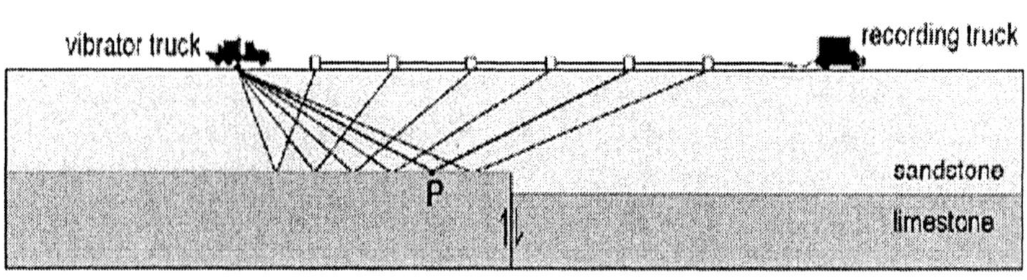

Figure 17.1. Schematic of the seismic method.

Seismic data require extensive processing to obtain an image of the subsurface, which is then carefully examined for potential accumulation points of hydrocarbons. Frequently, bright amplitudes or structural high curve shapes are indications of hydrocarbon accumulation points. For example, there are two bright spots in Figure 17.2 (around 10,000 feet underground). One is indicated with pressure anomaly and the other is next to where the sonic log is.

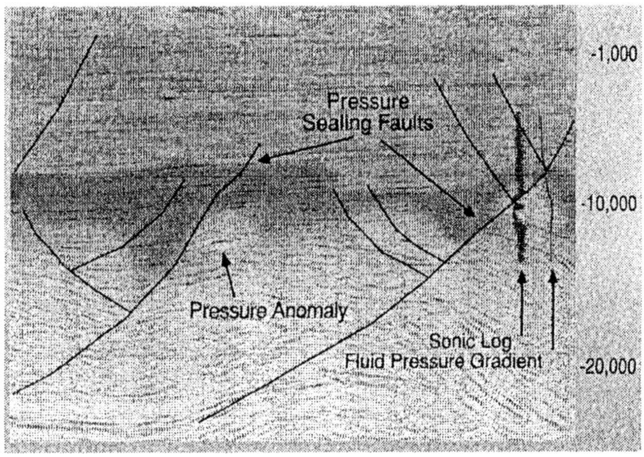

Figure 17.2. A seismic image with two bright spots.

Seismic bright spots are not sufficient evidence of the existence of hydrocarbons. To prove the presence of hydrocarbons, a well needs to be drilled. Current statistics shows that only one in three wells drilled encounters hydrocarbons; dry wells are not uncommon experiences in the oil exploration process.

Besides the presence of hydrocarbons, there are other criteria that are essential for a profitable oil production:

- First, the reservoir needs to be porous enough to contain large amounts of fluid.
- Second, these pores need to be mostly filled with hydrocarbons (instead of other substances, such as water).
- Lastly, these pores need to be connected with each other so that hydrocarbons can be extracted easily.

These three criteria are known as *porosity*, *hydrocarbon saturation*, and *permeability*. If these properties are uniform throughout the reservoir, measurements made in one exploratory well would be sufficient to determine the economic feasibility of the entire reservoir. Unfortunately, most reservoirs are heterogeneous. As a result, before deciding if an oil field is worthwhile, we need to estimate these properties for the whole area. The process of determining these spatially varying geological properties is known as reservoir characterization.

3. Reservoir Characterization

Reservoir characterization is a critical step in reservoir development and future production management. Knowing the details of a reservoir allows the simulation of different scenarios. The problem, however, is to define an accurate and suitable reservoir model including small-scale heterogeneity. Currently, the most abundant data about the reservoir, which is the seismic data, do not have enough resolution. The typical resolution of seismic data is on the order of 100 feet or more, which does not give enough detail of reservoir properties.

In contrast, well log data, which are collected by inserting sensing devices into an exploratory well, give an excellent description of the well at scales ranging from centimetres to hundreds of meters. However, due to its high cost, only a few well locations have log data. This scarce information is usually not sufficient to build a reservoir model that includes the small scale variations. In between wells, geological information needs to be estimated (see Figure 17.3).

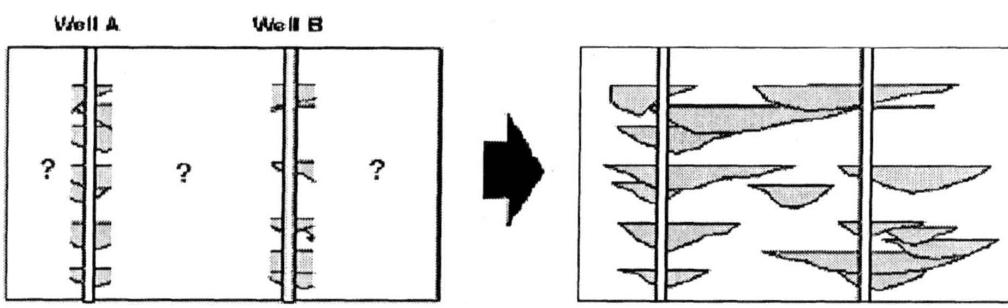

Figure 17.3: Geological information between wells needs to be filled to build a reservoir model.

A viable approach to construct a reservoir model is using both seismic and well log data. This approach first builds a framework using seismic scale and well log scale data, then fills in small details using statistical or soft computing methods. These methods are generally stochastic in nature, hence they generate many different realizations (reservoir models). This trait is actually an advantage, since the uncertainties of the different scenarios can be studied by generating a large number of equally possible reservoir models. Consequently, this approach of applying stochastic methods on seismic and well log data is widely used in reservoir characterization (Huang, Wong and Gedeon, 2000; Xie, 2001; Wong, Aminzadeh and Nikravesh, 2002; Nikravesh, Aminzadeh, and Zadeh, 2003). Currently, we are working on a research project that applies different stochastic methods to integrate seismic and well log data for reservoir characterization. This paper presents the part of the project that applies soft computing on well log data to model one of the reservoir properties: permeability.

3.1 Permeability Modelling

Permeability is a measure of fluid conductance in porous media. It is the most difficult reservoir property to estimate but of great importance to reservoir management decisions, such as drilling location. Two types of data were provided for this work: well log data and core permeability data. Well log data contain geological information, such as porosity and density. However, current logging technology is still unable to measure the permeability of a well. In order to obtain permeability information, rock samples (called cores) need to be removed from the exploratory wells and examined with various laboratory experiments. The permeability of the core is then calculated using Darcy's law (Scheidegger, 1974). The experimental method is described in (Yu and Lee, 2002).

It is commonly observed that permeability is related to other geological properties, although a theoretical proven mathematical equation to describe such a relationship does not exist. The objective of this work is to build a model that approximates such a relationship. This model can then be used to estimate reservoir permeability where well log data are available but core permeability data are not. In the future, it will also be used to estimate reservoir permeability where only seismic data are available. In this case, required well log properties are estimated from seismic data.

We carry out this modelling task using Genetic Programming (GP) (Koza, 1992) and Adaptive-Network-based Fuzzy Inference Systems (ANFIS) (Jang, 1993; Jang and Sun, 1997). We first describe the overall system (model) structure in Section 4. The modelling process is explained in Section 5.

4. The Hybrid GP-Fuzzy System

In general, developing a generic system that gives good permeability estimation for all types of reservoir is a difficult task. This is because permeability is highly dependant upon the rock formation (lithology) of a well. Moreover, different permeability ranges demonstrate different geological characteristics. We therefore adopt a divide-and-conquer approach to build the system:

- The first layer of the system identifies lithology group.
- The second layer of the system predicts permeability range.
- The last layer of the system estimates the permeability value.

Figure 17.4 shows the high-level structure of the 3-layer permeability estimation system.

The lithology analyzer identifies which of the five possible groups (sand, shaly sand, sandy shale, shale and high impedance sand) the rock belongs to. Within each group, a classifier predicts the range of permeability (high, medium or low). Depending on the range, a different fuzzy system is used to estimate the permeability value. In summary, given well log data inputs, the system analyzes the lithology, classifies the permeability range and then estimates the permeability value.

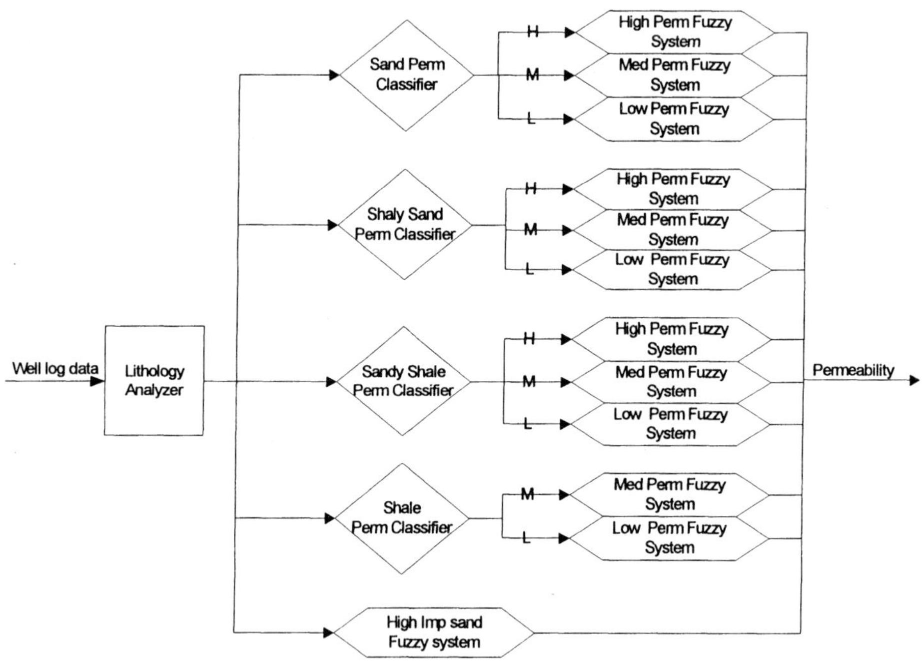

Figure 17.4.: The hybrid system high-level structure.

5. Modelling Process

We use four types of well log data to build the system. They are *porosity*, *density*, *velocity* and *gamma ray*. We do not use log data from some of the contemporary logging technologies, such as Nuclear Magnetic Resonance (NMR), because they are not always available across all well locations. In order to maximize future usage of the system, only common types of well data are used. Another advantage of this approach is that the established methodology can be applied to build models for other oil fields.

5.1 Data sets

We have log data from 5 wells; all of them have matching core permeability data. Table 17.1 gives the number of data samples in each lithology group and permeability range. These groups and ranges are established based on geological knowledge and the analysis of the data sets. The details will be explained in Section 5.2 and 5.3. The total number of data samples is 827.

Table 17.1. The number of data in each lithology groups and permeability ranges.

	Sand	Shaly Sand	Sandy Shale	Shale	High Impedance Sand
High Perm	148	81	83	-	43
Medium Perm	29	23	76	20	-
Low Perm	11	27	139	147	-
Total	188	131	298	167	43

5.2 Lithology Analyzer

One way to define lithology group is based on the percentage of shale in the rock. This information, called V-shale, can be computed using well log *gamma ray* data. Upon plotting V-shale vs. permeability (Figure 17.5), it is apparent that these two variables have an almost logarithmically linear relationship: the higher the V-shale value, the lower the permeability. The critical decision that needs to be made is how to partition these data into different lithology groups so that each group has similar geological properties in order to make it easier to model the data. An analysis of the s-curve suggests four lithology groups: sand, shaly sand, sandy shale and shale. The V-shale cut points for the four groups are 0.15, 0.4 and 0.75.

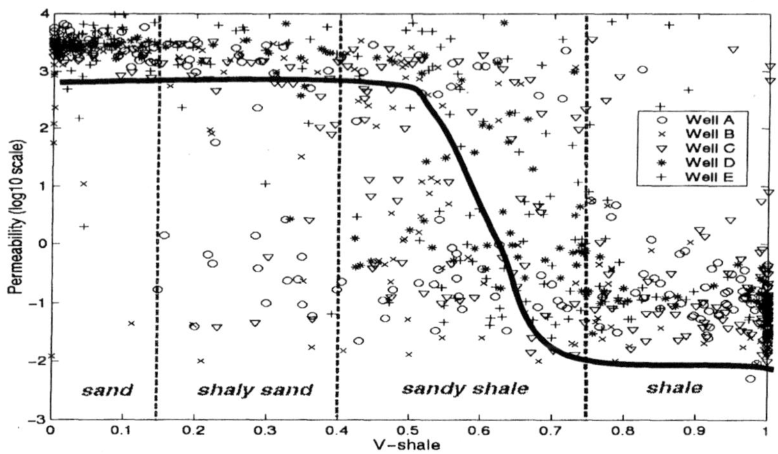

Figure 17.5: V-shale vs. permeability (log10 scale).

Among them, sandy shale data do not show any clear relationship pattern. The permeability in sandy shale is therefore the most difficult to estimate. It will be shown in Section 6 that this is indeed a challenge to our hybrid system.

High impedance sand, which has a *velocity* greater than 13,000, normally gives high permeability, regardless of the shale content. It is therefore made into a lithology group by itself. Moreover, it requires no classifier as it only has one permeability range: high (see Figure 17.4).

5.3 GP Classifiers

In each lithology group, we used GP symbolic regression to train the classifiers that predict permeability ranges (high, medium or low). The cut points of high, medium and low permeability are different in different lithology groups. For example, in sand, permeability greater than 2.5 is high and less than 1.0 is low. Anything in between is medium. Table 17.2 gives the permeability cut points for the four lithology groups. Note that shale has only one cut point since it has only two ranges: medium and low. Also, the values given are in log10 scale. This is a common practice in handling geological data.

Table 17.2. Cut points of high, medium and low permeability in 4 lithology groups.

Lithology Group	High vs. Med	Med vs. Low
Sand	3.0	2.0
Shaly Sand	2.5	1.0
Sandy Shale	2.5	1.0

Lithology Group	High vs. Med	Med vs. Low
Shale		1.0

The GP software we use, *Discipulus*(RML Technologies, 1998-2003), is only able to train binary classifiers. For lithology groups that have three possible classes, two classifiers are required. Figure 6 shows the operational flow using two classifiers to identify three permeability ranges.

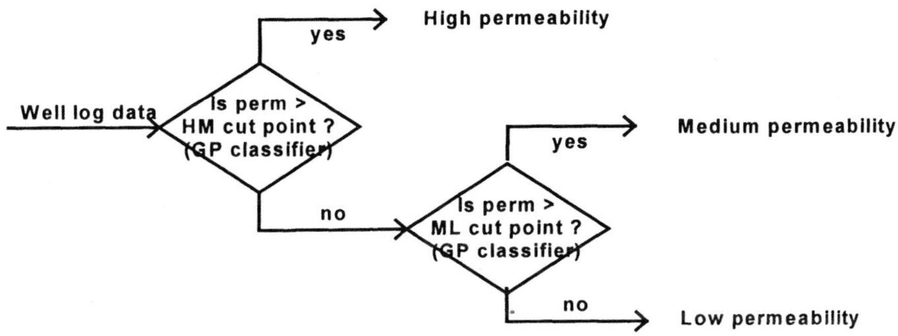

Figure 17.6: Using two classifiers to identify three permeability ranges.

Table 17.3 lists some of the GP parameters used to train the classifiers. The fitness is based on hit rate: the percentage of the training data that are correctly classified. During tournament selection, four candidates are randomly selected to compete for two winners. If two candidates are "tied" in their hit rates, the mean squared error measurement is used to select the winners. The "tied threshold" for mean squared error measurement is 0.01% in this work. If two classifiers are tied in both their hit rates and mean squared error measurements, one of them is selected as the winner randomly.

Table 17.3. Some of the GP parameters to evolve the classifiers.

Objective	Generate a classifier that distinguishes high vs. medium or medium vs. low permeability.
Functions	addition; subtraction; multiplication; division; abs; data transformation
Terminals	Porosity (ϕ), density (v), velocity ($1/\rho$) and constants.
Fitness	Hit rate then squared error fitness
Hit rate	The percentage of the training data that are correctly classified.
Selection	Tournament (4 candidates/2 winners)

There is another set of GP parameters that are not fixed but selected by the software for each run. These parameters include *population size*, *maximum program size*, *crossover* and *mutation rates*. In the first run, one set of values for these parameters were specified. When the run does not produce improved solution for a number of generations, this run is terminated and a new run starts with a new set of parameter values. The system records the best solution found throughout the runs. The best classifier found in the runs is selected as the final winner. In this work, each classifier is set to continue for 100 to 120 runs depending on the solution quality, i.e. more runs are made if the solution found so far is not satisfactory.

The software also assembles the best classifiers into teams. A team may consist of any odd number of classifiers. The whole team decision is the majority vote of the team members. Since the number of classifiers in a team is odd, there is always a winner (i.e. no tie). The following is an example of a team:

$f_1(\phi, v, \rho) = \ldots \rho * v - \phi^2 / \rho \ldots$
$f_2(\phi, v, \rho) = \ldots \rho^v - \phi^2 / \rho + v + 1.2 \ldots$
$f_3(\phi, v, \rho) = \ldots v * v - \phi^2 / \rho - (\phi + 2.3) \ldots$
cut point = 2.5;
Majority ((f_1 >*cut point*), (f_2 > *cut point*), (f_3 > *cut point*))

5.4 ANFIS Fuzzy Inference System

ANFIS is a fuzzy modelling tool that generates Takagi-Sugeno-Kang (TSK) type fuzzy inference systems (Sugeno, 1985; Sugeno and Kang, 1986; Sugeno and Tanaka, 1986) based on the given input and output data. A TSK fuzzy system has the following structure:

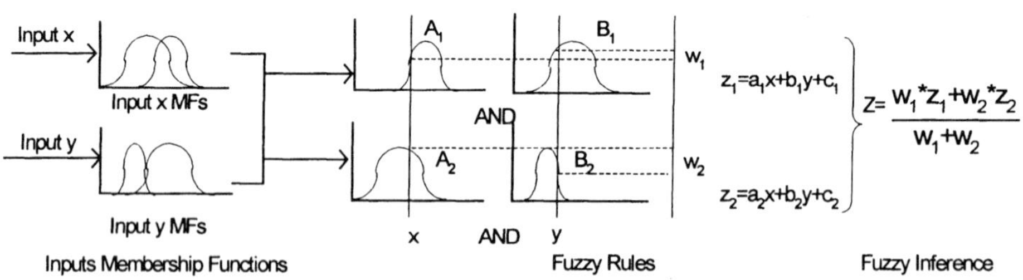

Figure 17.7: TSK fuzzy inference system structure.

The first component is a set of input membership functions (MFs). Membership functions map crisp inputs to linguistic values or labels. A membership function can have any shape, such as triangular, Gaussian and trapezoidal, as long as it varies between 0 and 1. In Figure 17.7, each curve defines a membership function. The transformation of a crisp input into degree (between 0 and 1) of match with one or more linguistic values is called "fuzzification".

Fuzzy rules are conditional statements in if-then format. The if-part consists of linguistic values and fuzzy operators (AND, OR, NOT). The then-part is a first order linear equation. The following is a TSK fuzzy rule:

*If porosity is high and density is low, permeability = -42572*porosity-53/velocity-115807*density + 260911.*

Fuzzy inference is a method to interpret the input values and, based on the fuzzy rules, assigns output values. In a TSK system, the output value is calculated based on the firing strength w_i of each rule. The formula is given in Figure 17.7.

TSK is an abstract system model and can be implemented in many different ways. One implementation is ANFIS, which use a feed-forward network architecture to implement the TSK system. Figure 17.8 is the equivalent ANFIS representation of the TSK system in Figure 17.7. Layer 1 corresponds to the input membership functions while layer 2 corresponds to the fuzzy rules. Layer 3, 4 and 5 correspond to the calculation of the final output z.

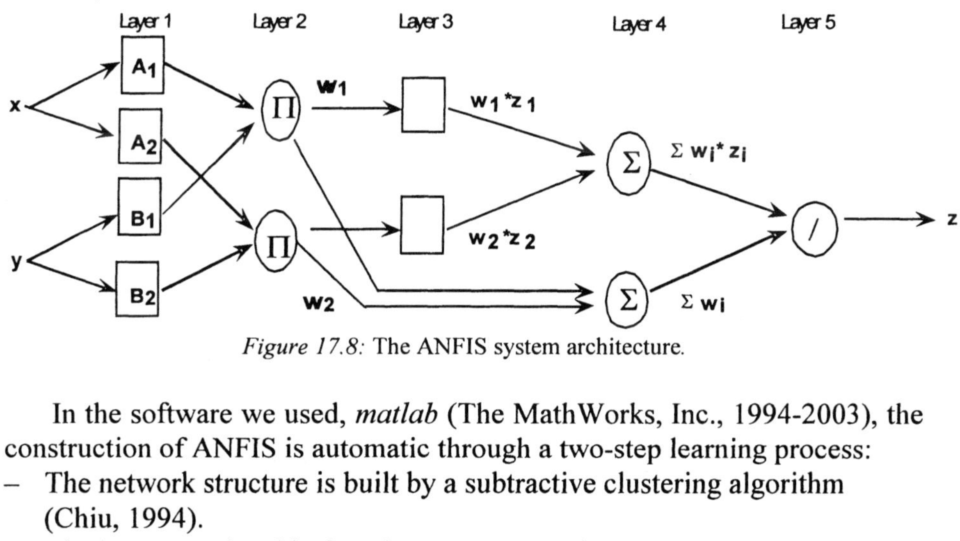

Figure 17.8: The ANFIS system architecture.

In the software we used, *matlab* (The MathWorks, Inc., 1994-2003), the construction of ANFIS is automatic through a two-step learning process:

- The network structure is built by a subtractive clustering algorithm (Chiu, 1994).
- The input membership function parameters (layer 1) are tuned by a back propagation algorithm while the output parameters (layer 3) are tuned by a least squares estimation.

The first learning step determines the number of fuzzy rules (layer 2). The clustering algorithm first partitions the data into groups and then generates a minimum number of rules to distinguish the fuzzy qualities associated with each of the groups. The shapes of the input membership functions (layer 1) are not learned but specified by users. Depending on the shapes of the input membership function selected, different parameters (e.g.

mean and standard deviation for Gaussian) are initialized. Meanwhile, the output parameters (e.g. the coefficients in the linear equations) are initialized.

The second learning step adjusts input and output parameters to minimize the error. More specifically, in the forward pass, training inputs go forward till layer 4 and the output parameters are identified by the least squared estimate. In the backward pass, the error rates propagate back and the input membership function parameters are updated by gradient decent.

The first learning step is a one-pass process. In contrast, the second learning step is an iterative one. Users can specify the "error tolerance" as the stopping criterion. The learning process stops after the learning error has reached the specified error tolerance.

6. Results

The accuracy of the GP classifiers is given in Table 17.4. As expected, sand and shale have simple, uniform property and are easy to model. In contrast, sandy shale has the most complex geological formation (discussed in Section 5.2). As a result, the classifiers have the lowest accuracy rate.

Table 17.4. Classification accuracy of the GP classifiers.

	Sand	Shaly Sand	Sandy Shale	Shale
High Permeability	93.24%	93.83%	67.47%	-
Med Permeability	82.76%	78.95%	78.95%	85%
Low Permeability	100%	88.89%	80.58%	96.64%
Total	92.02	90.08%	76.51%	97.01%

The TSK fuzzy systems give permeability estimations that are well correlated with the core permeability for all lithology groups. The coefficient of variation values (R^2) are between 0.9 and 0.95.

We combined the GP classifiers with the TSK systems then ran the overall system on the data set. Figure 17.9 shows the results. Among the 5 lithology groups, sandy shale data have the worst match between estimated and core permeability. This is due to the misclassification of permeability range by the GP classifiers (Table 17.4). Once the permeability range is misclassified, it is very hard for the TSK system to give correct permeability estimation. This indicates that the accuracy of the classifiers is of vital importance to the overall system performance. We are currently investigating different techniques to improve these classifiers including different classifier ensemble methods and Receiver Operational Characteristics analysis for better cut point selection.

Figure 17.10 gives the permeability estimations of the current system. It is clear that the hybrid GP-fuzzy system outperforms the current system on all 5 wells' data. We also made detailed performance comparisons with previous systems using data from only one well (well C). Figure 17.11 shows that the hybrid system (Perm-FIS) gives permeability estimations that match the core permeability (Perm-Core) better than the two previous works (Perm-Log is the current system and Perm-Vsh is the system used when only shale data is available). For example, the Core-Perm ("+" in Figure 17.10) between depth 6360 and 6340 feet are with high values ($> 10^3$). The value decreases between 6340 and 6335 feet and raises again at 6330 feet. The hybrid system gives permeability estimations that match this trend of high-low-high pattern very well. In contrast, Perm-Vsh gives permeability estimations around $10^{2.3}$ across this depth range, matching badly to the Core-Perm trend. Another previous work, Perm-Log, gives a better match than Perm-Vsh does. However, its estimations match poorly between the depth of 6340 and 6325 feet where Core-Perm drops. We are very encouraged by the results and convinced that this hybrid modelling approach is well suited for reservoir characterization.

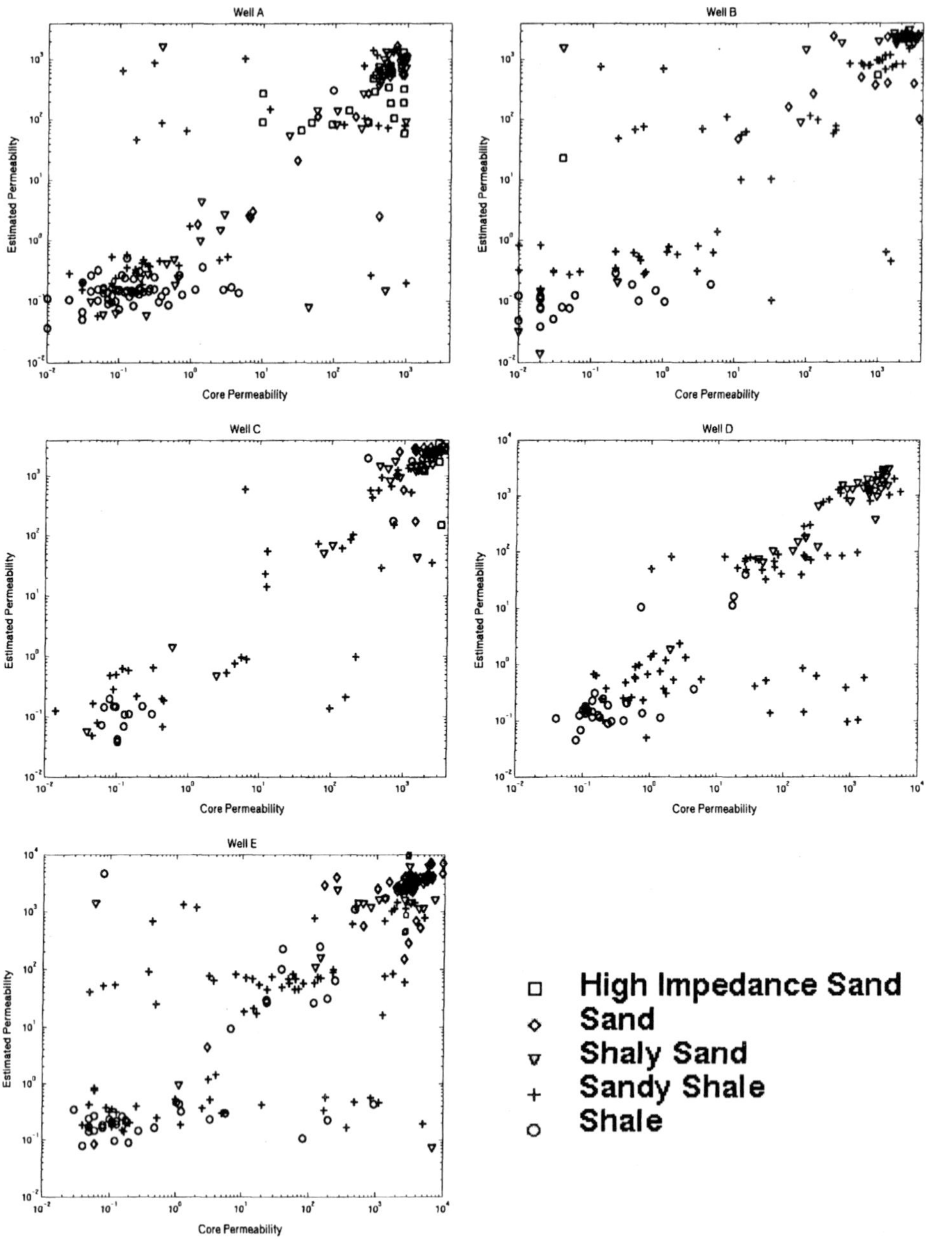

Figure 17.9: The overall hybrid system performance (estimations vs. core permeability).

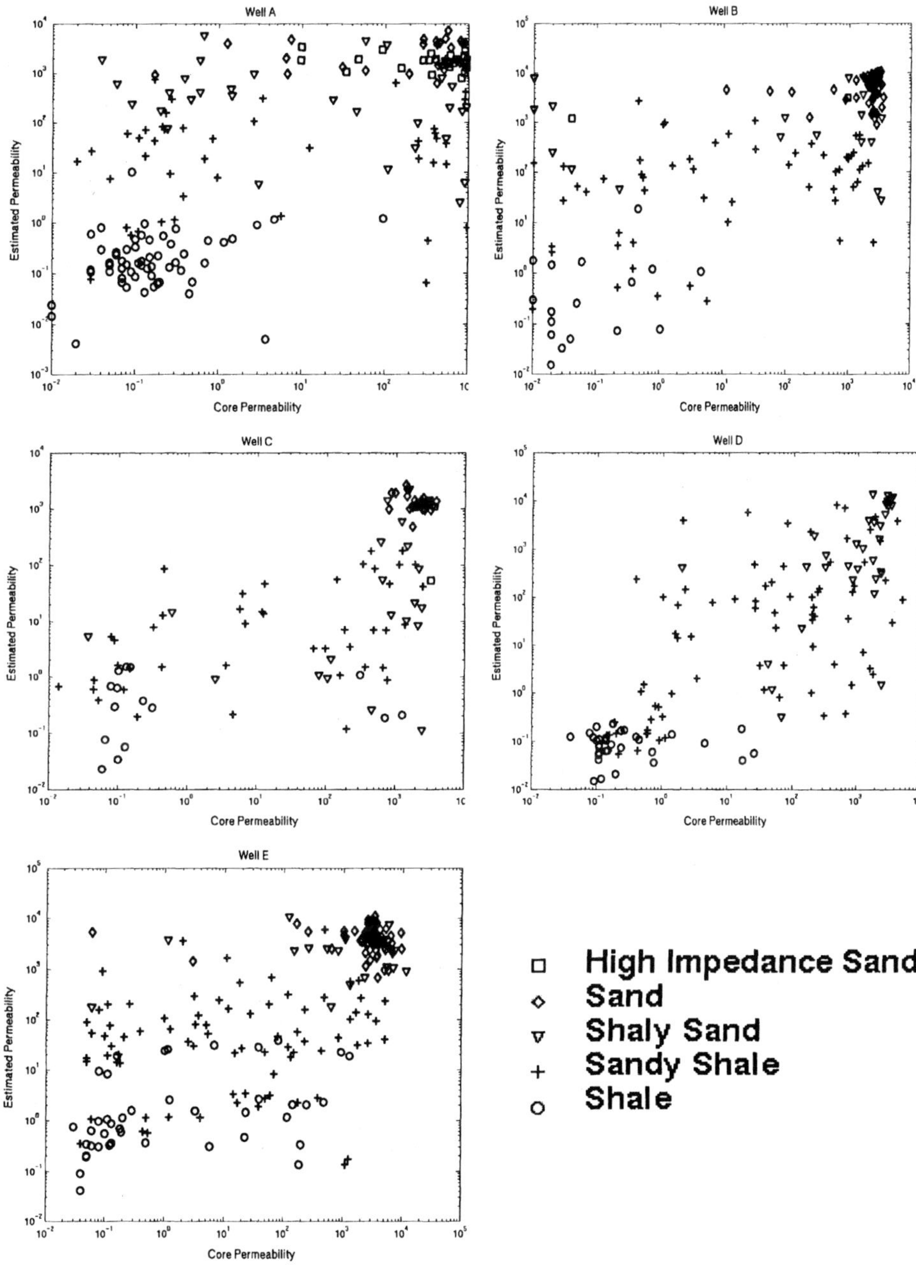

Figure 17.10: The current system performance (estimations vs. core permeability).

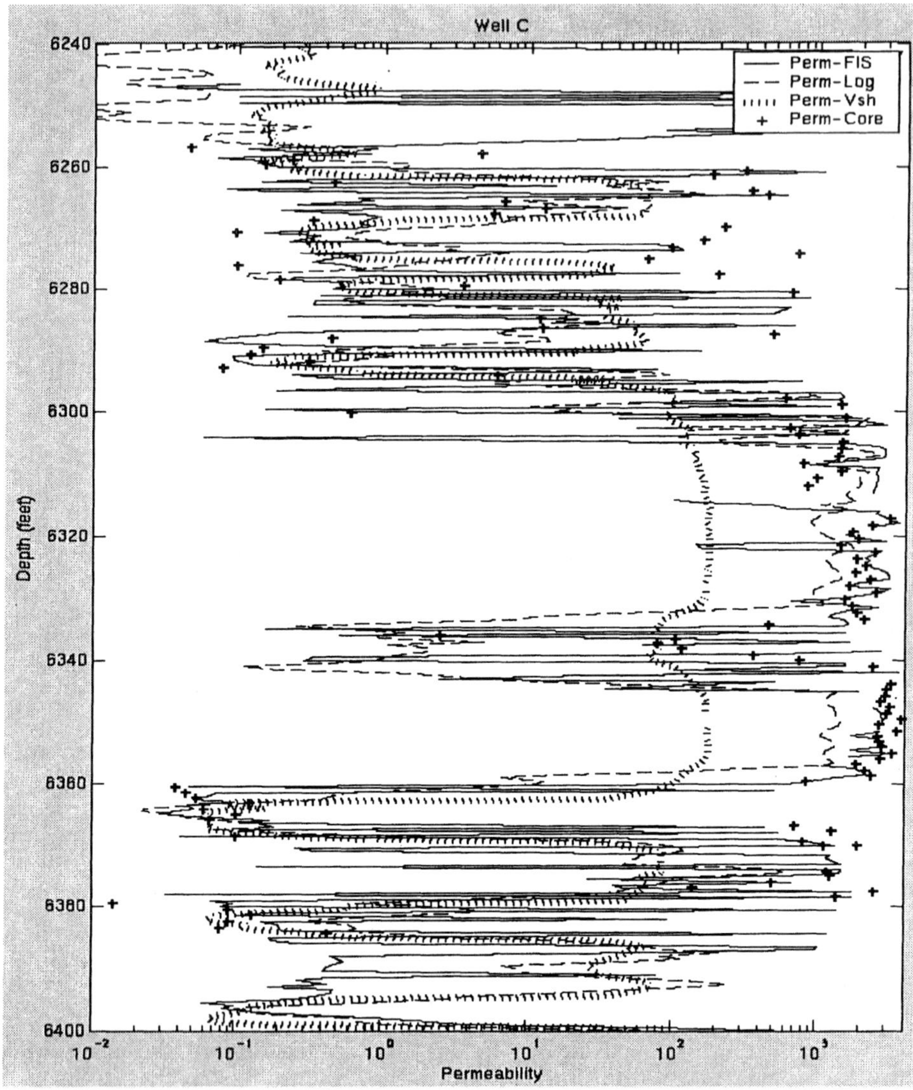

Figure 17. 11: Performance comparison of the hybrid system with two previous systems.

7. Discussions

Real world applications are results driven. Meanwhile, there is time pressure to deliver. Since the classification accuracy rates were mostly satisfactory, we made little effort in GP theoretical analysis. The only exception is the sandy shale classifiers because they have much lower accuracy than other lithology groups and the misclassification impairs the overall hybrid system performance.

One investigation we made was on the sensitivity of the mean squared error (MSE) tied threshold. Recall that the fitness measurement is based on hit rate rather than MSE (see Section 5.3). We have tested two MSE thresholds (1% and 0.1%) using cut point 1.0 to evolve the classifier that separates low from high/medium permeability. Table 17.5 and 17.6 give the results. Note that sandy shale classifiers have a different structure from that in Figure 17.6. It first separates low from high/medium, and then distinguishes high from medium. In this way, both classifiers have a balanced set of positive and negative samples for training (see Table 17.1).

Table 17.5. Confusion matrix for sandy shale classifier using MSE tied threshold 1%.

	Estimated Low	Estimated High/Medium	Total
Core Low	126 (90.65%)	13 (9.35%)	139
Core High/Medium	36 (22.64%)	123 (77.36%)	159
Total	162	136	298 (83.56%)

Table 17.6. Confusion matrix for sandy shale classifier using MSE tied threshold 0.1%.

	Estimated Low	Estimated High/Medium	Total
Core Low	116 (83.45%)	23 (16.55%)	139
Core High/Medium	21 (13.21%)	138 (86.79%)	159
Total	137	161	298 (85.23%)

With MSE tied threshold 1%, the evolved classifier identifies more data as low permeability (162 vs. 137) while the classifier built using threshold 0.1% identifies more data as high/medium permeability (161 vs. 136). This indicates that many evolved classifiers have the same hit rate and using the second fitness measurement (MSE) for selection can impact the quality of the classifiers. In this case, a smaller MSE threshold trains a better classifier.

What is a good sensitivity measure for an effective classifier? Generally, we prefer robust solutions, hence would prefer a medium range tie threshold. However, in this case, the accuracy of this classifier is very important to the overall system performance because it is the first classifier applied to the data. Once a data point is misclassified, the second classifier can not correct it and the ANFIS fuzzy system would not be able to give the correct permeability estimation. We therefore select the classifier that was trained using the smaller MSE threshold. However, we decide not to train new classifiers using MSE thresholds smaller than 0.1% in order to preserve the robustness of the classifiers.

Evaluation of alternative technologies is also very important in this project. We had tested various modelling technologies on sandy shale data. These include Decision Tree (C4.5) and Neural Networks. The results show that GP classifiers have the best accuracy rates.

Would theoretical work impact GP applications and vice versa? My view is yes. Most GP applications are multi-disciplinary projects, which provide the opportunity to investigate "what GP is doing" from the prospect of application domain. For example, while working on an engine oil blending project (Yu and Rutherford, 2001), a sub-equation with value between -1 and 1 was evolved during GP symbolic regression. The final viscosity increase of the engine oil is based on this value and adjusted by various additives used. With a thorough analysis, this might provide a different way to understand the GP process.

Theoretical works also help practitioners explaining GP to people who are not familiar with the technology. In the past few years, I have been working with researchers from many different fields. It is helpful to be able to discuss GP search process with them using fundamental theorems.

8. Conclusions

Optimization of oil production is becoming more important as the global energy supply is facing various challenges. We have presented a hybrid GP-fuzzy system for reservoir characterization and demonstrated its effectiveness on modelling reservoir permeability. The divide-and-conquer approach is very effective in dealing with the geological heterogeneity in a typical reservoir. This is shown in the exceptional performance of the hybrid systems: once the lithology group and permeability range are identified, estimation of permeability becomes more robust.

We are continuing this research in improving the classifiers, particular that of sandy shale. Meanwhile, we are testing the system on other well log data and investigating issues for future system deployment.

Acknowledgements

I would like to thank Bill Langdon, Mark Kotanchek, Conor Ryan and Terry Soule for their comments and suggestions. Thanks to Rick Riolo for encouraging me to write up the discussions section. We also wish to acknowledge ChevronTexaco Overseas Petroleum, for providing data and granting permission for these results to be published.

References

Chiu, S. (1994). A Cluster Estimation Method with Extension to Fuzzy Model Identification. In *Proceedings of Third IEEE International Conference on Fuzzy Systems*, pp. 1240-1244. IEEE Press.

Huang, Y., Wong, P. M., and Gedeon, T. D. (2000). "Permeability Prediction in Petroleum Reservoir using a Hybrid System." In *Soft Computing in Industrial Applications*, Y. Suzuki, S.J. Ovaska, T. Furuhashi, R. Roy and Y. Dote (eds.), Springer-Verlag, London, 437-446.

Jang, J-S R. (1993). ANFIS: Adaptive-Network-Based Fuzzy Inference System. *IEEE Transactions on Systems, Man and Cybernetics 23(3)*: 665-685.

Jang, J-S. R. and Sun, C-T. (1997). *Neuro-Fuzzy and Soft Computing: A Computational Approach to Learning and Machine Intelligence.* Prentice Hall.

Koza, J. (1992). *Genetic Programming: On the Programming of Computers by Means of Natural Selection.* Cambridge, MA: MIT Press.

The MathWorks, Inc. MATLAB. Natick, MA, 1994-2003.

Nikravesh, M., Aminzadeh, F. and Zadeh, L. A. (2003). *Intelligent Data Analysis for Oil Exploration, Developments in Petroleum Science*, Elsevier, in press.

RML Technologies, Inc. Discipulus 3.0, Littleton, CO, 1998-2003.

Scheidegger, A. E. (1974). *The Physics of Flow through Porous Media.* University of Toronto Press.

Sugeno, M. (1985). *Industrial Applications of Fuzzy Control*, Elsevier Science Pub. Co.

Sugeno, M. and Kang, G. T. (1986). Fuzzy Modeling and Control of Multilayer Incinerator. *Fuzzy Sets and Systems 18:* 329-346

Sugeno, M and Tanaka, K. (1991). Successive Identification of a Fuzzy Model and Its Applications to Prediction of Complex Systems. *Fuzzy Sets and Systems 42:* 315-334

Wong, P. M., Aminzadeh, F. and Nikravesh, M. (2002). *Soft Computing for Reservoir Characterization and Modeling.* Series Studies in Fuzziness and Soft Computing, Vol. 80, Physical Verlag, Springer.

Xie, D. (2001) Thin-bed Reservoir Characterization using Integrated 3-D Seismic and Well Log Data: a Case Study of the Central Boonsville Field, Fort Worth Basin, North-Central Texas, PhD Dissertation, Michigan Technological University.

Yu, T. and Lee, S. (2002). Evolving Cellular Automata to Model Fluid Flow in Porous Media. In *Proceedings of the 2002 NASA/DoD Conference on Evolvable Hardware*, pp. 210-217. IEEE Computer Society.

Yu, T. and Rutherford, J. (2001). Modelling Sparse Engine Test Data Using Genetic Programming. In *Proceedings of Fourth Workshop on Mining Scientific Datasets*, pp. 42-48. Chandrika Kamath.

Chapter 18

ENHANCED EMERGING MARKET STOCK SELECTION

A Genetic Programming Approach

Anjun Zhou

Advanced Research Center, State Street Global Advisors, Boston, MA 02110

Abstract: Emerging stock markets provide substantial opportunities for investors. The existing literature shows inconsistency in factor selection and model development in this area. This research exploits a cutting edge quantitative technique - genetic programming, to greatly enhance factor selection and explore nonlinear factor combination. The model developed using the genetic programming process is proven to be powerful, intuitive, robust and consistent.

Key words: genetic programming, emerging market, stock selection

Introduction

Emerging stock markets have for many years attracted attention from market participants due to their return potential and low correlation with developed markets (Harvey 1994). However, emerging markets have tended to generate a very volatile return stream. As a result, stock selection in emerging markets is even more challenging than in developed markets.

Given the task and potential reward, there have been a number of studies that attempt to predict stock returns in emerging markets. For example, Achour et al (1998a, 1998b) examined about 30 firm-specific factors for several emerging markets and concluded that their relative contributions to stock selection vary depending on economic situations and markts. Claessens et al (1998) examined a number of factors and suggested that sensitivity to the market movement (beta), size and trading volume are primary factors, while dividend yield and price-earning ratio (P/E) are secondary factors. Rouwenhorst (1999) found that momentum (trend-following factor) and size matter, value stocks outperforms growth stocks, and that there is no beta effect. Wright (1999) documented that emerging market returns have considerable correlation from one period to another, which makes the return from previous periods a good potential predictor. Many investment firms also have been conducting research in this area and publish the relative importance of factors depending on markets and time periods. Despite the breadth and depth of these studies, there has been little consensus on the best factors and methodology to predict stock returns in emerging markets.

Assuming a set of factors has been selected, the traditional way of building a model is to combine these factors in a linear fashion. This can be traced back to the history of financial research. First used as a convenient and simple tool, linear specification eventually grew to be a default assumption used to develop models without much justification or by imposing some assumptions. Classic examples are the CAPM and Arbitrage Pricing Model. A more recent example is the linear information model proposed by Ohlson (1995). In the Ohlson model, linearity follows from a particular specification of how excess profits are competed away over time but a small change in this specification destroys the linearity.

The linear specification can be convenient and sometimes intuitive, but it often fails to capture the complexity of the true relationships. To avoid this potential bias and search for the best possible emerging market stock model, we decided to turn to some unconventional methods. After searching and reviewing several potential candidates, we decided to adopt genetic programming to explore nonlinear factor combination and enhance factor selection.

Evolved from the genetic algorithm developed by Holland (1975) and later extended and advocated by Koza (1992), genetic programming has been applied to a number of scientific areas such as biology, medicine, computer science and information system. In the financial industry, however, genetic programming remains a rather cutting-edge methodology, which yields a handful of pioneer works. For example, Neely, Weller and Dittmar (1997) used genetic programming technique to identify technical trading rules in the foreign exchange markets. Chen and Yeh (1997) tested efficient market hypothesis with genetic programming method. Wang (2000) adopted genetic programming to enhance trading and hedging in equity spot and futures markets. Apoteker and Barthelemy (2000) demonstrated the value of genetic algorithms in building a early warning system for country risk assessment for emerging markets.

The purpose of this research is to combine traditional techniques with genetic programming to develop an emerging markets stock selection model that is intuitive, powerful, robust and consistent.

1. Data

Our emerging market portfolios use MSCI emerging market index as benchmark, one of the most widely used benchmarks by global portfolio managers. We focused our research on six major emerging markets in the MSCI universe: Brazil, Israel, Korea, Mexico, Taiwan and South Africa. In order to construct the potential factors that predict stock returns, we downloaded over 60 monthly data items from MSCI, Worldscope and IBES databases for the period January 1995 – December 2001, ranging from balance sheet information to earnings estimates to price volatility. Due to liquidity concerns, we selected companies with market capitalization of $100 million or more, which tend to be traded relatively easily.

2. Methodology

As mentioned in the introduction section, most stock selection models used by portfolio managers are linear. This is essentially more of a convention and convenience than anything else. Nonetheless, even if it is believed that a linear relationship is not appropriate, attempting to explore the true relationship between factors and returns would result in exhaustive searches through various factor combinations and modeling methods. This inevitably demands a daunting amount of time and effort. One way to meet this challenge is by adopting genetic programming technique.

Following the structure used in Neely, Weller, and Dittmar (1997), we designed GP process in three steps: training, selection and validation. The training step generates the initial best populations (master population). The selection step takes the resulted population from the training step and compares them against newly generated populations. And the validation step serves as an out-of-sample test. We used 50% of randomly selected sample for the training step, and 25% for each of the other two steps.

In our application, the genome trees are factor models. When it comes to function selections, there are many ways to construct a nonlinear factor model. For example, we started with allowing some mathematical functions such as exponential, square root, minimum and maximum, as well as relationship functions such as if… then…. The GP generated models often have complicated formula, which make it difficult for our portfolio managers to explain to clients in an intuitive way. As a compromise, we decided to use only four basic arithmetic operations: addition, subtraction, multiplication and division, along with factors and constants, to construct models. We also constrained the tree depth to be up to 6 and non-terminal nodes to have up to 2 child nodes.

Our objective function is to find the model that yields the best information ratio (IR), the risk-adjusted relative return against a benchmark, through the training, selection and validation steps.

We ran 10,000 generations, with the size of population of 1000 and the 1% best outcomes being copied to the master population. We assumed the probability of crossover operation as 95% and mutation operation as 5%. 80% of leaf nodes are factors and 20% constants. The GP process will terminate when any one of the following three conditions is met:

1. A convergence threshold of 0.99 is reached.
2. The selected number of generations is reached.
3. A specified objective value is found.

3. Stock Selection Model

As the first step, we examined over 40 factors in the areas of value, growth, momentum, earning estimate sentiment, profitability, debt information, and behavioral finance. We then performed comprehensive factor testing, using traditional measures such as information coefficients, return tricile spread[1] and turnover. Information coefficients are the rank correlation between each factor and 1-month forward returns. Return tricile spread is obtained by first ranking stocks by a factor and dividing these stocks into 3 tiles, where stocks in tile 1 has the highest level in the factor and those in tile 3 the lowest, then calculating the difference between average return of tile 1 and that of tile 3. Turnover for tile 1 is the percentage of stocks that bought and sold in/from tile 1 in two consecutive periods. It measures the frequency of stock changes in a portfolio.

Based on this analysis, we selected 15 promising factors as input to the genetic programming process. In doing so, we eliminated factors that do not perform well or directly intuitive. As a result, the GP outcomes are more likely to be successful both in performance and intuitiveness.

For proprietary reasons, we cannot show the model we actually developed for the portfolio management team. Instead, we demonstrate with one of the best performing models:

B/P * diffusion * forward E/P * growth yield,

[1] The reason to rank stocks into three tiles instead of the commonly used five or ten is due to limited observations that fall into a country and sector bin.

where B/P is book to price ratio, diffusion is time-weighted difference between up earnings revisions and down earnings revisions, forward E/P is rolling forward 12-month earning to price ratio, and growth yield is estimated change in earnings per share for the next 12 months. The final GP tree is shown in Figure 18-1.

We think this model is interesting for two reasons. First, it selects 4 factors from 15 potential factors, which cover the areas of value, growth and sentiment and keep a good balance across these areas. Second, the model combines the factors in a multiplicative fashion and yet remains relatively straightforward and intuitive.

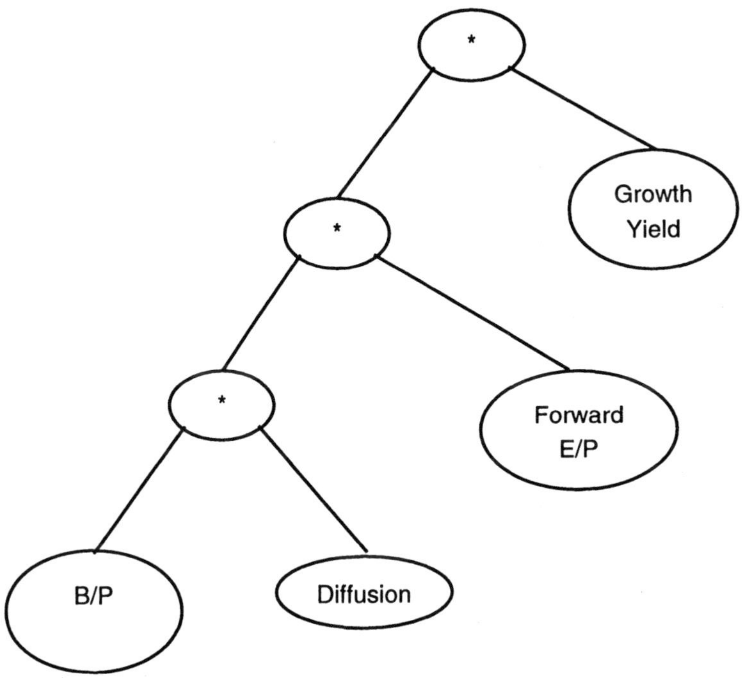

Figure 18-1. A Genetic Programming Tree

3.1 Linear versus Nonlinear Models

To understand the similarities and differences between a linear model and a nonlinear model, we used two hypothetical factors - Value and Growth, ranked them between 1 – 100 and generated two alphas, the forecast returns, as the following:

Linear Alpha = (Value + Growth) * 0.5

Nonlinear Alpha = $\sqrt{\text{Value} * \text{Growth}}$

We next selected the top 25% stocks based on each alpha and plotted them in Figure 18-2.

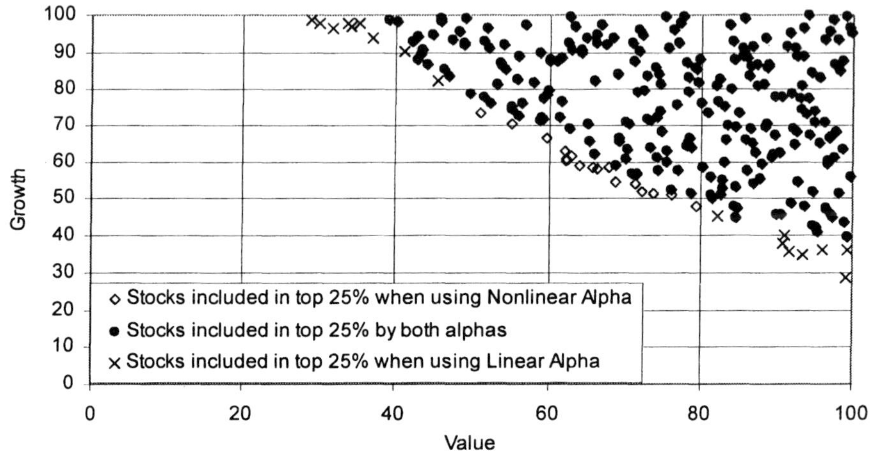

Figure 18-2. Linear versus Nonlinear Models

It is shown that, first of all, there is an overwhelming overlap for the stocks selected in the top 25% by both alphas. Secondly, the nonlinear model favors stocks with more consistent ratings across the factors and avoids those with stellar ratings on some factors and lousy ratings on others. This method helps to screen out companies with potential problem areas and single out companies that are strong overall. It is especially important in emerging markets when data is known not to be 100% reliable.

4. Results

4.1 Factors

From the 15 factors we input, the GP process selects a few that work well together. In this sense, the GP process essentially enhances factor selection.

Table 18-1. Factor Rank Correlation (1995 – 2001)

	B/P	Forward E/P	Diffusion	Growth Yield
B/P	1.00			
Forward E/P	0.37	1.00		
Diffusion	-0.06	0.01	1.00	
Growth Yield	0.19	0.33	0.09	1.00

Table 18-1 shows the Spearman rank correlation of the factors used in the proposed GP model. Rank correlation overcomes the potential bias caused by outliers, which are more a potential problem in emerging markets due to possible data errors. It is shown that all 4 factors are relatively weakly correlated, which makes each of them a unique contributor to predicting the stock returns. With B/P negatively correlated to diffusion, it adds diversification effect to the model.

Table 18-2 shows the in-sample test results for the 4 factors. In dividing in-sample and out-of-sample periods, instead of cutting off by a calendar date, we randomly selected 80% of observations as in sample, and the rest 20% as out of sample. The reason it is handled this way is that factor behavior is different from year to year. We prefer to focus on the consistency of factor performance, rather than relying on a certain period in making investment decisions. To meet our investment mandate, we performed tests on country and sector neutral basis.

Panel A of Table 18-2 provides information coefficients (IC) and turnover for tile 1. The five ICs are the rank correlation between the factor and 1-month and first, second, third, and fourth 3-month forward returns. This provides information on the speed and magnitude of decay in the predictive power of each factor.

It is shown that forward E/P has the strongest predictive power both in the short-term and near-term. Diffusion is a strong factor but with a higher turnover ratio. Panel B shows factor tricile spreads on both equal-weighted and cap-weighted basis. Tricile spreads are obtained by forming three long/short strategies using stocks in the relevant tiles. For example, to obtain the top-bottom tricile spread on a cap-weighted basis for factor diffusion, we ranked stocks by diffusion and put them into three tiles. We then calculated tile 1 return by adding return of the stocks that belong to tile 1, weighted by their market capitalization. Tile 3 return was calculated in a similar way. The tricile spread is thus the difference between tile 1 return and tile 3 return.

As shown, forward E/P is clearly the best of all in both counts. All factors but B/P have stronger power in the equal-weighted scenario than the cap-weighted scenario. The spreads from Buys and Sells strategies show that the effect of all factors on 1-month forward return is monotonic mainly in the equal-weighted scenario.

Table 18-2. Factor In-Sample Tests 1995 - 2001(Randomly selected 80% of observations)

A. Information Coefficients and Turnovers

Factors	ICs					Turnover
	1mo	1st 3mo	2nd 3mo	3rd 3mo	4th 3mo	Tile 1
Forward E/P	0.052	0.074	0.055	0.042	0.053	37.4%
Diffusion	0.026	0.040	0.024	-0.004	-0.011	58.2%
Growth Yield	0.024	0.016	0.000	-0.009	0.011	40.2%
B/P	0.008	0.000	-0.016	-0.006	-0.011	36.2%

B. Tricile Spreads

Factors	Tricile Spread (Eql Wgt)			Tricile Spread (Cap Wgt)		
	Top-Bottom	Buys*	Sells**	Top-Bottom	Buys*	Sells**
Forward E/P	13.51	3.90	9.61	9.05	-0.12	9.17
Diffusion	7.10	2.56	4.55	1.59	-2.57	4.15
Growth Yield	8.41	2.53	5.89	2.07	-2.06	4.14
B/P	6.50	2.83	3.67	6.64	3.08	3.55

* Buys: Top tile – Middle Tile
** Sells: Middle Tile – Bottom Tile
*** Tests on country and sector neutral basis.

4.2 The GP Model

To evaluate the performance of the proposed model (which we refer to as GP Model hereafter), we created a benchmark model, which is an equal-weighted three-factor linear model. The three factors largely overlap with those used in the GP Model and has over 50% weight in value. We used this model as a benchmark because it would be the kind of model that traditional portfolio managers use in picking stocks.

Table 18-3 provides in-sample test results for the GP Model and Benchmark Model, which once again was obtained by using randomly selected 80% of observations. It is shown that with a little higher turnover, the GP Model overall has strong ICs and tricile spread. The returns from Buys and Sells strategies confirm that the effect of the model on 1-month forward return in the three tiles is monotonic.

Table 18-3. Model In-Sample Tests 1995 - 2001(Randomly selected 80% of observations)

A. Information Coefficients and Turnovers

Models	ICs					Turnover
	1mo	1st 3mo	2nd 3mo	3rd 3mo	4th 3mo	Tile 1
GP	0.054	0.064	0.022	0.012	0.026	46.1%
Benchmark	0.046	0.062	0.023	0.019	0.020	44.5%

B. Tricile Spreads

Models	Tricile Spread (Eql Wgt)			Tricile Spread (Cap Wgt)		
	Top-Bottom	Buys*	Sells**	Top-Bottom	Buys*	Sells**
GP	16.54	6.16	10.37	10.39	2.84	7.55
Benchmark	15.29	6.01	9.28	11.04	2.85	8.20

* Buys: Top tile – Middle Tile
** Sells: Middle Tile – Bottom Tile
*** Tests on country and sector neutral basis.

Table 18-4 provides out-of-sample test results. The much higher turnover than in the in-sample test is due to the nonconsecutive periods, which resulted from the 20% randomly selected observations. As shown, overall the GP Model outperforms the Benchmark Model, particularly by cap-weighted tricile return spread.

Table 18-4. Model Out-of-Sample Tests 1995 - 2001

A. Information Coefficients and Turnovers

Models	ICs					Turnover
	1mo	1st 3mo	2nd 3mo	3rd 3mo	4th 3mo	Tile 1
GP	0.044	0.068	0.025	0.042	-0.005	92.0%
Benchmark	0.041	0.072	0.030	0.024	0.003	91.5%

B. Tricile Spreads

Models	Tricile Spread (Eql Wgt)			Tricile Spread (Cap Wgt)		
	Top-Bottom	Buys*	Sells**	Top-Bottom	Buys*	Sells**
GP	19.15	5.86	13.29	13.87	3.01	10.86
Benchmark	18.73	4.26	14.47	3.54	-3.32	6.87

* Buys: Top tile – Middle Tile
** Sells: Middle Tile – Bottom Tile
*** Tests on country and sector neutral basis.

To further confirm the power of the GP Model, we ran portfolio simulations and examined performance across markets and time periods. Table 18-5 provides the summarized results for the six major markets. As shown, the GP Model boasts on average 109 basis-point higher annual excess returns than the Benchmark Model. The turnover is smaller and industry bets are closer to the industry-neutral mandate. The "Cap" is the percentage change of log portfolio capitalization versus log benchmark capitalization. It shows a more consistent selection of stocks using the GP Model. Since these measures are essential to the portfolio managers, in a sense it is the simulation results that gave them confidence for the GP model.

Table 18-5. Simulation Results

		Benchmark	GP
Excess Returns	Minimum	-3.98%	-2.42%
	Mean	-0.01%	0.10%
	Median	-0.08%	0.10%
	Maximum	6.42%	5.03%
	StdDev	1.68%	1.07%
	IR	3.79%	10.40%
	Annualized	0.52%	1.61%
Industry Bets	Maximum	7.10	3.31
	Average	0.57	0.29
Cap	Minimum	-51.45	-23.20
	Mean	-19.90	0.05
	Median	-16.84	3.61
	Maximum	17.96	20.92
	StdDev	18.56	12.62
Turnover	Mean	2.72	1.83
	StdDev	5.68	3.53
	Minimum	0.00	0.00
	Maximum	39.69	24.36

* All results are monthly except Annualized Excess Returns

Figure 18-3 compares the year-by-year performance of the two models. It is shown that in most years the GP Model outperforms the Benchmark Model by a large margin. One exception is 1998, when the Israel equity market was in bad shape and the factor diffusion performed poorly.

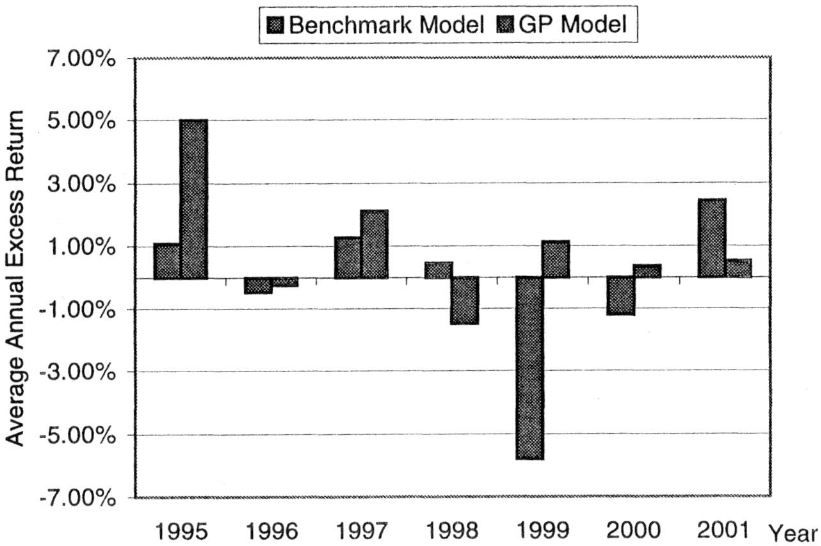

Figure 18-3. Performance by Year

5. Conclusions

Genetic programming revolutionizes the factor combination process and greatly enhances the factor selection process. The methodology has demonstrated tremendous power in emerging market stock selection. One potential concern is that the GP outcomes may not be intuitive and thus difficult to explain to clients. One way of dealing with it is, as in our case, to enforce the investment intuition throughout the process by initial factor screening and careful function selection. We believe that genetic programming will gain wider acceptance by the investment industry in the near future.

Acknowledgments

The author would like to thank Anthony Foley for his advocacy on the application of genetic programming and many valuable comments on this paper, Jennifer Sjostedt for setting up the research structure, and George Hoguet and emerging markets team for their support and helpful feedback. The views expressed are those of the author and do not necessarily reflect the opinions of State Street Global Advisors. All errors remain mine.

References

Achour, D., C. Harvey, G. Hopkins, and C. Lang. (1998a). Stock Selection in Emerging Markets: Portfolio Strategies for Malaysia, Mexico, and South Africa. *Emerging Markets Quarterly,* 38-91.

Achour, D., C. Harvey, G. Hopkins, and C. Lang. (1998b). Firm Characteristics and Investment Strategies in Africa: The Case of South Africa. Working Paper.

Apoteker, T., and S. Barthelemy. (2000). Genetic Algorithms and Financial Crises in Emerging Markets. Working Paper.

Chen, S, and C. Yeh. (1997). Toward a Computable Approach to the Efficient Market Hypothesis: An Application of Genetic Programming. *Journal of Economic Dynamics and Control 21 (6):* 1043-1063.

Claessens, S., Dasgupta, S., and J. Glen. (1998). The Cross-Section of Stock Returns: Evidence from the Emerging Markets. *Emerging Markets Quarterly,* 4-13.

Connor, G., and S. Sehgal. (2001). Tests of the Fama and French Model in India. Working paper.

Domowitz, I., Glen, J., and Madhavan, A. (1997). Market Segmentation and Stock Prices: Evidence from an Emerging Market. *Journal of Finance 52 (3):* 1059-1085.

Harvey, C. (1994). "Portfolio Enhancement Using Emerging Markets and Conditioning Information." *Portfolio Investment in Developing Countries.*

Holland, J. (1975). *Adaptation in Natural and Artificial Systems.* Ann Arbor: University of Michigan Press.

Koza, J. (1992). Genetic Programming: *On the Programming of Computers by means of Natural Selection.* Cambridge: MIT Press.

Neely, C., Weller, P., and Dittmar, R. (1997). Is Technical Analysis in the Foreign Exchange Market Profitable?: A Genetic Programming Approach. *Journal of Financial and Quantitative Analysis 32(4):* 405-426.

Olhson, J. (1995). Earnings, Book Values and Dividends in Equity Valuation. *Contemporary Accounting Research 11:* 661-687.

Rouwenhorst, K. (1999). Local Return Factors and Turnover in Emerging Stock Markets. Yale University Working Paper.

Serra, A. (2000). The Cross-Sectional Determinants of Returns: Evidence from Emerging Markets' Stocks. Working paper.

Wang, J. (2000). Trading and Hedging In S&P 500 Spot and Futures Markets Using Genetic Programming. *Journal of Futures Markets 20(10):* 911-942.

Wright, J. (1999). Long Memory in Emerging Market Stock Returns. Federal Reserve Board of Governors International Finance Discussion Papers; no. 650.

Chapter 19

THREE FUNDAMENTALS OF THE BIOLOGICAL GENETIC ALGORITHM

A Biologist's View of Biological Inspiration vs. Biology Envy

Stephen Freeland

Department of Biology, UMBC, Catonsville, MD 21250 email: freeland@umbc.edu

Abstract: Evolutionary computing began by lifting ideas from biological evolutionary theory into computer science, and continues to look toward new biological research findings for inspiration. However, an over enthusiastic 'biology envy' can only be to the detriment of both disciplines by masking the broader potential for two-way intellectual traffic of shared insights and analogizing from one another. Three fundamental features of biological evolution illustrate the potential range of intellectual flow between the two communities: particulate genes carry some subtle consequences for biological evolution that have not yet translated mainstream EC; the adaptive properties of the genetic code illustrate how both communities can contribute to a common understanding of appropriate evolutionary abstractions; finally, EC exploration of representational language seems pre-adapted to help biologists understand why life evolved a dichotomy of genotype and phenotype.

Key words: particulate genes; genetic code; phenotype; genotype; biology envy

1. Introduction

At present, we know of only two explanations that can account for complex (i.e. highly improbable) 'design' wherever it is found: conscious intelligence and natural selection. Biological explanations are necessarily limited to using the latter to explain what has spontaneously emerged, *ex nihilo*, during 4 billion years: any appeal to Intelligent Design steps beyond the remit of science. By contrast, evolutionary computing (EC) inherently relies on a degree of intelligent design to define a system of rules within which evolution takes place. Each community faces a different challenge:

the biological theorist is ultimately charged with explaining in detail why one 'run' of the algorithm turned out precisely as it did. The EC researcher is charged with designing rules that, once initiated, auto-generate adaptive solutions without compromising the spontaneity of the adaptive process.

These challenges are qualitatively different, and there is no explicit need for EC to emulate biology. In these proceedings, Ostrowski and Reynolds provide an excellent example of how a conscious rejection of biological analogy need not impede the creation of an efficient EC search algorithm. That said, the sophisticated adaptation evident within the biosphere is a natural inspiration for the intelligent design component of EC systems. Indeed, EC not only began by transposing fundamental concepts of natural selection into computer science (Holland, 1992) but continues to look towards biology for new ideas (e.g. Banzhaf, these proceedings). Thus, elsewhere in this volume, Spector's thought-provoking essay argues that further 'borrowing' from biology is both inevitable, as EC techniques continue to grow in complexity and desirable, as biological research continues to change our perception of the evolutionary process. As an evolutionary biologist, I do not disagree with this conclusion but extend it to include Goldberg's sentiment, expressed at the workshop, that over zealous borrowing could transform biological inspiration into a blind "biology envy" (Notably, biologists themselves have long decided that an analogous blind "physics envy" is unhelpful to their research, (Yates, 1978)). In this context, I introduce three fundamental features of biological evolution:

1. particulate genes and population genetics
2. the adaptive genetic code
3. the dichotomy of genotype and phenotype

Each phenomenon is chosen to represent a different point in the spectrum of possible relationships between computing and biological evolutionary theory. The first is chosen to ask whether current EC has fully transferred the basics of biological evolution. The second demonstrates how both biological and computational evolutionary theorists can contribute to common understanding of evolutionary abstractions. The third is chosen to illustrate a question of biological evolution that EC seems better suited to tackle than biology. Taken together, their relevance is not to showcase what biology can offer EC, but rather to explore the difference between usefully borrowing from (or feeding into) biological evolutionary theory, and the indiscriminate one-way intellectual traffic endemic to biology envy.

In principle, distinguishing biological inspiration from biology envy is straightforward. It is the difference between importing biological phenomena into EC because they improve the evolutionary heuristic (e.g. genetic drift), as opposed to importing phenomena simply because they have emerged during 4 billion years of evolution (e.g. walruses). However, this distinction also inherently implies equivalency: to qualify as inspiration, a 'useful' feature of biological evolution must also translate, functionally, into EC.

A plausible case in point is recombination. Exchange of genetic material between individuals appears near-ubiquitous in biological systems, and has long been regarded as an adaptive feature that increases genetic variation and thereby the power of natural selection (e.g Fisher, 1930). Within EC, the introduction of recombination has proved far more controversial: effects analogous to reductions in biological fitness caused by epistasis (the different phenotypic effects of a gene depending on its genomic context) often result, enough to elicit the comment during this workshop that:

> Our biological friends assure us that recombination is not a problem, but our experiences tell us otherwise. (*Goldberg, pers. comm.*)

This observation from EC may stimulate biologists to re-evaluate the perceived status of recombination as an adaptive phenomenon. However a different possibility is that the communities' observations are correct, but that biological systems somehow limit the potential for epistatic chaos with features that have not been incorporated into mainstream EC. In other words, recombination in the two systems may not be equivalent. One suggestion for such a feature relates to the population genetics of particulate genes.

2. Particulate Genes and Population Genetics

Today it is widely recognized that Darwin's theory of natural selection, as published, would not have worked. The problem lay not with what he added to our understanding, but rather in what he failed to address:

> The laws governing inheritance are quite unknown...Noone can say why the child often reverts in certain characters to the grandmother or grandfather. (Darwin, 1859)

Mainstream thinking of the time viewed the genetic essence of phenotype as a liquid that blended whenever male and female met to reproduce. It took the world's first professor of engineering, Fleeming Jenkin (1867), to point out the mathematical consequence of blending inheritance: a novel advantageous mutation arising in a sexually reproducing organism would dilute itself out of existence during the early stages of its spread through any population comprising more than a few individuals. This is a simple consequence of biparental inheritance. Mendels' theory of particulate genes (Mendel, 1866) replaced this flawed, analogue concept of blending inheritance with a digital system in which the advantageous version (allele) of a gene is either present or absent and biparental inheritance produces diploidy. Thus natural selection merely alters the proportions of alleles in a population, and an advantageous mutation can be selected into fixation (presence within 100% of individuals) without any loss in its fitness. Though much has been written about the Neo-Darwinian Synthesis that ensued from combining Mendelian genetics with Darwinian theory (e.g. Mayr and Provine, 1998), it largely amounts to biologists' gradual acceptance that the particulate nature of genes alone provided a solid foundation to build detailed, quantitative predictions about evolution. Indeed, derision of mathematical models of genes in populations as "bean bag genetics" (Mayr, 1963) overlooks the scope of logical deductions that follow from particulate genetics. They extend far beyond testable explanations for adaptive phenomena and into deeper, abstract concepts of biological evolution. For example, particulate genes introduce stochasticity into evolution. Because genes are either present or absent from any given genome, the genetic makeup of each new individual in a sexually reproducing population is a probabilistic outcome of which particular alleles it inherits from each parent. Unless offspring are infinite in number, their allele frequencies will not accurately mirror those of the parental generation, but instead will show some sampling error (genetic drift). The magnitude of this sampling error is inversely proportional to the size of a population. Wright (1932) noted that because real populations fluctuate in size, temporary reductions can briefly relax selection, potentially allowing gene pools to diffuse far enough away from local optima to find new destinations when population size recovers and selection reasserts itself. In effect, particulate genes in finite populations improve the evolutionary heuristic from a simple hill climbing algorithm to something closer to simulated annealing under a fluctuating temperature.

One final property of particulate genes operating in sexual populations is worthy of mention. In the large populations where natural selection works most effectively, any novel advantageous mutation that arises will only reach fixation over the course of multiple generations. During this spread, recombination and diploidy together ensure that the allele will temporarily find itself in many different genetic contexts. Classical population genetics (e.g. Fisher, 1930) and experimental EC systems (e.g. O'Reilly 1999) have focused on whether and how this context promotes selective pressure for gene linkage into "co-adapted gene complexes". A simpler observation is that a novel, advantageous allele's potential for negative epistatic effects is integral to its micro-evolutionary success. Probability will favor the fixation of alleles that are good 'team players' (i.e. reliably imbue their advantage regardless of genetic background, (see Dawkins, 1989). Many mainstream EC methods simplify the population genetics of new mutations (e.g. into tournaments), to expedite the adaptive process. This preserves non-blending inheritance and even genetic drift, but it is not clear that it incorporates basic difference between biology and EC contribute anything to our understanding of why recombination seems to play such different roles in the two systems?

3. The Adaptive Code Book

Molecular biology's Central Dogma connects genes to phenotype by stating that DNA is transcribed into RNA, which is then translated into protein.

The terms transcription and translation are quite literal: RNA is a chemical sister language to DNA. Both are polymers formed from an alphabet of four chemical letters (nucleotides), and transcription is nothing more than a process of complementing DNA, letter by letter, into RNA. It is the next step, translation, that profoundly influences biological evolution.

Proteins are also linear polymers of chemical letters, but they are drawn from a qualitatively different alphabet (amino acids) comprising 20 elements. Clearly no one-to-one mapping could produce a genetic code for translating nucleotides unambiguously into amino acids, and by 1966 it was known that the combinatorial set of possible nucleotide triplets forms a dictionary of 'codons' that each translate into a single amino acid meaning (Frisch, 1966). The initial surprise for evolutionary theory was to discover that something as fundamental as the code-book for life would exhibit a high degree of redundancy (an alphabet of 4 RNA letters permits 4x4x4=64 possible codons that map to one of only 20 amino acid meanings, Crick, 1966). Early interpretation fuelled arguments for Non-Darwinian evolution (King and Jukes, 1969): genetic variations that make no difference to the protein they encode must be invisible to selection and therefore governed solely by drift (Kimura, 1968). More recently, both computing (e.g. Banzhaf, 1994; Kargupta, 2001) and biological (e.g. Kauffman, 1993; Huynen *et al.*, 1996) evolutionary theory have started to place this coding neutrality in the bigger picture of the adaptive heuristic. Essentially, findings appear to mirror Wright's early arguments on the importance of genetic drift: redundancy in the code adds selectively neutral dimensions to the fitness landscape that renders adaptive algorithms more effective by increasing the connectedness of local optima.

At present, an analogous reinterpretation is underway for a different adaptive feature of the genetic code: the observation that biochemically similar amino acids are assigned to codons that differ by only a single nucleotide. Early speculations (e.g. Woese, 1965) that natural selection organized the genetic code so as to minimize the phenotypic impact of mutations have gained considerable evidential support as computer simulation enables exploration of theoretical codes that nature passed over (comprehensively reviewed in Freeland *et al.*, 2003). However, it seems likely that once again this phenomenon has more subtle effects in the broader context of the adaptive heuristic. We have suggested that an 'error minimizing code' may in fact maximize the probability that a random

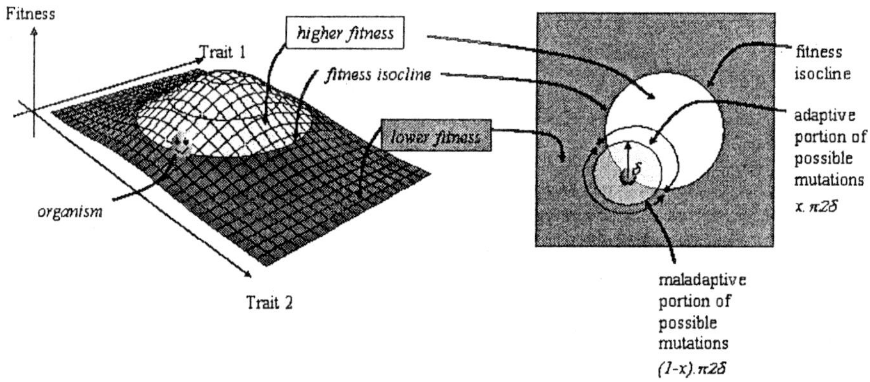

Figure 19.1. The fitness landscape for an organism of 2 phenotypic traits: (a) for any organism, we may define an isocline that connects all trait combinations of equal fitness; (b) (the fitness landscape from above): a random mutation of magnitude δ that has tradeoff effects on both traits defines a circle of radius δ around the organism. The probability that this mutation will improve fitness (i.e. that the organism will move within the white area) is inversely proportional to its magnitude (δ) (from Freeland *et al.*, 2002)

mutation produces an increase in fitness (Freeland, 2002) According to Fisher's (1930) Geometric Theory of gradualism (Fig. 19.1). Preliminary tests for this phenomenon reveal an even simpler influence: the error minimizing code smoothes the fitness landscape where a random genetic code would render it rugged (unpublished data). By clustering biochemically similar amino acids within mutational reach of one another it ensures that any selection towards a specific amino acid property (e.g. hydrophobicity) will be towards an interconnected region of the fitness landscape rather than to an isolated local optimum.

4. The Genotype/Phenotype Dichotomy

Implicit to the concept of an adaptive genetic code is a deeper question that remains largely unanswered by biology: why does all known life use two qualitatively different polymers, nucleic acids and proteins, with the associated need for translation? Current theories for the origin of this dichotomy focus on the discovery that RNA can act both as a genetic storage medium, and as a catalytic molecule. Within the most highly conserved core of metabolism, all known organisms are found to use RNA molecules in roles we normally attribute to proteins (White, 1976). Morevoer, *in vitro* experiments have begun to demonstrate a unexpectedly broad potential for RNA-enzyme ('ribozyme') catalysis (reviewed in Doudna *et al.* 2002). Put together, these observations have produced the RNA World hypothesis (Gesteland *et al.*, 1999) in which primordial, self-replicating RNA molecules eventually evolved metabolic protein clothing.

However, this answer to *how* the dichotomy evolved has largely eclipsed the question of *why* RNA evolved a qualitatively different representation for phenotype. A typical biological answer would be that the larger alphabet size of amino acids unleashed a greater catalytic diversity for the replicators (e.g. Szathmary and Maynard Smith, 1995), with an associated increase in metabolic sophistication that optimized self-replication. Interestingly, we know that nucleic acids are not limited to the 4 chemical letters we see today: natural metabolically active RNA's utilize a vast repertoire of post-transcriptional modifications (reviewed in Grosjean and Benne, 1998) and synthetic chemistry has demonstrated that multiple additional nucleotide letters can be added to the genetic alphabet even with today's cellular machinery (Picirilli *et al.* 1990). Furthermore, an increasing body of indirect evidence suggests that the protein alphabet itself underwent exactly the sort of evolutionary expansion early in life's history (Ronneberg *et al.* 2001).

Given the ubiquity of nucleic acid genotype and protein phenotype within life, biology is hard-pressed to assess the significance of evolving this 'representational language'. The choice of phrase is deliberate: clearly the EC community is far ahead of biology in formalizing the concept of representational language, and exploring what it means. Biology will gain when evolutionary programmers place our system within their findings, illustrating the potential for biological inspiration *from* EC.

5. CONCLUDING REMARKS

Three fundamental features of the biological genetic algorithm are: particulate genes, the adaptive genetic code and the dichotomy of genotype and phenotype. For particulate genes, I ask whether EC can gain from biology by considering the detailed dynamics by which a novel, advantageous allele invades a wild-type population. This naïve question serves to illustrate an important, general criterion for curtailing biology envy: does the EC representation of a biological phenomenon achieve true functional equivalency? It is unclear whether this question gets routinely and rigorously assessed when novel biological phenomena are brought into EC.

On a different front, research into adaptive features of the genetic code illustrates how EC and biological evolutionary research can and do contribute to a common understanding of the general evolutionary dynamic. Neutrality, redundancy and fitness-landscape smoothing are abstractions that apply unambiguously to both communities.

For the dichotomy of genotype and phenotype, I argue that biology has so far failed to even recognize a question that the EC community is actively addressing: how and why does representational language influence the evolutionary dynamic? A luxury of EC is to bend fundamental rules of evolution beyond anything biologically plausible, and thus to answer questions where biologists have assumptions. Wherever conceptual equivalency can be demonstrated, EC will penetrate biological evolutionary theory not by borrowing from it, but by escaping biological assumptions long enough to tell us why they exist.

Acknowledgements

Funding for this work came from Templeton Foundation grant 938-COS27, and from UMBC. I am indebted to Rick Riolo and Bill Worzel for inviting me to this workshop, to Lee Spector, Wolfgang Banzhaf, Duncan MacLean and all the workshop attendees for their warm introduction to the EC community and to Anjeanette Agro for persistent inspiration in a complex world.

References

Banzhaf, W. (1994). Genotype-phenotype mapping and neutral variations ---a case study in Genetic Programming. In Parallel Problem Solving From *Nature III*: 322-332. Davidor, Y., Schwefel, H.-P. and Männer, R., Springer: Berlin.

Banzhaf, W. *(these proceedings)*. "Artificial Regulatory Networks and Genetic Programming."

Crick, F. H. C. (1966). Codon-anticodon Pairing: The Wobble Hypothesis. *Journal of Molecular Biology 19*: 548-555.

Darwin, C. R. (1859). *On the Origin of Species by means of Natural Selection.* John Murray.

Dawkins, R. (1989). *The Selfish Gene.* 2nd edition, Oxford University Press; 84-86

Doudna, J. A. and Cech, T. R. (2002). The Chemical Repertoire of Natural Ribozymes. *Nature 418*: 222–228

Fisher, R. A. (1930). *The Genetical Theory of Natural Selection.* Oxford: Clarendon Press.

Freeland, S. J. (2002). The Genetic Code: An Adaptation for Adapting? *Journal of Genetic Programming and Evolvable Machines 3(2)*: 113-127

Freeland, S. J., Wu, T., and Keulmann, N. (2003). "The Case for an Error Minimizing Genetic Code." In *Origins of Life and Evolution of the Biosphere. (in press)*

Frisch, L. (ed.) (1966). The Genetic Code. In *Proceedings of Cold Spring Harbor Symposia on Quantitative Biology,* pp. 1-747.

Gesteland, R. F., Cech, T., and Atkins, J. F. (eds.) (1999). *The RNA World.* 2nd edition, Cold Spring Harbor Laboratory monograph series 37.

Grosjean, H. and Benne, R. (1998). *Modification and Editing of RNA.* American Society for Microbiology.

Holland, J. H. (1992). *Adaptation in Natural and Artificial Systems.* The MIT Press.

Huynen, M., Stadler, P., and Fontana, W. (1996). Smoothness Within Ruggedness: The Role of Neutrality in Adaptation. *Proc. Natl. Acad. Sci. 93*: 397-401.

Jenkin, F. (1867). (Review of) The Origin of Species. *North British Review 46*: 277-318.

Kargupta, H. A (2001). Striking Property of Genetic Code-like Transformations. *Complex Systems 13*: 1-32.

Kauffman, S. A. (1993). *The Origins of Order - Self-Organization and Selection in Evolution.* New York: Oxford University Press.

Kimura, M. (1968). Genetic Variability Maintained in a Finite Population due to Mutational Production of Neutral and Nearly Neutral Isoalleles. *Genetical Research 11*: 247–269.

King, J. L. and Jukes, T. H. (1969). Non-Darwinian Evolution. *Science 164*: 788–798.

Maynard, Smith J. and Szathmáry, E. (1995). *The Major Transitions in Evolution.* W.H. Freeman and Co.

Mayr, E. (1963). *Animal Species and Evolution.* Harvard University Press.

Mayr, E. and Provine, W. B. (1998). *The Evolutionary Synthesis: Perspectives on the Unification of Biology.* Harvard University Press.

Mendel, G. (1866). "Versuche über Pflanzen-Hybriden" Verhandlungen des naturforschenden Vereines, Abhandlungen Brünn 4:3-47. (English translation: available at http://www.mendelweb.org/MWGerText.html)

O'Reilly, U-M. (1999). Foundations of Genetic Programming: Effective Population Size, Linking and Mixing.

Ostrowski D. and Reynolds R.G. *(these proceedings).* "Using Software Engineering Knowledge to Drive Genetic Programming Design Using Cultural Algorithms."

Picirilli, J. A., Krauch, T., Moroney, S. E., and Benner, S. (1990). Enzymatic Incorporation of a New Base Pair into DNA and RNA Extends the Genetic Alphabet. *Nature 343*: 33-37.

Ronneberg, T. A., Landweber, L. F., and Freeland, S. J. (2001). Testing a Biosynthetic Theory of Genetic Code Evolution: Fact or Artefact? *Proc. Natl. Acad. Sci. 97(25)*: 13690-13695.

Spector, L. *(these proceedings).* "An Essay Concerning Human Understanding of Genetic Programming."

White, H. B. (1976). Coenzymes as Fossils of an Earlier Metabolic State. *Journal of Molecular Evolution 7.*

Woese, C. R. (1965). On the Evolution of the Genetic Code. *Proc. Natl. Acad. Sci. 54*: 1546–1552.

Wright, S. (1932). The Roles of Mutation, Inbreeding, Crossbreeding and Selection in Evolution. In *Proceedings of the 6th International Congress of Genetics 1*, 356-366.

Yates, F. E. (1978). Is Physics Envy the Curse of Biology? *American Journal of Physiology 234(1).*

Index